Health and
Environment and Quality Audits

Internal auditing is an essential tool for managing compliance and for initiating and driving continual improvement in any organization's systematic HSEQ performance.

Health and Safety, Environment and Quality Audits includes the latest health and safety, environmental and quality management system standards—ISO 9001:2015, ISO 14001:2015, and ISO 45001:2018. It delivers a powerful and proven approach to risk-based auditing of business-critical risk areas using ISO, or your organization's own management systems. It connects the 'PDCA' approach to implementing management systems with auditing by focusing on the organization's context and the needs and expectations of its interested parties. The novel approach leads HSEQ professionals and senior and line managers alike to concentrate on the most significant risks (*Big Rocks* and *Black Swans*) to their objectives. It provides a step-by-step route through *The Audit Adventure*™ to provide a high-level, future-focused audit opinion. The whole approach is aligned to the international standard guidance for auditing management systems, ISO 19011:2018.

With thousands of copies now sold, this unique guide to HSEQ and operations integrity auditing has become the standard work in the field over four editions, while securing bestseller status in Australasia, Europe, North America, and South Africa. It is essential reading for senior managers and auditors alike. It remains the 'go-to' title for those who aspire to drive a prosperous and thriving organization based on world-class HSEQ management and performance.

Dr Stephen Asbury is the author of seven books on safety, risk management, and decision-making for Taylor & Francis. He is Chartered Fellow of the Institution of Occupational Safety and Health (CFIOSH), an Emeritus Professional of the American Society of Safety Professionals (ASSP), and a Fellow of the Institute of Environmental Management and Assessment (FIEMA). He has almost 40 years' experience from assignments in over sixty countries on six continents.

Health and Safety, Environment and Quality Audits

A Risk-based Approach

Fourth Edition

Stephen Asbury

CRC Press is an imprint of the
Taylor & Francis Group, an **informa** business

Designed cover image: © Shutterstock

Fourth edition published 2024
by CRC Press
2385 NW Executive Center Drive, Suite 320, Boca Raton, FL 33431

and by CRC Press
4 Park Square, Milton Park, Abingdon, Oxon, OX14 4RN

CRC Press is an imprint of Taylor & Francis Group, LLC

Third edition published by Routledge 2018

Second edition published by Routledge 2014

First edition published by Butterworth Heinemann 2006

Library of Congress Cataloging-in-Publication Data
Names: Asbury, Stephen, author.
Title: Health and safety, environment and quality audits : a risk-based approach / Stephen Asbury.
Description: Fourth edition. | Boca Raton, FL : CRC Press, 2024. | Includes bibliographical references and index.
Identifiers: LCCN 2023020594 (print) | LCCN 2023020595 (ebook) | ISBN 9781032429083 (hbk) | ISBN 9781032427577 (pbk) | ISBN 9781003364849 (ebk) | ISBN 9781032583310 (ebook+)
Subjects: LCSH: Total quality management. | Organization. | Auditing. | MESH: Total Quality Management—standards | Management Audit—standards | Organizational Culture | Safety Management—standards
Classification: LCC HD62.15 .A845 2023 (print) | LCC HD62.15 (ebook) | DDC 658.5/62—dc23/eng/20230707
LC record available at https://lccn.loc.gov/2023020594
LC ebook record available at https://lccn.loc.gov/2023020595

ISBN: 978-1-032-42908-3 (hbk)
ISBN: 978-1-032-42757-7 (pbk)
ISBN: 978-1-003-36484-9 (ebk)
ISBN: 978-1-032-58331-0 (eBook+)

DOI: 10.1201/9781003364849

Typeset in Times
by Apex CoVantage, LLC

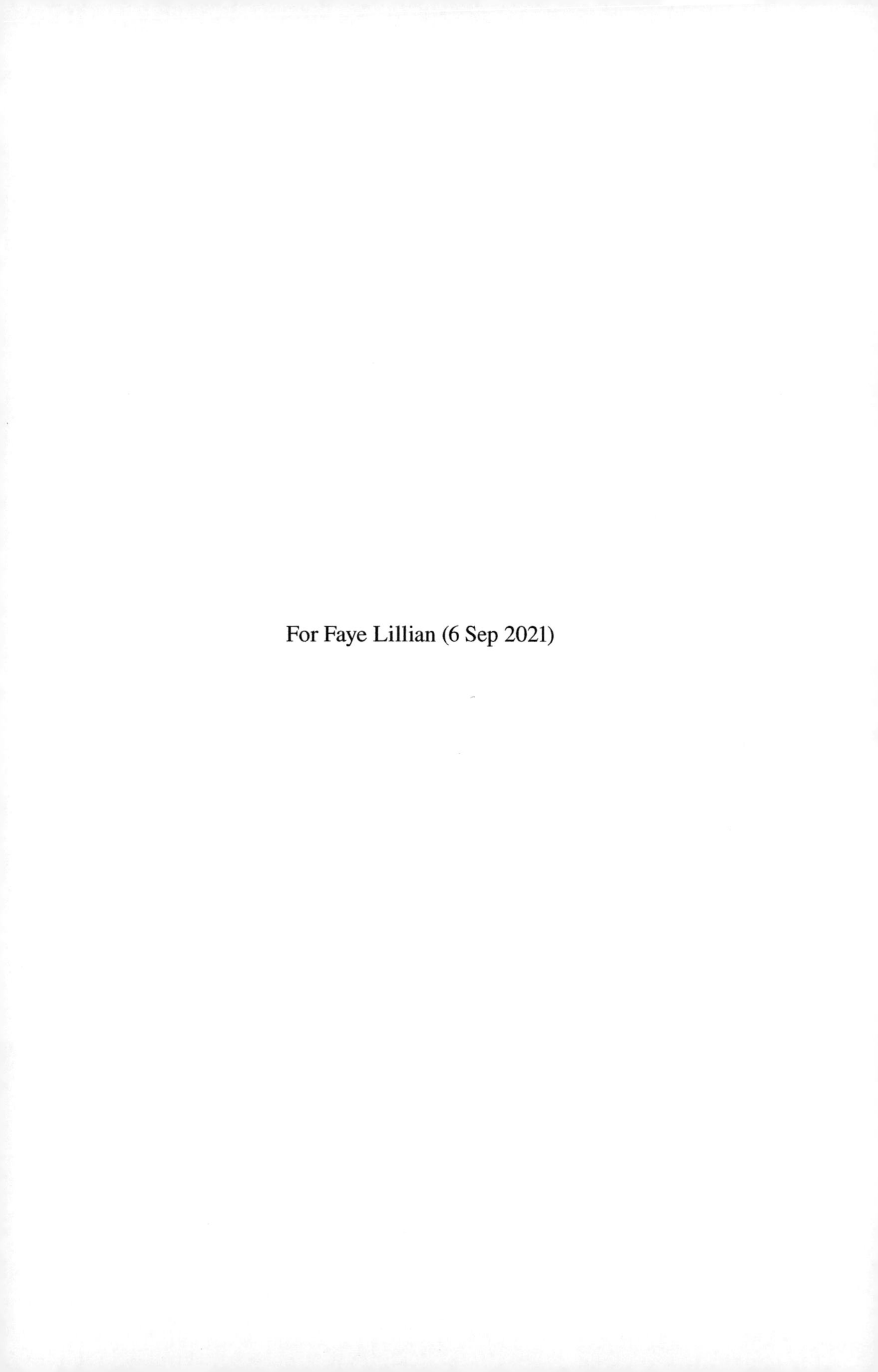

For Faye Lillian (6 Sep 2021)

Contents

INITIATE
CAN'T YOU SLEEP DEAR?
NO DEAR... I NEED TO HAVE AN AUDIT ADVENTURE SOMETIME SOON...
HOME SWEET COMFORT ZONE

Figures

Tables

Case Studies

About the Author

Dr Stephen Asbury is a four-time award-winner. He is the author of seven books on safety, risk management, and decision-making for Informa Taylor & Francis, and over 50 journal articles and conference presentations.

Stephen's first qualification was in law, following which he progressed to work as a safety and risk manager in the construction, technical plastics and polymers, and defence and automotive engineering sectors, before joining the London insurance market in 1995. Since 1999, he has founded and headed up two international consulting groups (CRS, and AllSafe Group).

In a career extending almost 40 years, he has been involved with the implementation and auditing of HSEQ, BCP, and fleet management systems at over 40,000 unique locations in over 60 countries on six continents, with organizations including Bombardier, Chevron, Coca-Cola, Marks and Spencer, McDonald's, Panama Canal Commission, RasGas, and the world's largest company, Saudi Aramco.

Stephen is a Chartered Fellow of the Institution of Occupational Safety and Health (CFIOSH), an Emeritus Professional of the American Society of Safety Professionals (ASSP), a Fellow of the Institute of Environmental Management and Assessment (FIEMA), and a Six Sigma Green Belt.

He was a statutory director of IOSH (1998–2003) and IOSH Services Limited (2004–08), a member of IOSH Council (1998–2012), and chaired several IOSH standing committees, including Professional Affairs (1998–2013). He played a critical role in securing the award of IOSH's Royal Charter in 2003 and in securing permission for IOSH to grant a personal Charter to competent members from 2005.

Stephen was awarded an MBA with Distinction by De Montfort University, Leicester, UK in 1995, and was conferred Doctor of Professional Studies (DProf) by Public Works by Middlesex University London in 2021.

Outside of work, he enjoys F1 motorsport, NBA basketball, and travelling.

Foreword

This fourth edition recognizes that we live in extraordinary times. Our organizations operate in complex and dynamic risk landscapes where near-term operational resilience is as important as reinventing business models, products, and services.

The good news is that a 'robust' business management system continues to provide the machinery for organizations to deploy policy, mitigate risk, and improve performance. A 'living' management system allows organizations to respond to extraordinary times at pace, building resilience and de-risking reinvention.

Audit must act, not to defend the status quo or mark process owners' homework, but to help shine a light on risk and performance and to drive principled improvement.

In these extraordinary times the philosophy, principles, and approaches to audit in Stephen's book are more important than ever.

Vincent Desmond
Chief Executive Officer
The International Register of Certificated Auditors (IRCA)
and the Chartered Quality Institute (CQI)
London, UK

Endorsements

In my Presidential year, Stephen Asbury worked closely with me to re-enforce the IOSH CPD scheme, and together, we had considerable success. This success was by and large due to the passion and tenacity of someone who, for many years, has remained a stalwart volunteer for IOSH and other organisations, an advocate for the complete ethos and understanding of the structure and control of HSEQ-MS.

Now in its fourth edition, Stephens's thoughts on structure in control and auditing of HSEQ-MS have been expressed through this book since 2007. It is inspiring and wonderful that his approach continues to guide health and safety and other risk practitioners on how to deliver meaningful business risk management, efficiency, and continual improvement. In everything he does, Stephen remains true to PDCA roots, easily seen through this wonderful fourth edition.

As chair of IOSH Professional Committee for more than twelve years, he led the creation of the current IOSH membership structure, CPD, IPD, and the Code of Conduct. This structure allowed IOSH to achieve its Royal Charter in 2003 and its permission to grant an individual Charter to its competent members from 2005. A prolific IOSH member, mentor, professional, and inspiring person, Dr Stephen will continue to assist through these nuggets of experience in the narrative of his books, and this one is no exception.

It gives me great pleasure to endorse this new edition of how YOU can assist your employer or client deliver exceptional HSEQ-MS success.

Jimmy Quinn CFIOSH
IOSH President, 2020–21

Competent auditing is a vital function of well-run organisations. It provides assurance to boards and senior management that appropriate controls and governance arrangements are in place, to both manage environmental impacts effectively and support performance improvement.

Now, more than ever, it is crucial for organisations and broader society to manage their relationship with the environment, not only to reduce the impacts they have but also to create new opportunities for development and growth.

I very much welcome this book. It will be a great help to auditors in delivering assurance and value to business.

Sarah Mukherjee MBE
Chief Executive
Institute of Environmental Management and Assessment (IEMA)

As a past General Manager for Royal Dutch Shell, my time spent doing HSE audits provided some of the most rewarding experiences in my career. There is no better way to learn about a business than by asking questions, seeking evidence, and prioritising the findings against the risks. However, carrying out an audit brings with it the responsibility to follow the process.

Stephen Asbury is probably the best instructor that I have come across, and he certainly received the highest level of feedback for the courses that he delivered for the PetroSkills oil and gas training alliance.

Stephen brings enthusiasm, ability to communicate, and an understanding of the subject that comes through in his writing. If you have an opportunity to participate in an audit, seize it, and enjoy.

Dr Adrian Hearle
former Shell Head of Distribution, and former Director HSE, PetroSkills

Health and safety management is an integral part of business risk management, with auditing being an essential component for helping ensure efficacy and continual improvement. Audits should not be dreaded or adversarial, but regarded as opportunities for organizations to learn and for their auditors to share good practices. The international adoption of ISO 45001 is a timely reminder of the value of structure in establishing control of health and safety risks.

Stephen Asbury's excellent book, now in its fourth edition, can assist employers and prospective and practising auditors to better understand their respective roles and also the potential value to the organization of a well-designed and conducted audit undertaken by a competent auditor or audit team.

Rob Strange, OBE
Chief Executive (2001–2013) The Institution of Occupational Safety and Health (IOSH) Leicester, UK

Check is a cornerstone of the *Plan Do Check Act* cycle, which is fundamental to an occupational health and safety management system. The audit element of the management system is a very valuable part. This is the only real way you will know if what you have planned is actually being implemented and working as it should.

An audit allows you to identify opportunities to implement improvements to make the system and the organization run better and improve its performance. Think about how your car runs:

While you are driving, you check your speed and fuel; this is like checking your incident, illness, and lost-time statistics. You also perform inspections of your car's essentials, like oil level, water levels, tyre pressure, and tread depth. This is like your own safety inspections. But to ensure that the car is running as efficiently as it should and that key components are not in need of replacement, you have a service by a competent mechanic. These days it is likely to mean a computer-based diagnostic analysis of the whole car's systems. This analysis will identify any adjustments or opportunities to improve performance.

An audit is so much more than looking at your key performance indicators. It is a holistic review and analysis of your management system and its performance that will allow you to identify areas to improve that performance.

Phil Bates
former member of ISO/PC 283 Working Group on ISO 45001

Stephen Asbury and I have been associated for over twenty-five years. Back then, he was Royal Insurance's risk engineer assigned to our account, and we conducted many audits together in Europe and here in the US.

Audits have increasingly become an essential part of doing business and have not only been embraced by our management but built into the educational structure of McDonalds and our Hamburger University. Safety and the protection of our customers and employees are the highest priority.

Risk-based audits play a major role in allowing us to provide that protection, and I am pleased to endorse Stephen's methodology presented within the fourth edition of this extremely popular book.

Jim Marshall
Director, Insurance & Safety (retired)
McDonald's Corporation, Oak Brook, Illinois, USA

Auditing is an essential component of effectively implemented management systems—it provides assurance to management, enables an opportunity to alert to shortfalls, and, where appropriate, advises management on actions to be taken.

This book, *Health & Safety, Environment and Quality Audits: A Risk-based Approach*, offers a unique and extremely clear overview of *The Audit Adventure*™ which will be invaluable to those who are involved with auditing, whether as an auditor or those who are audited. This risk-based approach described herein is consistent with ISO 19011:2018 and all the ISO Annex SL-based management system standards. It provides not only the background to auditing but outlines each stage of an excellent auditing process with tips, real-life examples, and case studies. This new edition updates the status of all the related standards and includes a range of new case studies to illustrate points made in the text.

This new fourth edition is therefore ideal reading for students taking auditor training and/or developing their auditing skills or simply updating their knowledge. Furthermore, it provides outstanding additional reading for those undertaking a wide range of health, safety, environmental and quality courses, ranging from the NEBOSH General Certificate to postgraduate qualifications, or for anyone who needs to clearly understand the concepts of the audit process.

Dr Jonathan Backhouse, MA, MRes, CFIOSH, MIFireE

Stephen is world renowned for his contribution to the field of health, safety and environment assurance, and risk-based auditing. I was privileged to have worked with him in South Africa, Europe, and many parts of Asia to sincerely share his strong qualities of dedication, perseverance, and such fun to work with. He takes pain to complete his tasks with aplomb, is a great team player, orchestrator yet an excellent mastermind. His penchant for detail and customer satisfaction is worthy of emulation.

This book *HSEQ Audits* succinctly traces the logic of the effective risk-based audit approach, with a culmination of years of continuous improvement in the art and science of auditing. I recommend Stephen and his approaches to auditing to any organization wanting to improve their risk management or health, safety, and environment management systems.

Dato Lokman Awang, DIMP, MBA(Fin), CMIIA, MICG, BAppSc (hons) (Mining)
Managing Director, Proactive Control Sdn Bhd. Kuala Lumpur, Malaysia

Maintaining control in a very large and complex organization of many divisions and many sites such as ours requires thoughtful structure in control systems. Over the years, we have learned to drive improvement into our systems by learning positively from our experiences—actively and reactively. Our commitment to validate our competence and continual improvement is driven by our senior management and satisfies our customers' compliance requirements, so we have maintained SHE and Q systems for many years.

There is always a possible danger that some sites might try to do the bare minimum (or less), ramping up their control only when an external audit draws close. And so, this is where our internal audit programme fits. It is designed to regularly, reliably, and thoroughly assess the performance of our management systems and controls to assure and assist our divisions and sites to deliver against their business objectives.

Stephen Asbury has provided management systems training to all our senior, division, and site managers. It was extremely well received. This book captures the essence of the 'Asbury live' risk-based auditor training event, and I am pleased to recommend it to you.

Ian Kempson
SHEQ Manager
Bunzl Catering & Hospitality Division

My greatest challenges have come from implementing HSE programmes in the emerging markets of the Far East, Africa, Eastern Europe, and Eurasia, where Stephen Asbury first provided the foundation of my assurance programmes.

In my experience, an HSE practitioner requires the skills to positively influence the top management of a business from a position of strength, credibility, and neutrality. In turn, management teams at all levels must be willing to use audit information "intelligently" for sustainable change to occur. For audit results to be utilised effectively, it is my view that they are best regarded in the same way as a business uses a profit and loss account.

In this fourth edition of this excellent book, Stephen has updated his comprehensive insight into the effective tools needed to develop and sustain an effective assurance programme delivering that elusive "value add" to any organization.

Fred Alderson CMIOSH
Present and past positions: HSE Manager, Belhaven Brewery (Greene King)
HSE Manager, The Scottish Salmon Company
HSE Manager, Britvic Soft Drinks
Vice President Global Operational Risk, Deutsche Post DHL
Head of Loss Control, Coca-Cola Hellenic
Counter-intelligence Operative, RAF

It is hard to believe that the fourth edition of this book is soon to be published. Having read all three previous editions over the years and personally used many of the tools and techniques provided, it is a given that the fourth will again provide us all with simply the best audit advice and guidance available from a master practitioner.

Mike Hann
Health and Safety Manager
Mayflower Theatre, Southampton, UK

Abbreviations

ACM asbestos-containing material
AFWP audit finding working paper
ALARA as low as reasonably achievable (also ALARP and SFRP)
ALARP as low as reasonably practicable
BCF business control framework (also MSS)
CFC chlorofluorocarbon
CFSI counterfeit, fraudulent, and suspect items
CoCo Criteria of Control Board
COSO Committee of Sponsoring Organizations of the Treadway Commission
CSR corporate social responsibility
EMS environmental management system
E-SEAP eliminate, substitute, engineering, administrative, PPE (the hierarchy of control from ISO 45001:2018, clause 8.1.2)
EU European Union
FDA Food and Drug Administration
FOFO fear of finding out (also FOBFO, fear of being found out)
G7 Group of Seven
G20 Group of Twenty
GATT General Agreement on Tariffs and Trade
GDP gross domestic product
HAZOP hazard and operability study
HSEQ-MS health, safety, environment, and quality management system (also H&S-MS, HSE-MS, and similar variants)
IH industrial hygiene
ILO International Labour Organization
IMF International Monetary Fund
IPPC integrated pollution prevention and control
ISO International Organization for Standardization
KPI key performance indicator
MOC management of change
MSA management self-assessment
MSS management system standard (also BCF)
OBO operated by others
OECD Organization for Economic Co-operation and Development
OH occupational health
OH&S-MS occupational health and safety management system
OI operations integrity
PCAOB Public Company Accounting Oversight Board
PDCA Plan-Do-Check-Act (also PDSA, 'Study')

PELR	political, economic, legal, resources
PEST	political, economic, social, technical
PML	possible maximum loss
PPE	personal protective equipment
QMS	quality management system
RAM	risk assessment matrix
RPI	retail price index
SCBA	self-contained breathing apparatus
SFRP	so far as is reasonably practicable (from the UK Health and Safety at Work, etc. Act 1974)
SHEQ	safety, health, environment, quality (also HSEQ, QEHS, etc.)
SKATE	skills, knowledge, attitude, training, experience
SMART	specific, measurable, achievable, right (or realistic), timely (or timed)
SSA	safety self-audit
SWOT	strengths, weaknesses, opportunities, threats
ToR	terms of reference (aka 'remit')
UN	United Nations
VUCA	volatility, uncertainty, complexity, ambiguity
WTO	World Trade Organization

Preface to the Fourth Edition

> *Several historical indicators of global recessions are already flashing warnings. The global economy is now in its steepest slowdown following a post-recession recovery since 1970. Global consumer confidence has already suffered a much sharper decline than in the run-up to previous global recessions . . .*
>
> —World Bank (2022)

Austerity, and Asylum. Brexit, and Bird flu. China (PRC and ROC). Cost of living, Corruption, Climate, and CO^2. Drugs, Debt, and Demographics. Energy prices, and Electric vehicles. Food supply, Food banks, and Furlough. Gas, and George Floyd. Homeworking, Hacking, Homelessness and Hatred. Inflation, and Interest rates. Jeff (Bezos, Amazon; online sales vs High Street decline). Knife crime. Loss and Damage. Mental health and Migration. NHS. Obesity. Population (8bn on 15 November 2022), Pollution, Pandemic, and Partygate. Quantitative easing, and Qatar (FIFA World Cup). Russia, Recession, and Racism. Strikes, and Slavery. Trussonomics, and Trump. Ukraine. Vladimir (Putin). Wade v Roe, and Whataboutism. ftX (collapse of crypto). Youth capacity (of 15–24s to make change in the world). Zaporizhzhia nuclear power station.

My A-Z picks out some features of the global chaos since the third edition, published in October 2018. I heard someone had called this a period of *VUCA* (volatility, uncertainty, complexity, ambiguity). Either way, a perfect storm in the macro environment. Geopolitical and economic chaos in 120 words. So many events, so much change, in so little time. The world is heading for recession (World Bank, 2022) as consumer confidence falls, and central banks hike interest rates to combat the old enemy of inflation. The effects of climate change ravage our planet as we seek to overcome a desperate inheritance for our children—a test of international resolve to steer us all away from disaster. And an illegal war in Europe has reignited the old east-west Cold War tensions. I could go on, but I know you get the picture. Many organizations find themselves in a mess.

And so, against this depressing backdrop, health and safety, environment, and quality management systems (HSEQ-MS) are fading away, dying in organizations, due to more important, more pressing matters, right?

Err, no.

Every year, the International Organization for Standardization (ISO) conducts a survey of certifications to its management system standards. The latest data available*, published in September 2023 (ISO, 2023a), is for the year to 31 December 2022. It shows the following:

- ISO 9001:2015 (ISO, 2015a) had 1,265,216 certificates covering 1,666,172 sites issued by certification bodies that have been accredited by members of the IAF (International Accreditation Forum), representing a 17% increase in certificates, and a 15% increase in sites on the previous year (2021 = 1,077,884 certificates and 1,447,080 sites).

- ISO 14001:2015 (ISO, 2015b) had 529,853 certificates covering 744,428 sites issued, representing a 26% increase in certificates, and a 22% increase in sites on the previous year (2021 = 420,433 certificates and 610,924 sites).
- ISO 45001:2018 (ISO, 2018a) had 397,339 certificates covering 512,069 sites issued, representing a 35% increase in certificates, and a 38% increase in sites on the previous year (2021 = 294,420 certificates and 369,897 sites).

**New data is published in September each year for the preceding year, and readers are encouraged to track the adoption of certifications in future years on the ISO website www.iso.org*

Me? I anticipate continued growth in certified (and non-certified) adoption, and we have stable standards for the foreseeable future:

- ISO 9001:2015 was reviewed and confirmed in 2021 and remains the current standard. It will not be reviewed again before 2026.
- ISO 14001:2015 was also reviewed and confirmed in 2021 and remains the current standard. It will not be reviewed again before 2026.
- ISO 45001:2018 was reviewed and confirmed in 2022 and remains the current standard. It will not be reviewed again before 2027.

I have prepared new illustrations which summarize ISO 9001:2015, ISO 14001:2015, and ISO 45001:2018 as they are—see Figures P.1, P.2, and P.3.

In Chapter 2, I'll also share a new illustration of an integrated HSEQ management system model based upon this 'holy trinity' of management standards. My thanks to Rocky Vega for her assistance with these illustrations.

Proper and meaningful implementation of HSEQ management system standards (MSS)—whether they are ISO standards, your sector's, your client's, or your own organization's standards—require competent auditing if the benefits are to be maximised. Clause 9.2 of all ISO MSS mandate an internal audit process. That's where this book has fitted in since its first edition in 2007. Back then, it intentionally aligned with the first ISO guidance standard for management system auditing, ISO 19011:2002, and it continues to align with the latest edition of that standard today.

ISO 19011:2018 (ISO, 2018b) is the current version, and this in the final stages of a planned review commenced in 2022. At the time of writing, it was at international harmonized stage code 90.60 *Completion of Main Action/Close of Review* (ISO, 2023). The results of this review are expected shortly, and I anticipate nil or minimal changes. It will not be reviewed again before 2028.

In the preface to the first edition (which you'll find herein on pages xlvii–xlviii), I asked why anybody might write a book about auditing. I answered saying that we live in a world where enterprises of all types, sizes, and sectors must be able to prove to both those inside and outside their organizations that they are being managed in a way which is consistently acceptable to all of society. This remains true today.

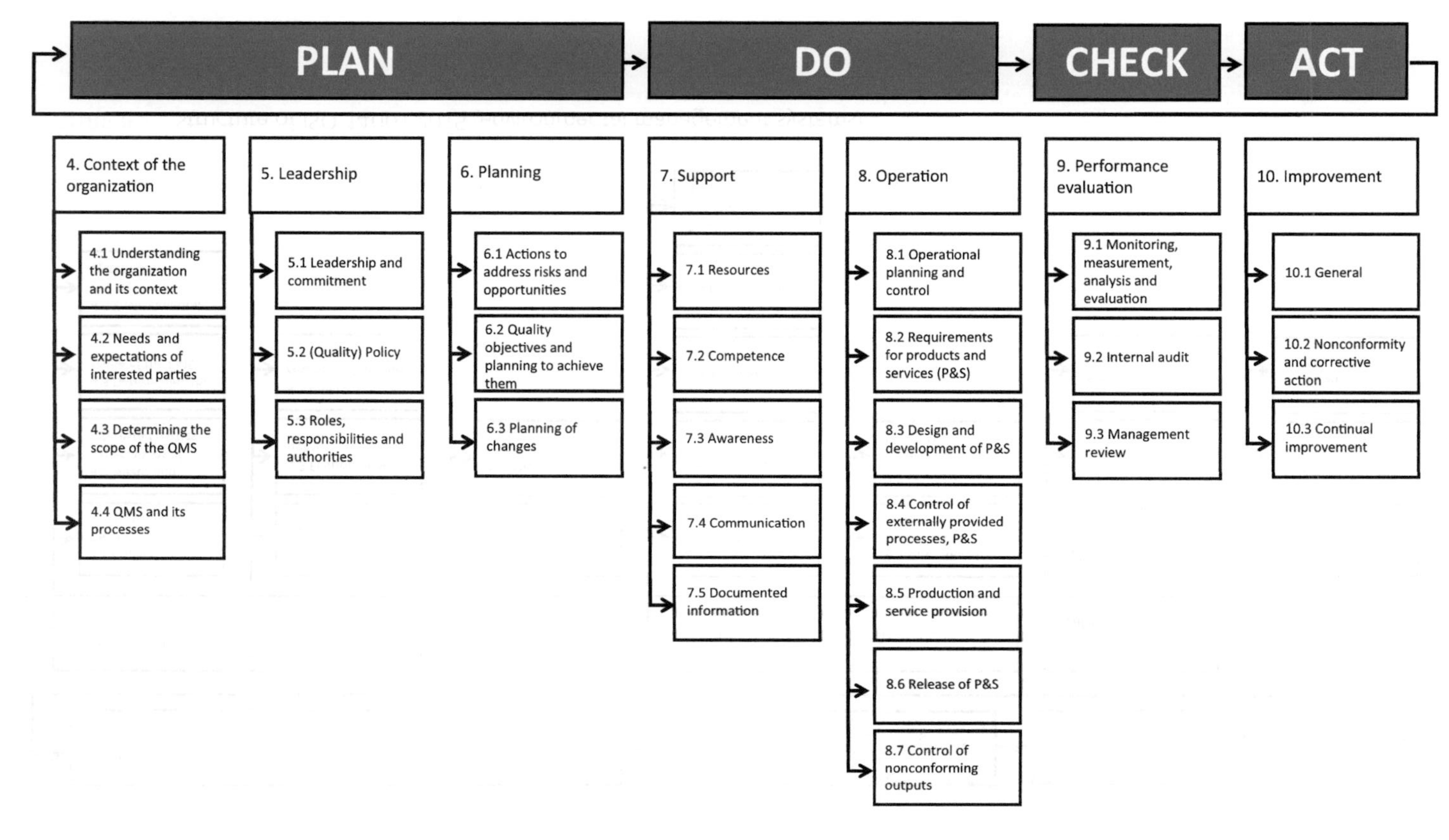

FIGURE P.1 Structure of ISO 9001:2015 (quality management systems).

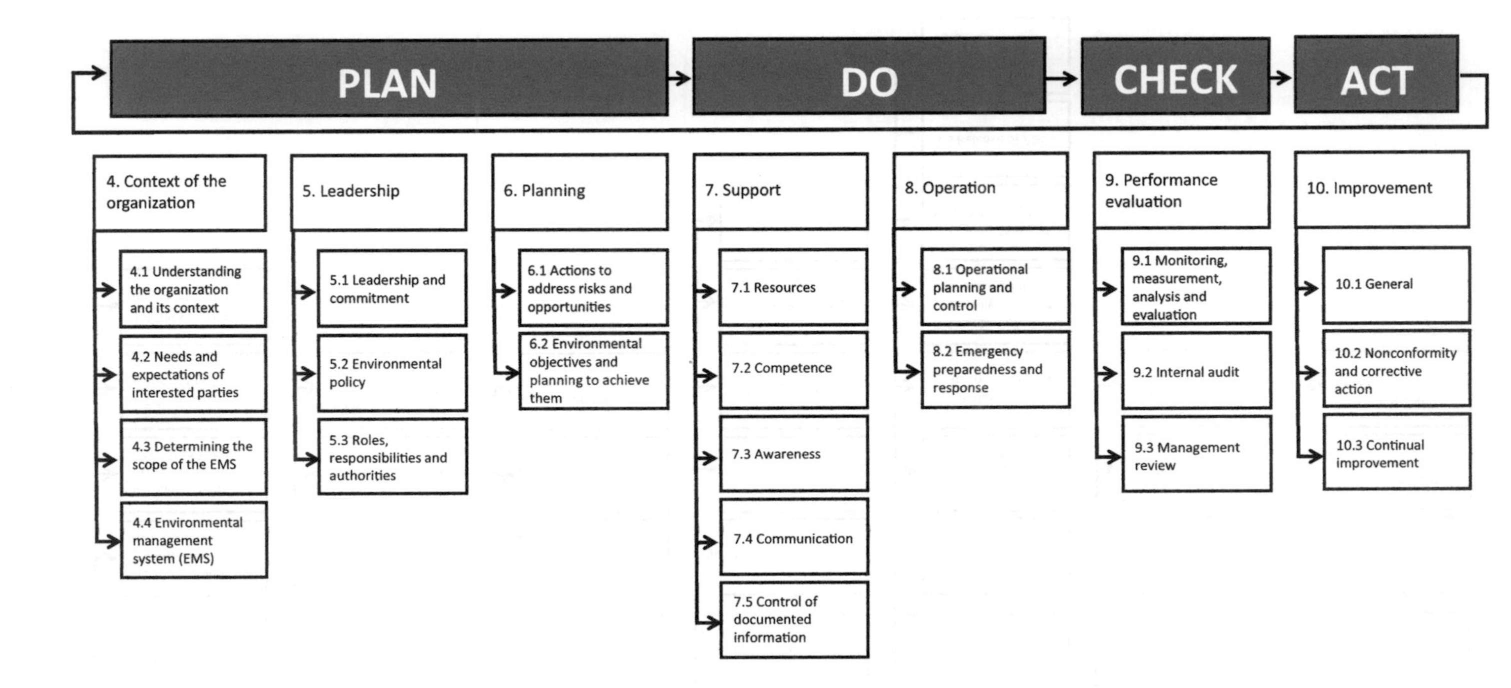

FIGURE P.2 Structure of ISO 14001:2015 (environmental management systems).

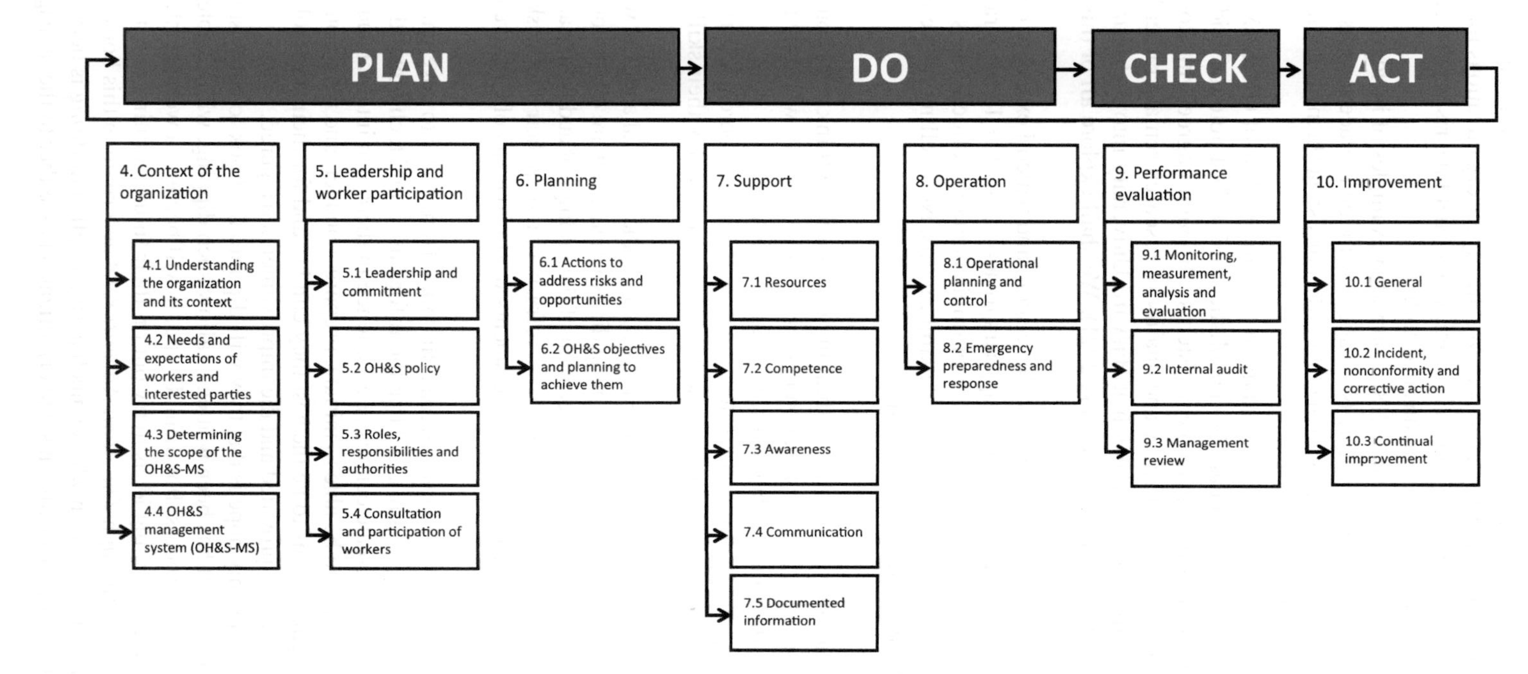

FIGURE P.3 Structure of ISO 45001:2018 (occupational health and safety management systems).

In the preface to the second edition (2013), I expanded that thinking as an expectation of improved, 'better performance'. We all expect organizations that introduce hazards into our global and local societies, and that take risks to be successful, to properly control them. That also remains true today.

In the preface to the third edition (2018), I provided examples of incident, data, and human rights losses. I had done so before (see page xliv in the Preface to the second edition, and page xlviii in the Preface to the first). My consistent theme remains that it is time to manage and audit HSEQ better.

Since the first edition of this book and the systematic review of OH&S-MS between 1887–2004 by Robson et al. (both published in 2007), I note the evidence of the improved effectiveness of management systems, although I agree with Robson et al. (ibid.) and others that the quantity and quality of research remain rather limited. I am encouraged by the ongoing research at Harvard University into the *treatment effect* of adopting MSS (Viswanathan et al., 2021). We shall explore all of this in Chapter 2.

Increased adoption of HSEQ-MS as we have seen (and which I expect to continue as top-tier organizations push their tier 1 suppliers to adopt, and tier 1s encourage tier 2s, etc.) may count for little if good standards of implementation effectiveness cannot be achieved. Better auditing can initiate and maintain this. Now selling in five-figure quantities around the world, this book will:

- support senior managers who are thinking (or should be thinking) about setting up an internal audit function in their organization, or who may be questioning the value of their *existing* audit function;
- shine a bright light onto a field-tested, practice-hardened approach to risk-based auditing for those who may like to develop and deploy their skills as an internal auditor in the future; and
- improve the effectiveness of seasoned HSEQ and other internal auditors, who may already have management systems audit experience but are disillusioned with the style, process, results, or reception of the findings of audits they are asked to conduct. Or they may wish to improve, refresh, or top up their current skills—this new edition includes learning assessments worth 30 hours of CPD credit.

My hope is that the first two groups will read this book from cover to cover (and watch my *Microlearning*™ presentations), and that the information and techniques they learn will inspire them to create centres of excellence in their own internal auditing departments. I want them to be able to initiate, prepare, conduct, and report on audits which help their organizations to be the best they can be, and for their stakeholders to truly esteem the assurance provided and the improvements triggered.

For the third group, my hope is that they will dip into the book to contrast with and add to their practice. It has been written to allow such dipping, with Chapter 5 summarizing the whole process. For them, I hope, it will become a well-thumbed source, with useful tips and challenging ideas to try out on their future auditing assignments.

Along our journey through *The Audit Adventure*™ described in this book, you will have the opportunity to reflect on why so much activity called *auditing* is being done today with so little benefit accruing in some organizations—either to the managers of the entities being audited or to those people who expect every entity to be run by superheroes and paragons of virtue.

I always look forward to building on these ideas and sharing new experiences in future editions. I also try to support those interested in management systems and the people I'll call *Audit Adventurers* through the book's companion website at https://routledgetextbooks.com/textbooks/_author/asbury/

There, you'll find a host of useful materials for your use, including the following:

- Microlearning™ presentations of each book chapter (each 8–14 minutes' duration)
- Learning assessments on each book chapter*
- *The Audit Adventure™* video tutorial
- Documents for use in your own audits
- Sixteen OH&S-MS Toolkits for ISO 45001:2018
- Example MSS frameworks
- Articles and papers of interest
- Websites and contacts of interest

* Many HSEQ and audit professional bodies require their members to participate in continuing professional development (CPD) to show they are keeping their knowledge up to date and developing new skills in a structured manner. These per-chapter learning assessments, presented as ten multiple-choice tests, provide you with a recognized CPD opportunity. Successful completion of each assessment will confer three hours of CPD—30 hours CPD for all ten of the chapter assessments. This CPD opportunity is formally recognized by IEMA as follows:

> *IEMA is pleased to endorse the CPD learning content in this book and see it as a valuable contribution to our members' ongoing demonstration of their personal development.*

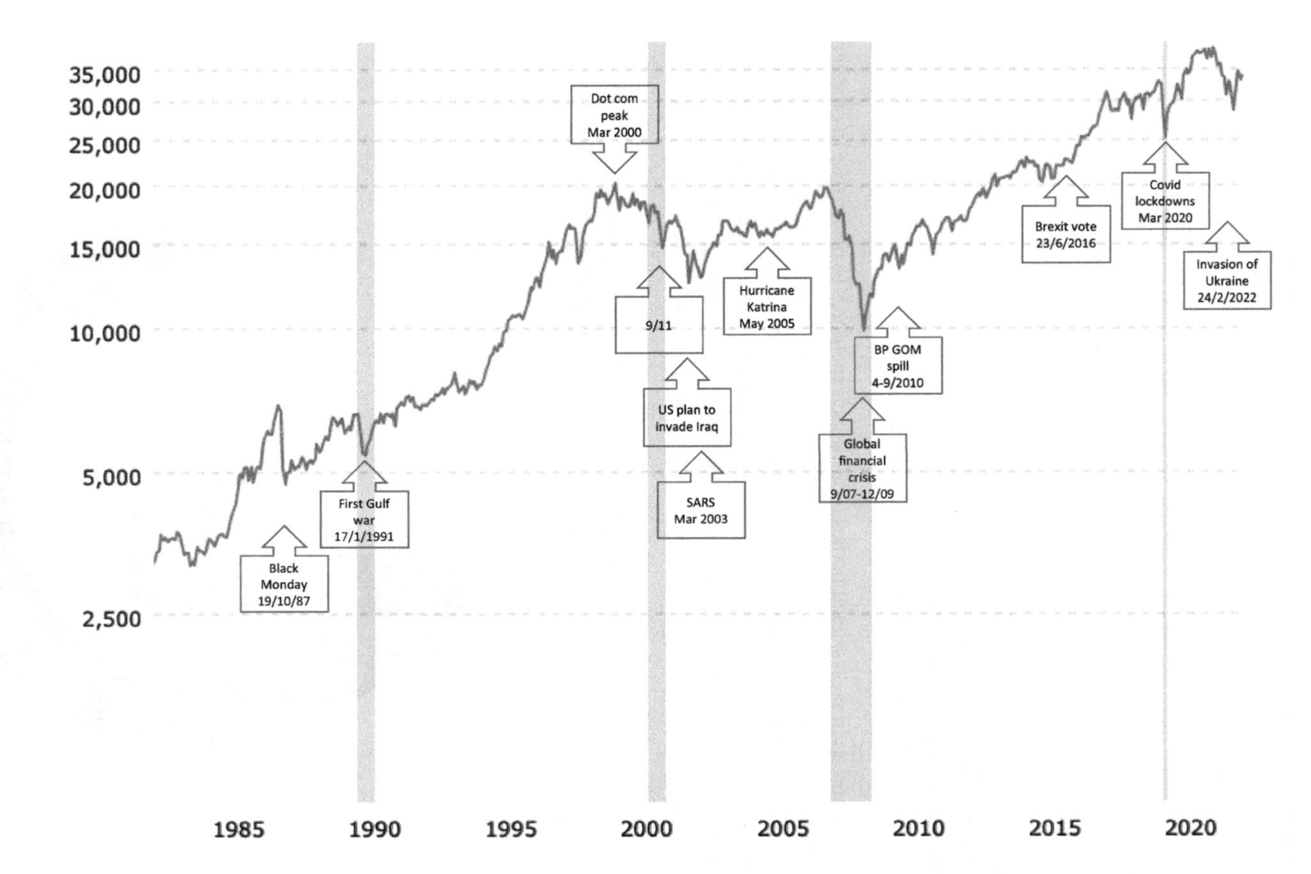

FIGURE P.4 Timeline of example world events shown along the Dow Jones Industrial Average (DJIA) from January 1983 to January 2023 (40 years). Graph data used with kind permission of Macrotrends LLC.

A guide to all the eBook+ and other online content is provided in Appendix 4.
You can also keep up to date with risk management news, views, and solutions by following me on X (Twitter) @Stephen_Asbury.
You'll also find me easily on LinkedIn, and I'll be pleased to accept your connection request.

As three times previously, I remain keen to share the ideas and experiences of auditors using the methodology presented in this book in future editions. Your comments, stories, tips, and ideas are welcomed and can be sent to me at stephen@stephenasbury.com. I promise to namecheck any that I use in future editions.

Global and national economic recovery will happen. It has before, as shown in Figure P.4. The timeline is the 40 years from 1983 to date, approximately one working lifetime. Recovery and growth will happen, and effective HSEQ/risk management, with effective risk-based auditing, can only hasten and support this.

Together we can, and we will, win the battle against ineffective auditing!

Preface to the Third Edition

> Every 15 seconds, somewhere in the world, one worker dies and another 153 have a work-related accident.
>
> —ILO, 2016

In just ten years, this book has become the bestselling book in the world on risk-based HSEQ-MS auditing. A good question might be *why?* It may be because over 15,000 people have attended the live *Asbury* auditing class and generally found the approach it commends to be both interesting and helpful to their practice. However, there are probably several other answers:

Firstly, it has kept up to date with the developments in management system standards—particularly those related to HSEQ. It charts the evolution of management system thinking from ancient China, through the work of Shewhart and Deming, and US defence standard MIL-Q-9858 in the 1950s to the *numbered standards* we know today—the trilogy of ISO 9001, ISO 14001, ISO 45001; and it considers other systems based on or influenced by these. The book's continued reference to the PDCA approach was subsequently adopted in 2012 by the International Organisation for Standardization (ISO) in its framework for management standards, Annex SL.

Secondly, it provides a straightforward, reliable, and repeatable approach for those who wish to adopt a risk-based auditing process in their organizations (as many have). *The Audit Adventure*™ method presented herein has tracked and mirrored the evolution of the guidance for auditing management systems: ISO 19011. When that standard was last published in 2011, there were eleven management system standards, but that number has since grown significantly to thirty-nine, with twelve others presently in development. Accordingly, a further revision to this standard is expected in mid-to-late 2018. This latest revision has been written with that in mind.

Thirdly, the book is a very practical source of helpful information, with over 50 case studies illustrating major points in the text, and dozens of tips learned from over 1000 HSEQ audits conducted by the author over the last 30 years.

Despite all the progress, we still kill people at work. The International Labour Organization (ILO, 2016) says that every 15 seconds, somewhere in the world, one worker dies, and another 153 have a work-related accident. In each of the two earlier editions, I have provided a world map showing some examples of catastrophic HSEQ-related losses since the last edition. I could have done the same again in this edition—the Savar building collapse in Bangladesh in 2013 (1129 killed), the Lac Megantic derailment in Canada in 2013 (47 killed and thirty buildings destroyed), the Soma mine disaster in Turkey (301 miners killed), the Tianjin port explosions in China in 2015

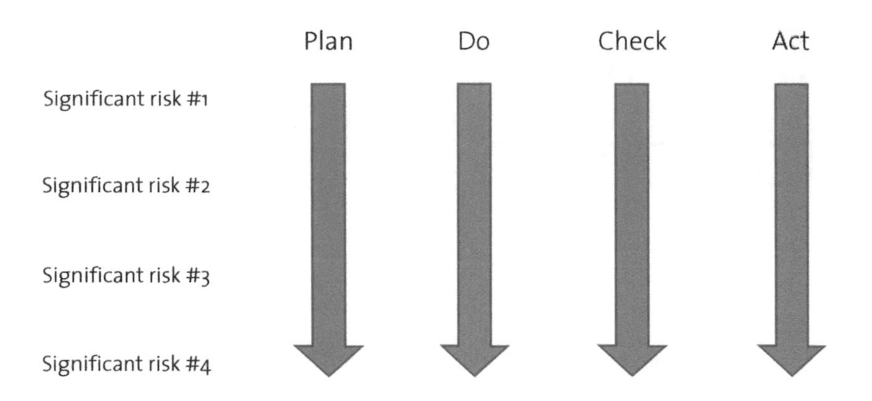

FIGURE P.5 Silos: How management systems are sometimes implemented and audited.

(173 killed), and the Gazipur boiler explosion in Bangladesh in 2016, which killed twenty-three. And I could have added some of the newer types of loss, such as the data breach at Equifax in 2017 (loss of personal information for 134 million customers) or those relating to workers' rights in the *gig economy*, for example Uber (2016). On a different writing day, I could have selected a dozen different examples for you to think about and, if you wished, to research further.

The bottom line remains that we must learn how to manage and audit HSEQ better. Figure P.5 shows a major part of the challenge we are to overcome. I worry that too many management systems seem to be more about creating paperwork than actually doing anything to mitigate risks. Expressed simply, too many organizations prepare and file job descriptions (and audit these job descriptions) or fill in and file risk assessment forms (and audit these risk assessments) in a silo-type (vertical) approach, rather than using management systems as they are intended *through* a (horizontal) continuity of planning, doing, checking, and acting to improve (PDCA). You'll be delighted to know that this book provides you and your organization with a highly effective and highly implementable solution.

This new edition of this book has been structured to be of interest to three broad sets of readers:

1 Senior managers who are thinking (or should be thinking) about setting up an internal audit function in their organization, or who may be questioning the value of their existing internal audit function.
2 Those who might like to develop their skills as an internal auditor in the future.
3 Seasoned HSEQ and other internal auditors who may already have risk-based or management system auditing experience—perhaps they are disillusioned with the style, process, and reception of the audits they are presently being asked to do—and wish to improve, refresh, or top up their skills.

Preface to the Second Edition

In the preface to the first edition of this book, back in 2007, I asked you to ponder why anyone might wish to write a book about auditing. I believe the answer to this question remains as straightforward now as it was back then. The expectation of internal and external stakeholders is still that organizations should be able to demonstrate acceptable standards of risk management. The pressure for this has, if anything, increased in the last six years—we all expect and demand better performance.

Let's be clear what we mean by *better performance* here. We expect organizations that introduce hazards into our global and local societies, and that take risks in order to be successful, to properly control them. The greater those risks, the more control we reasonably expect. Lawmakers call this approach to risks ALARP—as low as reasonably practicable. But we can express this more simply. We're happy to pay a fair price for the goods and services, and we don't like it when organizations kill their workers, their customers, or the public. We don't want them to pollute our lungs or the environment. Or lose our personal data. Or blow up the city. Employees expect to keep their jobs, get paid, and build their skills and careers. Suppliers wish to prosper over the years with their partners. And investors want their money back, with growth in their capital.

We expect senior managers to keep an all-seeing eye on their external environment, set their business objectives in the context of that environment, and then deal with the significant risks—the *Big Rocks*—that might impact on those objectives and the requirements of society at large. And, for all of us to be assured of management's proper governance and probity, we expect them to initiate independent audits of the management systems at agreed intervals, maintaining control where it works and taking corrective or improvement actions where these are found to be necessary.

Taken together, we call this *operations integrity* (OI). Operations integrity addresses all aspects of an organization's business, including security, which can impact its safety, health, or environmental performance. And, despite all the auditing done, there is a critical failure somewhere in the world almost every day, almost every week. Some examples are shown in Figure P.6, but this is by no means a definitive list. On a different writing day, I could have selected a dozen different examples for you to think about and, if you wished, to research further.

Facilities and assets that have sustained losses have invariably been audited. I have noticed that one of the common conclusions of many disaster enquiries is that the auditing of the management systems was defective. The problem with many audits is that they tend to be conducted at too low a level, with low-level understanding of the business and its context and low-level reporting of the findings—trivial matters unnecessarily

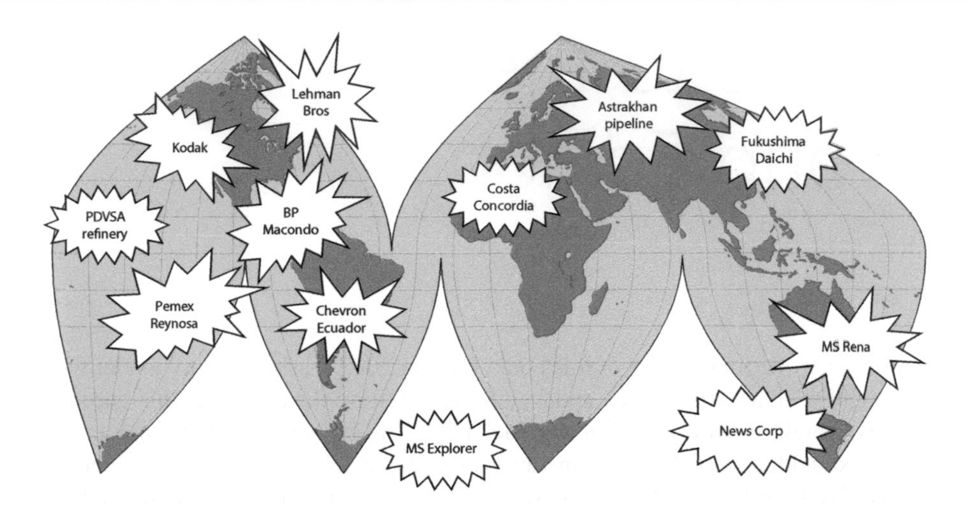

FIGURE P.6 A dozen examples of corporate failings, 2007 to date.

escalated, or significant matters reported out of context or lost amongst trivia. Too many audits are historically focused, on observed hazards, instead of future-focused, on proper control of critical operations. It's all too easy for an audit team to take the low-hanging fruit of personal protective equipment not being worn or training records being misfiled without focusing on what really matters to the organization and to society. And it's much easier for an audit team to report good news to management than bad. And if a management team sees auditing only as a means of providing themselves with assurance that things are as they should be, then this is what they are likely to be given. As Hopkins (2009) says, leaders who want to pinpoint unrecognized problems that may be lurking below the surface need to avoid any suggestion that they are asking for assurance; they need to be suspicious of audit reports that suggest all is well.

> When we asked senior management why they didn't know about many of the failings uncovered by the enquiry, one of them said, 'I knew everything was alright because I never got any reports of things being wrong'. In my experience, there is always news on safety and some of it will be bad news. Continuous good news—you worry.
>
> From a video lecture on the 1988 Piper Alpha disaster

Better auditors, with better auditing methodologies, challenge asset managers to demonstrate that their operations integrity management systems (OIMS) are working as intended. They provide assurance where these systems work, and ring alert bells when they do not. They regard problems as an indication of a defect in the management system. Auditing at its best uncovers both particular issues and the system defects which have allowed these issues to occur.

The second edition of this book set out to show how Health and Safety, Environment, and Quality (HSEQ) and other internal auditors can help management to avoid such failures and losses of integrity. It updates the first edition with reference to the latest international HSEQ and auditing standards, and provides over twenty new case studies and lots of new tips for effective auditing practice.

The work of writers and auditors of ISO management system standards (MSS), as well as those responsible for their implementation, will be significantly changed because of the publication of Annex SL (ISO, 2012a, previously 'ISO Guide 83') of the *Consolidated ISO Supplement of the ISO/IEC Directives*. The ISO has produced this annex with the objective of delivering consistent and compatible management system standards. In future, all new MSS will have the same overall 'look and feel' thanks to Annex SL. Current MSS will migrate to the new format during their next revision. This includes ISO 14001, which is presently being revised and is due for publication by 2015. The migration has, however, already started. ISO 22301:2012 was developed using a draft version of Annex SL, and ISO 27001 has been produced using the published version. The ISO 9001 requirements document has also started its revision cycle and will be developed using Annex SL. There is much change in the air, and this book addresses the need for its readers to be better informed.

Preface to the First Edition

Why, you might ask, would anybody wish to write a book about auditing? The answer is very simple. Today, we live in a world where enterprises of all types, sizes, and sectors must be able to prove to those both inside and outside their organizations that they are being managed in a way which is consistently acceptable to all of society.

In the main, enterprises have lost people's trust to carry out their activities relying purely on their owners', directors', or managers' word that everything is being done properly. Even when directors explain in great detail what their policies, guidelines, and standards are with regard to how they intend to carry out their activities, that may still not be good enough.

In the last ten to fifteen years, people outside—and often inside—all types of organizations have demanded demonstrable proof as to the extent to which enterprises are meeting their self-proclaimed standards. And over the same period, many groups claiming to represent interested people in society have persuaded enterprises to involve or engage them. There is no turning back.

The level of management performance needed to ensure that entities stand a chance of meeting these continually increasing levels of expectation is competing head-on with the level of management performance needed to create commercial success.

I believe that the conundrum of how to get the same individuals to achieve both goals simultaneously can be solved if entities create a function to carry out effective management system auditing.

'Corporate governance' and 'social responsibility' are the expressions used today to describe the governmental, legal, and societal reaction to these simultaneous phenomena of lack of trust and huge expectation.

There is a major challenge to agreeing a global approach because historically the US attitude to regulation has adopted a 'rule-book mentality', which means that when anything contravenes the prescribed letter of the law, organizations and officers are sued and possibly prosecuted. Meanwhile the UK and many international standards of accounting, auditing, ethics, and corporate governance essentially are 'principles based', which means that you really have to think about the *spirit* of the standard or rule—what is it expecting to achieve?—rather than just 'ticking boxes' as soon as you can show compliance with the 'letter' of the standard or rule.

The accountancy profession, particularly those elements authorized to carry out statutory audits, was affected for many years by what is often referred to as *the expectation gap*. This *gap* was the difference between the layman's perception of the type and extent of work that went into an audit and the actual work which was required by law. A statutory audit results in the auditor giving either an unqualified audit opinion so that the reader can impute that the entity's financial statements reflect

Ahold (Netherlands)
Aural Mining (Romania)
Barings Bank (UK/Singapore)
BCCI (UK/India)
Buncefield oil terminal (UK)
Cable & Wireless (Hong Kong)
Chernobyl reactor (Ukraine)
Esso/Exxon Longford gas plant (Australia)
Occidental Caledonia, Piper Alpha (UK)
Parmalat (Italy)
Resona Bank (Japan)
Shell Brent Spar (UK)
Shell Reserves (UK/International)
Union Carbide, Bhopal (India)

FIGURE P.7 Major non-US business control failings.

a *true and fair view*, or, on the contrary, an audit opinion that indicates the extent to which the statements are not true or not fair. It was as recent as 1990 in the UK in *Caparo Industries vs Dickman* that external statutory auditors were reminded by the justice system that they needed to manage this expectation gap rather better than before, because they owed a duty of care to other parties who may suffer an economic loss by relying upon their statutory audit opinion.

The resultant debate about the extent of external auditors' legal liability has been going on ever since, with a variety of ideas being put forward for mitigation in many jurisdictions across the world. A significant recent development has occurred in the USA with the creation of the Public Company Accounting Oversight Board (PCAOB) as the guardian angel of investors in US securities markets and charged with the responsibility to ensure that public company financial statements are audited according to the highest standards of quality, independence, and ethics.

The PCAOB was established by legislation known as the Sarbanes Oxley Act, which came into effect on 30 July 2002 as a response to the massive lack of trust and loss of confidence in the US capital markets caused by a litany of major corporate failures—immortalized by Enron and its auditor Arthur Andersen, Tyco, WorldCom-MCI, HealthSouth, Global Crossing, and Adelphia.

Many non-US regulatory bodies were already in place to protect investors, improve audit quality, and ensure effective and efficient regulation of firms. However, business control failings in entities of all types and sizes have occurred throughout the world—in Europe, Japan, Australia, Asia, Africa, South America, and Russia. Some examples are shown in Figure P.7. They will continue to happen because of the failure of some senior managers to either believe in the benefits of, or put sufficient priority on, implementing an effective business control framework or personally defer to them in their own behaviours and actions.

Corporate failure of varying kinds affects varying groups of stakeholders. Some of the most visible are major technical failures when people are killed and communities knocked sideways—such as the accidents in the North Sea (such as *Piper Alpha*), at the Longford gas plant in Australia, at BP Texas City in the USA, on the railways, and at Buncefield in the UK.

Acknowledgements

Each time I write a book, I am granted this pleasant opportunity to thank the great and the good people who have helped me, have influenced my thinking, or guided my career. Many of the names which follow I have thanked publicly before, while some have been helpful more recently. But really, I appreciate all your support.

My wife Susan remains the backbone of my career—'behind every successful businessman is an exhausted woman' is her favourite quote. I cannot do what I do without her.

I'm so proud of my daughter Kimberley for her academic, professional, and personal achievements. She makes me so proud to be her dad. With her husband, Alex, she also made me a grandfather for the first time in 2021. This book is dedicated to Faye.

There has been a supportive group around me professionally for much of my career. My thanks are due always to Michael Farmer, John Fawkes, Dr Alex Grieve, Hazel Harvey, Dr Adrian Hearle, Dr Andy Cope, the late Dr Arthur Rothwell, and Andrew Ure. Thank you to Dr Alan Page and Professor Hemda Garelick who guided and mentored me through my doctorate.

Several other people have helped this book with contributions large and small. My thanks to Judy Cahill at The W. Edwards Deming Institute, Rocky Vega who helped with my ISO illustrations, Macrotrends for the use of their data, and James Pomeroy for reopening an interesting conversation about plausible deniability. I renew my thanks to the family of the late Paul Richardson for *The Audit Adventure*™ illustrations which beautifully complement my text. My editorial and production team at Taylor & Francis and Apex CoVantage—especially my new editor James Hobbs, Kirsty Hardwick, Kari Budyk, and Balaji Karuppanan—who brought all the parts together.

David Wilkie, Ian Kempson, and Wayne Hemmings have been especially helpful in recent years. Thank you, too, to each of my training course participants from around the UK and the world for showing up, listening to my messages, and making what we might otherwise call 'work' such a great pleasure.

Finally, I express my love for my late parents, Alan and Betty. You crop up in my thoughts every day. I hope you're still proud of me.

Introduction

It remains true that too many audits result in an audible sigh of relief, or a scream of frustration, from the auditee who has been told that they have *passed* or *failed* the process. Just like that, a binary opinion has been dispensed that, in the worst cases, derails careers. Or the weekend with friends and family is cancelled while some *fix* is thrown together and implemented. Or maybe the auditee survives. Until the next cycle of audits comes around. They might even get to enjoy their weekend . . .

Many readers may believe that this is an outdated perception, but regrettably it is not. The problem is growing, not shrinking. Every year, technological advances make processes more complicated, and every year, management reacts to the need to be able to demonstrate compliance with ever-increasing governance requirements, such as changes to legislation or to the small print of a new swathe of head office or contractual provisions, or by doing more and more compliance auditing. Not long ago, I was invited to view a prototype *flying car*, shown in Figure I.1. Technology will continue to change organizations, and we must think ahead to be prepared for this. As you may imagine, the stakeholders on that project, including its regulators, will demand all manner of compliance checks before it flies over your house.

But hang on a minute. Why do we need to do all this compliance auditing? Put simply, it is because most directors, managers, and supervisors are overburdened just keeping their boat afloat and heading in roughly the right direction. So auditors are used as a safety net or a punchbag, in the sure knowledge that something will be overlooked somewhere. And then at least, we'll have someone to blame.

As a result, literally millions of hours of audits are carried out just in case somebody or something does not do their job properly. Forms will be filled in. Boxes will be ticked. Audits are seen as a necessary evil, because the audit plans say that we need to keep records to show that absolutely everything has been checked.

And so audit is unwelcome. It is dumbed down and rushed to get it out of the way so we can get on with the *real work* of making sausages, driving trucks, building *flying cars*, or whatever else being the reason we exist as an organization. This condescending view of the value of auditing has a knock-on effect, in that its effectiveness is seen more in terms of *doing the audit* rather than asking and answering difficult questions about the arrangements for preventing losses and verifying proper preparedness to recover from an incident should one occur.

My belief, shared with you in these pages, is that internal audit can and should engage the brightest of people in independently reviewing the application of selected risk-critical processes, activities, and services (including those in the design, assembly, and operation of flying cars) within an organization from a variety of viewpoints. There are other people and processes for checking compliance with the rules, and we should let them do their job.

Internal audit is thus perfectly placed to reinforce or challenge the way an organization is being managed; and whatever the type of audit, commercial or technical, the results should demand the respect and attention of senior management.

DOI: 10.1201/9781003364849-1

FIGURE I.1 Prototype *flying car.*

My book works through each of the main steps established by ISO 19011:2018 *Guidelines for auditing management systems* that enable internal auditors to deliver the exceptional quality of audit that will make a difference to a department, a site, the organization, and to society as a whole. This approach works for auditing all structured means of control, including management systems, and those which are HSEQ-related. And it gives auditors the opportunity to contribute to the success of their organization, to help to make them great.

> Keep away from people who try to belittle your ambitions. Small people always do that, but the really great make you feel that you, too, can become great.
>
> Mark Twain

THE AUDIT ADVENTURE™

The steps necessary for delivery of this exceptional quality of audit are encapsulated within a learning model called *The Audit Adventure*™. The origination of this model (Asbury, 2005) was part of preparations for the first edition of this book. Since 2007, it has been thoroughly tuned, and with the benefit of over sixteen years' refinement, it has dramatically improved the understanding of participants in auditor training courses as well as of people reading and using the book.

The Audit Adventure™ takes the approach to auditing to a high(er) level and makes it more future-focused by replicating very closely the movement dynamics and the thoughts of an auditor following and applying the internal audit approach and methodology currently being used (and taught) by the author's company and international clients to guide them through the five major auditing steps described in ISO 19011:2018:

- Initiating an audit culture (Chapter 3)
- Preparing audit activities (Chapter 6)
- Conducting the audit (Chapter 7)
- Concluding the audit (Chapter 9)
- Writing the audit report and following up (Chapter 10)

Figure I.2 shows Poldhu beach in Cornwall, a beautiful National Trust property located in the south-west of the UK. Figure I.3 illustrates a flattened and simplified Poldhu beach, representing the dynamics of *The Audit Adventure*™, with the hilltop car park and the scenic and interesting route down to the beach on the left, the granular detail of the sandy beach in the centre, and Mrs Rimington's cream tea emporium located on the opposite peak. The elevation above sea level at each point of the journey represents the level of detail involved in the audit: at the start, a high-level view of the business environment

FIGURE I.2 The beautiful beach and cove, Poldhu, Cornwall, UK.

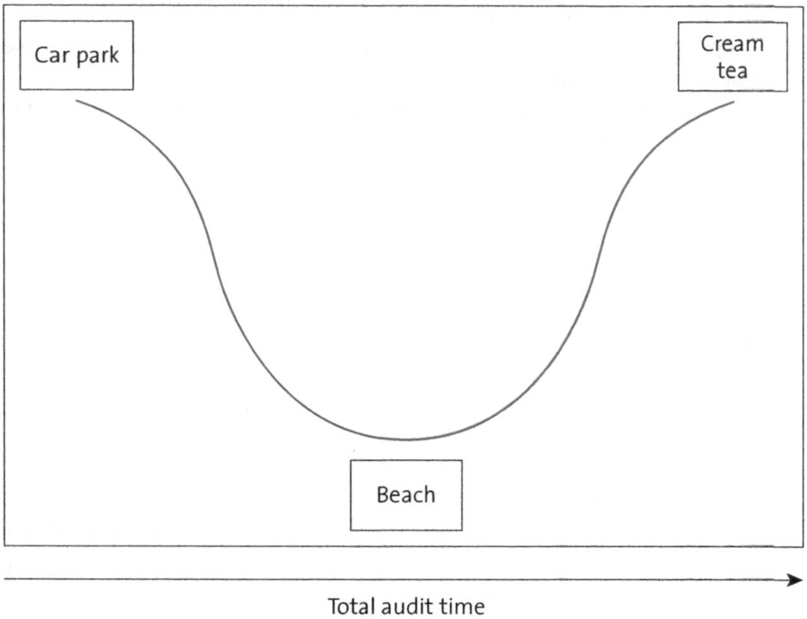

Car park

Cream tea

Beach

Total audit time

FIGURE I.3 *The Audit Adventure*™: A flattened and simplified dynamic.

provides the *Context* (ISO MSS clause 4) for the organization to be audited. You can see all the way to the horizon from here. It is at this level that senior management operates, sets corporate objectives, and communicates its intentions to stakeholders. It is right, then, that auditors commence their work in the same elevated place.

> The ship cannot be steered from the engine room. Navigation must come from the bridge, where there is vision.
>
> Dr Stephen Asbury

Operations integrity (OI) controls are often detailed, with (rightly) multiple layers of protection around the most critical and most valuable assets. A controlled descent to the beach, using the steps or one of the windy paths, represents the auditors' selection of significant activities, processes, or services for the work plan of a single audit. They are guided by the route they choose towards the detail of the controls necessary to achieve the barriers, or layers, of control required. As our *Audit Adventurers* find beach flora, fauna, and mineralia large and small, they seek and verify with *Gemba* attitude* the necessary elements of effective control through their detailed and probing questions.

**Gemba* attitude—going to the source to check the facts for yourself, to arrive at informed decisions (*gemba* means 'place' in Japanese).

The left-to-right arrow at the bottom of Figure I.3 follows the progress and stages of *The Audit Adventure*™. The steepness of the pathway to the beach, with the anticipation of what's there to be seen, creates a feeling of time flying by (Figure I.4), as the inevitable acceleration of the body due to the forces of gravity reminds the auditor that there is no going back.

FIGURE I.4 Time: It flies by . . .

Exhausted yet exhilarated from their explorations, the *Audit Adventurers* are extremely motivated to head for the best tea shop in Cornwall, owned by Mrs Rimington. This, of course, means making the ascent up the other side, away from the beach and its granular detail and towards the opposite peak. This is where management resides, and the auditors must speak once again the language of their auditee. It's high level, and the view from the tea shop is quite breathtaking.

> On a clear day, you can see France from up here!
>
> —Mrs Rimington, café proprietor

As they take their final exhausted steps and wait for tea and cakes to be served, the *Audit Adventurers* deliver their high-level, future-focused audit report to an audience that welcomes and values it.

A-FACTORS

To help you cut useful and memorable steps into the trails and pathways of *The Audit Adventure*™, you will find ninety *A-Factors* spread throughout the chapters of this book. They can be referred to as *Asbury* or *Auditing Factors* as you wish.

FIGURE I.5 The Audit Adventure™.

Along with dozens of case studies and scores of practical tips, these comprise your secret weapons and golden nuggets of information. They crystallize or summarize the major points for auditors from the main text. These *A-Factors* are also gathered in Appendix 1. Many readers of the earlier editions said that these were *worth the price of the book on their own.*

CASE STUDIES

You will find over 60 case studies, including five new ones, in this fourth edition. Through these cases, I use my experiences from real-life audits, conducted by audit teams I have led over the last thirty-five years, to powerfully illustrate to you some of the points made in the text as well as providing examples of auditing excellence.

TIPS

Finally, the book provides a generous serving of tips, tools, and techniques on how to conduct and deliver audits successfully. Also included are some tips and ideas on what *not* to do, including my thoughts on the instance from 1997 when I was escorted from an auditee's site by security.

To conclude this introduction, I'd like you to take a look at Figure I.5; I look forward, in the following pages, to taking you on your first *Audit Adventure™*.

Context of the Organization

1

> The organization shall determine external and internal issues that are relevant to its purpose and that affect its ability to achieve the intended outcomes of its OH&S management system [including] the relevant needs and expectations of workers and other interested parties [or stakeholders].
>
> —From ISO 45001, Clause 4, Context of the Organization

> The value of a company can be derived from adding the value of all future dividends written down to net present value. Therefore, a reasoned view of the future is essential.
>
> —Dr Stephen Asbury

INTRODUCTION

> For a summary of this chapter, you can listen to the author's Microlearning™ presentation, which you'll find on the book's Companion Website. Then, delve deeper by reading on for further details on the Companion Website. (See Appendix 4.)

Sometimes, when I speak at a conference or at another meeting of HSEQ or audit professionals, a participant tells me that they cannot get any *face time* with their top management. That they're not invited to board meetings. They feel excluded from high-level decision-making, etc. It frustrates them that they're thus compelled to work at an operational, day-to-day level. And they want to know why . . .

There might be many answers. I don't know them all, or your directors personally, but there are some things I DO know. Anyone who can help an organization to achieve its corporate objectives is VERY welcome in the boardroom. That's who gets hired. Lord Sugar hires those people (on the BBC TV show *The Apprentice*). And he 'fires' the others.

These days, if you're an HSEQ manager, it is no longer acceptable (if it ever was?) to just drone on about work procedures, incidents, lost time, and PPE. As well as being a subject matter expert, you MUST be able to speak the language of the boardroom (and subsequently deliver your strategy). And now ISO Annex SL (ISO, 2012a)—which you'll read about in this book—has brought understanding the organization and its context onto centre stage as regards designing, implementing, and auditing any management system (aka structured means of control) including ISO 9001:2015, ISO 14001:2015, and

DOI: 10.1201/9781003364849-2

ISO 45001:2018. To be successful with your H&S/E/Q management systems, you simply MUST understand the business environment—your organization's context—as well as the specialist HSEQ stuff. And so this is where we will begin our learning adventure.

Chapter 1 provides your condensed MBA and economics degree(!). It explains the critical factors of the business environment (*Context*) that will help you, the reader, to better understand the 'stuff' that your directors and board members must contend with. Obviously, I cannot cover the whole world in the 43 pages of Chapter 1, but following my approach and thinking will assist you to parallel my process in your own county and country. After building this understanding of the context of your organization, you'll be ready for Chapter 2, which introduces you to the evolution, present, and future of management system standards (MSS). Then Chapters 3–10 will take you on a journey—from start to finish—on how to conduct a risk-based management system audit that I predict will have you warmly welcomed into your organization's boardroom. As your organization transforms its approach to using HSEQ (and other) MSS, you too will be transformed. Let's get started . . .

CONSTANT WHITE WATER

It makes me smile wryly when I hear that it will be easier to do 'x', and we'll have more time to do 'y' when *we have concluded the changes we're going through at the moment.* I don't expect the rate of change anywhere to let up anytime soon. It reminds me of a river flowing downstream—smooth running, with occasional rocks, weirs, and rapids. Except the rapids are getting closer and closer. I think we should brace for constant white water.

Yes, times are a-changing, and change is speeding up, not slowing down. We must get used to it. Just a few years ago, when we made a purchase, we had the latest technology in our hands. Now we're holding obsolete stock for much of the time, as a new model just launched.

Who could have foreseen the changes in the world in the first twenty-odd years of the twenty-first century? The growth in influence, trade, yet distrust of China. Angst in the Eurozone, as the PIIGS (Portuguese, Irish, Italian, Greek, and Spanish) economies were hammered by the twin evils of underperformance and crippling public debt. The election and rejection of Trump. Global lockdowns to prevent the spread of SARS-CoV-2 (Covid-19). Quantitative easing. War in Europe. Skyrocketing interest rates, fuel prices, and the costs of living. Uncontrolled illegal migration. And for UK readers, Brexit (a portmanteau of 'British exit' from the European Union).

On 23 June 2016, the UK voted to leave the European Union (51.89% for and 48.11% against on a 72.2% voter turnout). It had been a member since January 1973. Following the resignation of Prime Minister David Cameron, the new British government led by Theresa May notified the EU of its intentions to withdraw on 29 March 2017, and withdrawal was scheduled for 29 March 2019. In the June 2017 general election, the Conservative party lost its majority, and political deadlock led to three extensions of the UK's *article 50* process.

The deadlock was resolved after the December 2019 general election in which new Prime Minister Boris Johnson led his party to an overall majority of 80 seats. The British parliament ratified a withdrawal agreement, and the UK left the EU at 11pm on 31 January 2020. This began a transition period ending 31 December 2020.

Following withdrawal, EU laws and the European Court of Justice no longer have primacy over British laws (except in select areas related to Northern Ireland, that is, European Single Market re goods, and the EU Customs Union). The European Union (Withdrawal) Act 2018 retained relevant EU laws as domestic laws, which the UK can amend or repeal. The Retained EU Law (Revocation and Reform) Bill, known as *REUL*, received Royal Assent on 29 June 2023 to become an Act of Parliament. The Act revokes 587 instruments of REUL which were retained by the *Withdrawal Act* in 2018, all of which will be repealed on 31 December 2023 (Joyston-Bechal, 2022). It also repeals the principle of supremacy of EU law and facilitates domestic courts to depart from retained EU case law. It seems that deregulation is coming to the UK.

Other changes following Brexit include the following examples. We encourage readers to read further on themes that may affect them and/or their organizations.

- Economic effects on the UK and the EU27, including changes to trade/imports/exports/duties
- Border impacts between Northern Ireland and the Republic of Ireland
- Possibility of a second Scottish independence referendum (following the one held in 2014)
- British firms offshoring to the EU
- UK no longer a shareholder in the European Investment Bank
- UK left the Common Agricultural Policy (CAP)
- UK left the Common Fisheries Policy (CFP)
- UK left the European Common Aviation Area (ECAA)
- British courts no longer bound by decisions of the EU Court of Justice
- British universities lose EU research funding
- Freedom of movement ended, reducing immigration to the UK (including reduced numbers of employees in the labour market, for example, LGV drivers, and NHS workers)

Wherever in the world you live and work, it is the *Context* where your organizations' leaders are asked to conduct their operations and their business—on shifting sands, in the fog, and facing unprecedented international change and competition. They need a telescope to see what's coming over the horizon, and a microscope to understand it when it arrives. In these new markets, for example:

- In 2001, China produced just 1% of the world's solar cells, while the US made 27%.
- In 2011, China produced 56% of global production; the US, a mere 3%.
- In 2023, China will produce more than 80% of photo-voltaic (PV) solar cells.
- By 2025, according to the International Energy Agency (IEA), this will be 95%.

Michael Schuman (2012a), Prasad (2022)

> Over the past quarter-century, two large forces have swept the world: globalization and the information revolution. . . . They have given us consumer goods and services that were unimaginable at cheap prices. But these forces make it much easier to produce economic growth by using machines or workers in lower-wage countries. Hiring high-wage workers—that is, workers in Western countries—becomes a last resort.
>
> Fareed Zakaria (2012)

Ben Silberman (2013:31), CEO of online pinboard *Pinterest*, says he is looking forward to 'Apps that let strangers help each other . . . by using crowd-sourced data. I think we're going to see a lot more of that.' What used to take years to research is now available in a nanosecond.

Organizations do not exist or operate in a vacuum. Senior managers are compelled to understand their business environment and take proper account of it as they set their business objectives. Times are turbulent, and managers must look carefully if they are to see and to predict what's coming. Management in the final three quarters of the twenty-first century (and beyond) will be tough still.

UNDERSTANDING THE BUSINESS ENVIRONMENT

> I think the most important CEO task is defining the course that the business will take over the next five years or so years. You have to have the ability to see what the business environment might be like a long way out, not just over the coming months. You need to be able to set a broad direction, and also to take particular decisions along the way that make that broad direction unfold correctly.
>
> —Chris Corrigan, Australian businessman (1946–)

The business environment (aka *Context*) concerns the current and future characteristics of the place where the organization, its supply chain, and its competitors operate. It's where its customers consume their goods and services. And it's where society observes organizations' practices. It is important that we understand the business environment, as it sets the context for understanding the needs and expectations of interested parties (aka stakeholders) and determines the necessary *Scope* for any management system. Management systems such as ISO 9001:2015, ISO 14001:2015, ISO 45001:2018, and countless organizations' own MSS which often mirror these, were developed using Annex SL (see Chapter 2). Clause 4 requires this comprehensive understanding of the context. It is also the understanding of context that your senior management (should) have.

However you see it, organizations of just about every type—no matter their global (in)significance, or how they differ in detail—are concerned with transforming inputs to outputs. They do this within a business environment which affects—and in turn is affected by—the conduct of their activities. This environment is increasingly complex, dynamic, and volatile. As a participant on one of our training courses said recently, 'Change is the only constant these days.'

If we can begin to understand this environment, and its possible or likely effects upon our organizations, it will not only assist us in understanding the practice of 'business' in its entirety, it will also help auditors to understand their own roles amidst it.

As you will see, auditing is concerned with seeking to provide a level of confirmation (or assurance) that an organization has addressed reasonably foreseeable *significant* risks to its operations and to the achievement of its objectives with a suitable framework (or series of interconnected frameworks) for internal control. When the selected framework is consistently and reliably implemented, and when management intentionally maintains a healthy state of unease about its operations (rather than merely assuming that everything is well as they have not heard otherwise), success, however measured, often follows.

Along with providing assurance for the present, auditing should also provide a future-focused view of the suitability of this control framework(s) for the changing business environments of the months and years ahead. Information about the future is generally much more valuable to managers than information about the present or the past. You'll read more about this later.

As I said, this chapter takes us on a short journey through the key elements of understanding business environments and organizational contexts, before moving on to summarize the role of managers as regards positioning their organizations for success. Finally, it answers the question *What is risk*?

Tools for Understanding the Business Environment

PEST

As any management student or management textbook will assuredly affirm, there are many tools and techniques available for analyzing business environments. A common tool for analyzing external environments is PEST, with its derivatives:

- PEST (Political, Economic, Social, Technical)
- PESTEL (PEST plus Environmental, Legal)
- PEST-CM (PEST plus Customers, Markets)
- STEEP (Social, Technical, Economic, Environmental, Political)

Figure 1.1 shows a representation of how the results of a PEST analysis may be recorded.

SWOT

Similarly, there is a common tool for recording the significant features of the internal environment and their interactions with the external environment. A SWOT analysis considers the internal strengths and weaknesses, along with potential opportunities and threats to its success.

Strengths (S) and weaknesses (W) are generally internally focused ('we are good at/not so good at . . .'), while the opportunities (O) and threats (T) arise from developing situations or conditions which are external to the organization ('there is an opportunity in the market if we could develop a product that could . . .').

Political/legal (P)	Economic/competitive (E)
Social/demographic (S)	**Technical/infrastructure (T)**

Figure 1.1 Basic PEST recording tool.

A format for this analysis would be very similar to the presentation of Figure 1.1, with SWOT replacing PEST as the box headers.

These analyses are excellent starting points for determining external and internal issues that are relevant to the organization's purpose and which affect its ability to achieve the intended outcomes of its strategy/vision/mission/objectives. We'll now examine both at overview level, prior to progressing to greater detail for external environments (as these are more common for many organizations). As necessary, I refer you to my Further Reading suggestions on pages 367–9 for more substantive texts covering the *business environment.*

Overview of the External Environment

Wherever you are right now, look all around. Business activity likely surrounds you. It is everywhere, starting each morning with the generation and distribution of the electricity we use to heat water for our breakfast tea. 'Business' is, however, quite difficult to categorize precisely—it probably concerns activities of trade (buying and selling), profit (either making one, or existing as a not-for-profit entity) and provision of services (whether governmental, charitable, religious, or otherwise), among many others.

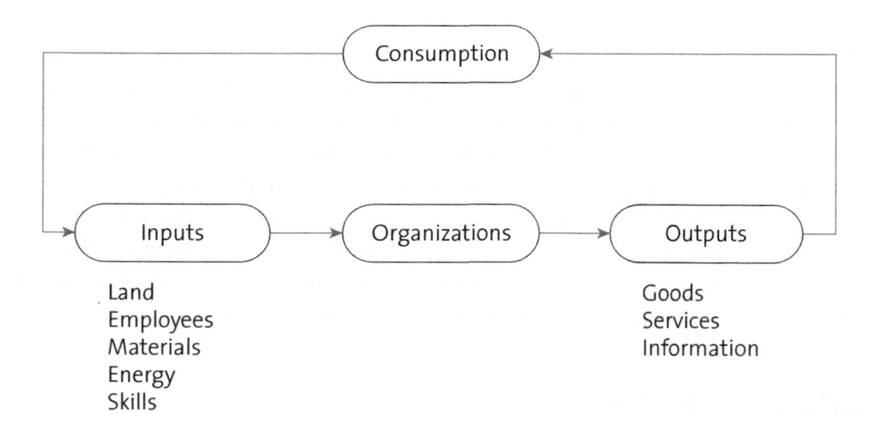

FIGURE 1.2 Simple schematic for the transformation of inputs to outputs.

A reasonable definition of 'business' is:

> occupation or trade, work to be done, or matters to be attended to.
>
> *Concise Oxford English Dictionary* 2008

An alternative, business-focused, definition is given by Peter Drucker (2013):

> Business has only two basic functions—marketing and innovation.

Throughout this book, I take the broadest possible view of the term, and I encourage readers to think of 'business activity' as it concerns their own undertakings, or those of the organizations they work with or know well.

Organizations have inputs (following on from my earlier examples, nuclear power, wind turbines, solar arrays, or gas), some process or activity adding value (manufacturing, delivery, conversion), and, finally, outputs (the goods or services, and their waste). The common feature of all organizations is the transformation of inputs to outputs. This is summarized in Figure 1.2.

Simply put, organizations of all types require land, labour, and capital resources—classically known by economists as the *factors of production*. Specifically, organizations require talented people with great ideas; a source of financial support for the enterprise; suitable buildings or accommodation for the process, activity, or service; a supply of materials; committed workers; satisfied customers; repeat business, and so on.

In accordance with the anticipated needs of their target consumers, these factors are combined to deliver the planned output (goods, services, or information). In successful organizations, this is a cyclical activity, as shown in Figure 1.2. An output which is consumed by the customers generates revenue and an appetite for the acquisition of new inputs, and a reward, financial or otherwise, for the financiers.

I'll relate this to a couple of simple scenarios.

Scenario 1—a shop

An entrepreneur, supported financially by parents or by a bank, purchases retail premises and some opening stock. The entrepreneur hires staff and advertises when the shop will open, with its unique selling points (e.g., that it is convenient or has locally sourced products). Customers visit and make purchases. With the revenue, the entrepreneur purchases new stock to replace that sold and makes the agreed repayments to the bank or reimburses the parents. If the entrepreneur has planned wisely, there may be a small profit as a reward, which will be inputted elsewhere in the economy (perhaps as purchases or savings).

Scenario 2—a charity

A registered charity seeks to raise disaster relief funds. It hires premises, engages staff for a call-receiving centre, and advertises its need for donations. Donations are made, and these are divided to pay for the premises, staff wages, and its advertising. Any residue is donated to relieve suffering in the disaster zone. If the charity has planned wisely, this residue will be sufficiently large to allow further advertising based on the success of the initial phase. Salaries paid to staff become inputs elsewhere in the economy.

A-FACTOR 1

Organizations are concerned with transforming inputs to outputs. Inputs create outputs, and outputs create inputs.

Business Organizations in Their Environment

These simple scenarios exampled here may be, of course, rather more complicated in the operational reality of business and commerce. Organizations are inseparably intertwined with the outside world—with their external environment.

This business environment—where all organizations conduct their enterprises—comprises a wide range of influences. These include the following:

- The prevailing political climate, the macroeconomic situation, the legal framework, and the (sociocultural) context as it relates to technology, education, entertainment and sport, religion, and even the prevalence of organized crime.
- The availability of resources (scarcity), whether potential customers are willing to trade, and the activities of any competitors.

The factors in the former group tend to have a slowly developing and general influence upon the enterprise, whereas the ones in the latter group represent day-to-day operational influences.

General Influences

These factors are discussed in greater detail later in this chapter, but a short overview is provided here to highlight some of the key external influences on businesses.

Politics

Different types of governments have different political aspirations and manipulate economies to these ends. This manipulation will tend to influence the business environment. For example, in Europe there is a significant political aspiration to combine national trade in an international trading block called the European Union. Not everyone likes this (e.g., Brexit). Governments are generally large organizations and employers of large numbers of people in public services.

Macroeconomics

Governments try to create (although sometimes they destroy) macroeconomic climates which are conducive to investment. Policies aimed at creating high or low levels of public sector borrowing, higher or lower levels of employment, and higher or lower levels of inflation are examples of how governments can intervene. Fiscal policies release (or restrict or withdraw) public sector spending, and other policies promote (or discourage) the creation of jobs.

Legal Influences

There is a framework of laws and regulations—some well-developed, some less so—in all countries. These laws define the relationships between the state, organizations, and individual citizens. In some territories, for example in the US, there is also an interrelationship between local (state) laws and national (federal) laws. Similarly, in Europe there is implementation of many legal requirements from federal level (EU directives) at country level (domestic legislation). Like the macroeconomic climate, legal influences can be viewed as being connected to the political perspective.

Sociocultural Influences

Demand and thus supply are driven by social and cultural factors. The demand for electronic goods increases where homes have access to mains electricity or batteries. The supply of locally produced textiles reduces when markets move overseas to take advantage of lower labour rates. Many people believe that cell phones were one of the most effective advancements in history at lifting people out of poverty, as they can revolutionize the average person's access to banking and person-to-person financial opportunities.

Technology

It seems that the speed of technological advance now is near-exponential, as anyone who has recently purchased a new mobile telephone, television, computer, or downloaded an app may have noticed. The willingness of organizations to invest in new technologies depends on their attitudes to the external market but is generally seen as a key to the success of an enterprise (or a country) over its peers.

Day-to-Day Operational Influences

Resources

Organizations rely upon their suppliers for resources. Likewise, the success of a supplier organization is sometimes dependent upon its customer. Accordingly, the operation of organizations is intertwined. They must attend to contracts, pricing agreements, delivery lead times, disputes, and contingencies as a part of the continuity from input to output.

I personally like the concept of the *Chinese Contract* introduced by Charles Handy in *The Empty Raincoat* (1994). This is an agreement between two parties so finely balanced that neither is advantaged at the cost of the other. Thus, neither are likely to walk away from the agreement.

Customers

Customers are vital to all organizations and their employees—they make paydays possible! The proven ability to meet or exceed current requirements (and to anticipate future requirements) as to price, quality, and delivery on time are the hallmarks of successful organizations. 'The customer is king' is loudly proclaimed by many organizations—while 'Our customers get in the way of our real work' may be whispered in some workplaces!

In well-led, total-quality organizations, employees tend to be more customer-oriented than elsewhere. Delivery of products, services, and information is becoming increasingly market-led, and organizing a business to satisfy the newly emerging needs of customers remains a vital requirement for longevity of trade.

I also like the metaphor expressed in *Who Moved My Cheese* (Johnson, 1999). It connotes a nimbleness and adaptability to customers that I have tried to apply to my own organization and commend to yours. It is an enlightening read for those passionate about thriving over the longer time frame.

Competitors

'Winning' or 'losing' in commercial environments often concerns one party's performance relative to another's. The 'other' is one or more competitors who may desire to provide to the same customers as you do with lower cost, higher quality, or differentiated goods and services.

Competition from overseas, where overheads may be lower, may be seen as particularly 'unfair'. Governments sometimes try to protect local markets with trade barriers such as import duties and taxes. Innovation by competitors can render your products and services obsolete. How an organization responds to its competitors (for example deciding on the time for aggressive product development or defensive pricing) may be a significant indicator of its future success.

A-FACTOR 2

Significant risks to organizations' objectives commonly arise in the external business environment with which they are inseparably intertwined. Senior managers should take proper account of this connection if they are to manage their enterprises successfully.

CASE STUDY

EVEN GIANTS CAN FALL

The long-established Eastman Kodak Company was a dominant player in the sale of photographic equipment until the late twentieth century. In 1976, for instance, its US sales of cameras and film represented 85 per cent and 90 per cent of the market, respectively. The company's financial model, based on selling cameras cheaply and targeting related consumables such as photographic paper and film for profit, worked well.

The company failed to respond, however, to the emergence of commercial digital technology in the 1990s, and as its sales fell away it began to run into difficulties. Kodak has not reported a profit since 2007, and in January 2012 it filed for Chapter 11 bankruptcy protection.

On 3 September 2013, it emerged from bankruptcy having sold many of its patents and technologies to its competitors including Apple, Google, and Microsoft.

LEARNING

Keeping an eye on the external business environment and on the competition—in this case largely Fujifilm—is very important.

OVERVIEW OF THE INTERNAL ENVIRONMENT

Organizations decide how to operate to meet their objectives. A common theme running through any analysis of internal environments is 'management' and the style in which it is conducted which creates a 'culture' in the workplace. Management concerns both the roles fulfilled by the individuals who manage an enterprise, and the process—the *management system*—by which the enterprise operate to meet its objectives.

At this stage, I should stress the interaction between internal and external environments. If an enterprise is to remain successful, attention needs to be paid by senior managers to balancing the competing environmental influences, by adapting to cope with any new circumstances facing the organization and then being ready to institute further changes as and when necessary.

Organizational Theory and Management

There are three main categories of organizational theory:

- Classic
- HR approach
- Systems-based

CASE STUDY

BUSINESS ENVIRONMENTS CAN BE VERY TURBULENT

A large UK-based entertainment organization had approximately 250 themed bars and nightclubs providing evening entertainment and dancing for adults. A cursory review of its business environment highlighted the following features to be managed if its business objectives were to be met:

Externally—an extension to licensing hours, a ban on smoking in pubs and clubs, a legal intent to reduce noise exposure levels, increased use of illegal substances, media focus on late-night town-centre disorder, and a rise in underage drinking.

Internally—retention of key staff as they progress with age, marriage, and family commitments, from 'happy to work at night' to finding working at night less acceptable; some parts of the business were for sale, and maintaining staff morale could be a challenge.

LEARNING

It is important to regularly reappraise the prevailing business conditions, as they can change very quickly.

Classic

Writers such as F.W. Taylor (1856–1915) viewed organizations as formal structures established to achieve objectives under the direction of top management. Taylor believed that management was responsible for 'scientific management'—methods related to the design of work, such as work study, that could be applied to improve production.

HR Approach

An alternative approach to the classic, formal organization is the HR approach, which emphasizes the importance of people in workplaces. The famous *Hawthorne Experiments*, conducted in the US between 1924 and 1932, showed that individuals at work were part of informal as well as formal structures, and that group influences were fundamental to understanding individual behaviours. Thus, influencing human behaviour is critical to enhancing organizations' effectiveness.

Systems-based

The approach, described earlier, to converting inputs to outputs and outputs to inputs (along with all the associated subsystems) produces systems-based organizational theories. Modern views of organizations focus on such systems-based approaches, in which management is a highly critical subsystem directing the enterprise towards its objectives.

Some of these *management systems* are externally verified and/or certified, for example Investors in People (for human resource development) in the UK, or ISO 14001:2015 (for environmental management systems) internationally.

Whichever organizational theory is preferred, an organizational structure to deliver it in practice is highly desirable.

Organization Structures

In all organizations—even sole traders, where, for example, a spouse assists with invoicing and 'the books'—there is a division of effort in pursuit of objectives. The resultant pattern of relationships is commonly known as the organization structure. This structure provides how the work is planned, communicated, carried out, supervised, and, when necessary, corrected.

A main feature of most organization structures is an embedded hierarchy along the lines shown in Figure 1.3.

There are five main approaches to organizational structures:

- By product/service
- By geographical location
- Functional
- Matrix organization/project team
- Virtual

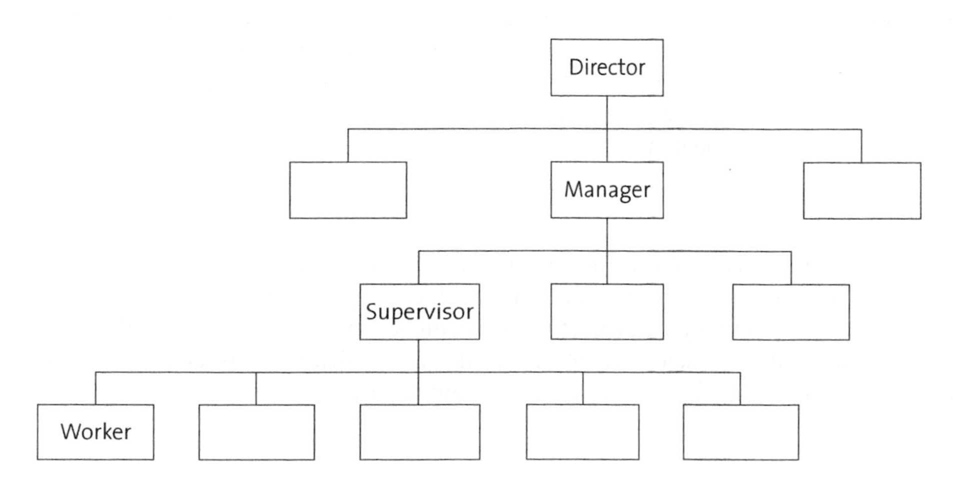

FIGURE 1.3 Example of a classic hierarchical organization chart.

By Product/Service

A high-street store, for example, may have the following departments, each with specialist staff to focus on the needs of customers—the structure follows the sales process:

- Menswear
- Ladieswear
- Home and garden
- Grocery

By Geographical Location

For example, a double-glazing company may structure regionally (North, South, East, West of territory/country) to provide a local contact point to customers and locally based management and installers.

Functional

A manufacturing organization may structure functionally—the structure follows the activity of the enterprise:

- Procurement/goods in/raw materials inventory
- Production
- Finished goods warehousing
- Sales and marketing
- Personnel/Human capital

Matrix Organization/Project Team

I'll give an example of this type of organizational structure. A Grand Prix motor race takes place annually at a national venue such as the Circuit de Catalunya, on the northern outskirts of Barcelona, Spain. A project team is brought together each year to combine in a matrix the skills of full-time travelling race management with local suppliers of accommodation, ticketing, catering, parking, waste disposal, and so on. It is disbanded after the event, until brought together again for the following year's event.

Virtual

Online auctions, such as eBay or Vinted, rely on a loosely connected web of buyers, sellers, and advertisers to achieve their business objectives—a formal organization chart is virtually invisible, but they are instinctive to use, and everyone seems to know their role.

In reality, some characteristics of each of these organizational structures could be present in a single organization to meet its present needs.

A-FACTOR 3

The structure of an organization is a means to an end, not an end in itself.

THE EXTERNAL ENVIRONMENT IN GREATER DETAIL

I will use another common business framework known as PELR (Political, Economic, Legal, and Resources) to build your learning about external environments in greater detail.

The Political Environment

Business activity takes place locally, nationally, and internationally. It is inevitable that governments will be involved in regulating and/or taxing these activities in some way. Markets for many products and services are globalizing, with governments around the world acting to remove (or add) barriers to trade. Understanding the basics of political systems, institutions, and processes provides organizations with greater opportunities to align themselves, and thus to achieve their business objectives.

Politics

A good question to ask audit practitioners reading this book is 'What is politics?' An attempt is made here to answer from this perspective.

The style and nature of any country's political system will tend to be underpinned by its historical and social values, national identity, and political philosophies. Revolutions come and go, while political evolution—incremental rather than radical change—is the norm, particularly in democratic, developed countries. This tends to bring some degree of stability to the business environment in these countries.

The two extremes of political systems:

- Totalitarian
- Democratic

Totalitarian Systems

Whether arising from the power of monarchy, from military conquest (as is sometimes called a 'junta'), or from elections, a totalitarian government will tend to act to restrict or prohibit political participation by others. The style of government tends to be rigid enforcement of rules and oppression of opposition.

> The Arab Spring replaced the harsh order of hated dictators with a flowering of neophyte democracies. But these governments—with weak mandates, ever shifting loyalties and poor security forces—have made the region a more chaotic and unstable place.
>
> Ghosh (2012)

Democratic Systems

Exemplified by free and fair elections at regular intervals and freedom of speech and media, democratic systems provide more balanced governments, with matters discussed and solutions accepted by all participants, even if they disagreed in the first place.

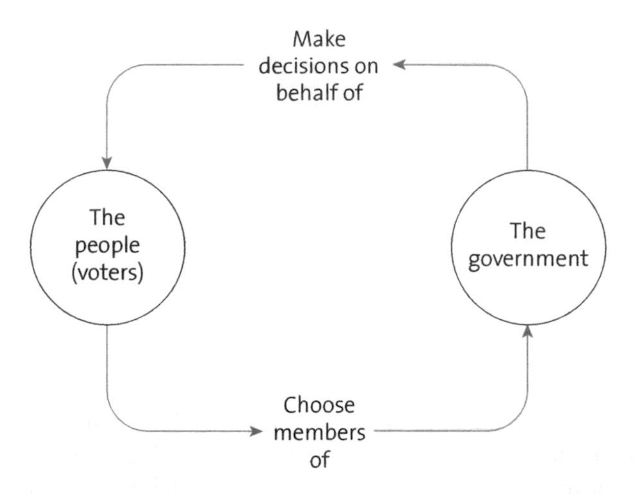

FIGURE 1.4 A model for democratic government.

A model for democratic government is shown in Figure 1.4.

To be recognized as a democratic government, more is needed than a transparent election process. The process should provide that the wishes of the electorate—in terms of the majority of votes cast—are reflected in the final result.

This point makes for an interesting debate where, for example, a 'first-past-the-post' system is in place. In the UK, only a simple majority of votes over other candidates is needed to be considered elected in a regional constituency. At national level, only a simple majority of candidates elected over other political groups (or parties) is needed for the party leader to be asked by the king to form a government in a representative assembly. As a result, it is often the case that the winner has less than 50 per cent of the total number of votes cast. Alternatives to this approach include proportional representation, where the aim is to produce political representation in proportion to number of votes cast. Overall, a first-past-the-post system tends to produce majority government (although it did not in the UK in 2010 or 2017), while proportional representation tends to produce coalition government. The significance of this is that majority governments tend to implement their manifestos (the pre-election 'sales pitch' to the electorate), while coalition governments tend to develop laws through negotiation and compromise with their governmental partners.

It is beyond the scope of this book to examine in any greater detail the political systems of further countries, and thus we move on to consider the basic functions of government.

The Functions of Government

The process of governing a country involves three main functions—making decisions, implementing those decisions, and enforcing compliance through a system of courts:

- Lawmaker
- Law implementer
- Law enforcer

Lawmaker

Governing involves taking major decisions which affect the lives and environments of organizations and individuals. Elected governments in a democratic system hold the power to make laws, and there is usually a series of checks and balances, sometimes including a bicameral legislature (that is, an upper and lower house) and other established processes, to ensure that law-making is proportionate and fair.

Law Implementer

A government has responsibility for putting laws into effect. Day-to-day administration is carried out by non-elected officials called civil servants whose major role is implementing public policy. While politicians may come and go, civil servants have permanent career positions. They are expected to act in a non-partisan way, and this allows for continuity of governance, for example, when one government cedes power to the next one.

Law Enforcer

The third arm of a government is a judiciary and a system of courts. It is a hallmark of democratic systems that there is separation between the law enforcement role of the judiciary and the other two main functions. An independent judicial system, free and able to challenge the government and review its decisions, provides a further check and balance to a democratic government, and it can protect citizens from a state that might have become too powerful.

Auditors need an appreciation of how political factors can impact upon auditees—for example, how laws are initiated, developed, implemented, and enforced (and sometimes, repealed). A review of legal compliance will be necessary in some auditing assignments.

Trans-frontier Government

As noted earlier, political influence is not restricted to national boundaries. International groupings such as the United Nations (UN), the Group of Twenty (G20) and the Group of Seven (G7; formerly G6, and G8), the World Trade Organization (WTO), and the European Union (EU) add far-reaching dynamics to an external environment and have an increasingly profound influence.

United Nations

The United Nations (UN) has 193 member states. Its aims are to facilitate cooperation in international law, international security, economic development, social progress, human rights, and world peace. It was founded in 1945, following World War II, to replace the League of Nations and to stop wars by providing a platform for dialogue.

G20

The Group of Twenty Finance Ministers and Central Bank Governors is a meeting among governments of twenty major global economies—nineteen countries and the European Union—representing almost 90 per cent of global GDP, 80 per cent of international global trade, and two-thirds of the world's population.

The group was first established in 1999, while summit meetings among the heads of government of the member economies have taken place since 2008. In 2009, G20 announced that it would take over from the G8 as the main forum for international economic cooperation.

G7

Established in 1975, the G7 group of countries—comprising the USA, Japan, Germany, France, Italy, Canada, and the United Kingdom, which together represent over 50 per cent of the world's economy—meets regularly to discuss matters of mutual interest. Known as 'economic summits', these meetings attract significant interest from media and protesters alike. Between 1997 and 2014, it was known as G8, reflecting Russia's inclusion, then expulsion following its invasion and annexation of Crimea.

'G8 + 5' meetings, involving Brazil, China, India, Mexico, and South Africa in addition to the original eight countries, have also taken place. It is unlikely that the G8 + 5 will meet in its current form including the presence of Russia. It had been suggested that the G7 may expand permanently to include these 'Outreach Five' economies, but agreement amongst the seven has not been found.

World Trade Organization

The World Trade Organization (WTO) was formed in 1995 to supersede the General Agreement on Tariffs and Trade (GATT), which had been formed in 1947 to assist with re-establishing trade at the end of World War II. With a large membership—and many other countries indicating that they wish to join—the WTO is charged with opening up global trading within a framework of agreed rules.

The WTO has certainly proven successful in promoting international trade. When China joined in 2001, for example, according to the UN, it was rewarded with over $710 billion in direct investment from international firms (Schuman, 2012b).

European Union

The European Union (EU) was, by 2013, a grouping of 28 European nations. It was established in 1958 by the Treaty of Rome, with six original members (West Germany, France, Italy, Netherlands, Belgium, and Luxembourg). Progressive enlargement in 1973 (UK, Denmark, Eire), 1981 (Greece), 1986 (Spain and Portugal), and 1995 (Austria, Finland, Sweden) was further magnified in May 2004, when ten new members were admitted (Cyprus, Czech Republic, Estonia, Hungary, Latvia, Lithuania, Malta, Poland, Slovakia, and Slovenia). Bulgaria and Romania were admitted in January 2007, and Croatia in July 2013.

In a referendum, UK people voted to leave the EU on 23 June 2016. Exit negotiations led to the UK's departure on 31 January 2020, leaving twenty-seven (EU27) in the grouping.

Turkey has been an official candidate for accession to the EU since 1999, but the negotiation process has been slow, and it is expected to be several years before membership is likely. There are presently seven other recognized candidates for future membership: North Macedonia (since 2005), Montenegro (since 2010), Serbia (since 2012), Albania (since 2014), and Moldova, Ukraine, and Bosnia and Herzegovina (since 2022).

The aim of the founding Treaty of Rome was to create a common market to promote trade and bring national economies closer together. This convergence led to the creation of a single market when the then-members signed the Single European Act, effective 31 December 1992, comprising the following features:

- Reduction in legal and other obstacles to cross-border travel and trade
- Harmonization of technical/safety standards
- Convergence on excise duties
- Mutual recognition of qualifications

Other treaty measures provided for the following:

- Monetary union, which was achieved on 1 January 1999, with the Euro presently used by twenty of the twenty-seven member states of the EU
- A social chapter to protect workers (including their health and safety)
- Common foreign and security policies

Putting in place such a significant series of changes has required literally hundreds of new laws and regulations in the member states. For completeness, a summary of the architecture of the EU, in terms of its principal bodies, follows. These bodies are the following:

- The European Parliament
- The Council of the European Union
- The European Commission
- The European Court of Justice

The European Parliament 'Europarl' or the 'EP', as it is commonly known, is a body of 705 members (following the withdrawal of the UK from the EU), directly elected every five years since 1979. It has representation from each member state based on the size of its population. These serve the second largest democratic electorate in the world (after India) and the largest transnational democratic electorate. Much of the work is undertaken by specialist committees, which make recommendations to full meetings held in Strasbourg.

The Council of the European Union The ultimate decision-making body of the EU comprises one minister from each member state. The presidency of the Council rotates between members on a six-monthly basis. It is responsible for major policy decisions, which are published as regulations, directives, recommendations, or opinions.

The European Commission The European Commission is the executive body of the EU—its 'civil service'—and the guardian of the treaties. It has offices in Brussels and Luxembourg and comprises twenty-seven commissioners and a staff of around 32,000 civil servants drawn from all member states.

The European Court of Justice The European Court of Justice sits in Luxembourg to pass judgement on the interpretation of EU laws. The Court can set aside measures adopted by the Commission, the Council, or governments of member states which are incompatible with the treaties. Decisions are binding upon the member states.

Auditors need an understanding of the political environment in which their audit work is to take place so that they can contextualize the possible risks areas on which they will later base their audit opinion.

I found McCormick (2015) particularly insightful on the finer details of the EU; details for this title are in the References section. I found Katz (2007) useful for reviewing the institutions of the US, although it is a few years old now. There are similar titles for other territories.

The Economic Environment

During November 1985, the US dollar (USD) and the British pound sterling (GBP) were trading roughly at parity (1:1) in London. At the time of writing, the exchange rate between these two great currencies is around 1.2:1. In 2007, when the first edition of this book was published, it was 2.1:1. In a world where trade is often international, where you operate and how you pay and get paid can have a significant effect on the price of raw materials and/or the value achieved from your sales invoices (more than doubling or halving either of these, in my example). And, of course, there are innumerable territories where this inflation/deflation ratio is much more significant. As markets globalize, the successes of organizations in different trading economies becomes increasingly interconnected. The economics of business is an important external factor to be considered and understood if an organization is to achieve its objectives.

Scarcity

Scarcity is based upon the relationship between consumers' 'wants' and the resources available (the input-output-consumption-input of Figure 1.2) to satisfy these wants. Consumers' 'wants' are said to be insatiable, while resources are inevitably finite. Thus, choices must be made concerning priorities. For example, does a society want better healthcare, or better education? More nurses, or more firefighters?

CASE STUDY

SCARCITY AND INELASTICITY

The publisher of an annual sports almanac undertook a strategic review with the aim of increasing company profitability. It confirmed that it was the only publication of its type in the market, and that this scarcity provided a near-certain inelasticity in demand from its customers. It increased its sale price and, right away, its profitability.

LEARNING

Much can be learned by an organization from carefully considering the scarcity of and the demand for its outputs.

In practice, scarcity is managed by several factors, including the following:

- Price—for example, diamonds are more expensive than rocks
- Rationing—for example, tickets to the cup final sell out
- Queuing—for example, there is a waiting list to see an eye surgeon

Price is deliberately at the top of this list and often sorts out the other two; supply and demand are invariably set in the marketplace.

Economic Systems

An important distinction in economic theory is between economic systems which are centrally planned and those which operate under market conditions.

Centrally Planned Economies

This type of economy (also known as a command economy) is generally associated with socialist economies such as Cuba, the Soviet Union 1917–91, and China until the late 1970s. The main production decisions are taken by a central authority. Characteristics of this type of economy:

- State control of resources
- State control of priority for use of resources
- Targets for production to balance supply and demand set by the state
- Prices set by the state

Free-market Economies

More common in the early years of the twenty-first century are free-market economies, where prices determine the allocation of resources. Characteristics of this type of economy:

- Privately owned resources, hence owners can choose how and when to consume them
- Privately owned organizations operating free from state intervention
- The customer is king—consumers choose how, when, and with whom to spend their money

A-FACTOR 4

Recognize that, ultimately, market forces tell organizations—if they are listening carefully—what to produce (quality), when to produce it (delivery on time), and the sustainable price. These as objectives should be reflected in the business plan.

Macroeconomics

Macroeconomic theory concerns an economy as a whole. It deals with such matters as overall levels of employment, the rate of inflation (a retail price index RPI is commonly used to measure price changes), and the annual rate of growth of output GDP of an economy.

A simple economy comprises cyclical flows of money (and other financial instruments) between organizations and consumers, as:

- organizations provide income to workers' households (salaries), and
- households spend salaries on organizations' products and services.

This cyclical flow shows that the fortunes of organizations are connected to the spending decisions of consumers—customers need to spend if organizations are to prosper. Thus, in any economy levels of income, output, expenditure, and employment are interrelated.

Recession occurs when these macroeconomic indicators move negatively, and growth occurs when these indicators move positively. Government (or central banks) use tools such as interest rates to encourage or suppress consumer activity to promote growth aligned to its own objectives for the economy. Similarly, increasing taxation to raise money for public expenditure injects additional income into this circular flow of money.

External economic factors also influence the spending decisions of consumers, for example, increases in the price of fuel at service stations, caused by many nations' determination not to purchase Russian crude oil following its invasion of Ukraine in 2022.

Of course, the economy is much more complicated than this short section can possibly reflect. Everything affects everything else, and nothing can replace local analysis at the time information is needed. Understanding how the macroeconomy works helps organizations to set and achieve their business objectives. It helps auditors to understand how business decisions are made (for example, in which years training budgets might be cut, and in which years they might grow), and thus to focus their work planning on foreseeable risks.

The Role of Financial Institutions

In a developed market economy, there will be financial institutions whose role is to channel funds from those willing and able to lend to those wishing to borrow. These intermediaries include private banks, state banks, and world banks.

Private banks generally lend to private customers on negotiated terms (usually based on the estimated level of risk in the transaction), gaining financial return from interest and other payments.

A state bank (such as the Deutsche Bundesbank in Germany) is a critical element in a country's financial system. It exercises control of over the domestic banking sector and sets monetary policy according to the needs of the economy and/or the geographical region.

World banks include the following:

- International Monetary Fund (IMF)
- Organization for Economic Co-operation and Development (OECD)
- The World Bank

IMF

Established in 1945, the role of the IMF is to provide a pool of international funds to promote growth in world trade. It also involves itself in assisting developing economies with debt problems.

OECD

An intergovernmental forum established in 1961 in which 38 member countries meet to discuss world economies. Its decisions are not binding, but its research is used by G7, G20, the IMF, and other bodies.

The World Bank

Established in 1944, the World Bank provides loans and technical assistance to developing countries, with the aim of reducing poverty.

The Legal Environment

Laws impact on many areas of activity conducted by all organizations. These include minimum employment conditions (including those relating to health and safety), sales contracts for supply of goods and services, taxation, and environmental discharges to air, land, and water. Penalty frameworks (known in the UK as sentencing guidelines) exist for those who are judged to have broken the law. These include fines and periods of imprisonment.

This section provides an overview of legal frameworks, written by a non-legal practitioner for non-legal practitioners.

Classification and Sources of Law

Laws have evolved over many years—in the UK, they are said to date to the conquest by William I (the Conqueror) in the year 1066. The essence of a law, from a general perspective, is that in return for the protection provided to an individual or organization by the law, the actions of that individual or organization are constrained. Thus, laws exist to regulate the behaviours of individuals and organizations, and collectively they set out the minimum standards of conduct desired by society at large in the territory to which they relate.

Laws are commonly derived from historic customs and practices, written laws which have passed through the political law-making process, and the cumulative judicial decisions of the courts. In civil law, lower courts are obliged to follow the decisions—the *ratio decidendi*—of higher courts.

As governments come and go, laws are enacted, amended, and repealed. Thus, the law in many countries is dynamic, and an auditor will need an appreciation of the legal framework that applies to their auditee's organization.

Law can be defined in many ways. The main features within a typical legal framework are summarized here.

Criminal Law

Criminal laws relate to that which has been prohibited by legislation or statute in the interests of society at large, and which is punished by the state on behalf of the people, upon conviction in a criminal court by a judge or a jury. Fines and/or imprisonment are possible punishments. In England and Wales, the starting point for all criminal prosecutions is before a lay judge in a magistrate's court. More serious cases are referred to a Crown Court for judgement by a lay jury and for sentencing by a professional judge.

TIP

ILO publishes a database of international OH&S legislation which I found useful, although sometimes not completely up to date. Known as 'LEGOSH', it can be found at www.ilo.org/dyn/legosh/en/f?p=14100.

Civil Law and Tort

Civil laws concern matters between individuals. A 'tort' is a civil wrong. Common torts include negligence, defamation, and trespass. An award of financial damages or an order of the court (perhaps requiring something to happen or for something to stop) are typical outcomes. In England and Wales, civil judgements may be made in a county court, and those where the remedy may be higher in value will be judged in the high court.

Both the criminal and the civil law systems have superior courts, with a right of appeal in the UK to the Court of Appeal, and ultimately the Supreme Court, the highest court for domestic purposes since it assumed the judicial functions of the House of Lords in 2009.

Public Law

Another useful distinction in law is between public and private law. Public law comprises those laws concerning the state, whether relating to national or international matters, or to the relationship between the state and an individual—for example tax laws.

Private Law

Private law comprises those laws concerning the relationships between individuals, such as family, property, trust, and contract.

International Law

The world seems to be becoming a smaller and more interrelated, interconnected place. There is an increasing tendency for nations to accede to international laws and treaties.

ILO

An agency of the United Nations, the International Labour Organization is concerned with social justice as it relates to the workplace. Its stated aims include promotion of employment rights and opportunities and social protection.

TIP

See the ILO website at www.ilo.org/global/lang--en/index.htm.

The Montreal Protocol

The Montreal Protocol is the agreed international framework that bans, except in specific circumstances, the manufacture of CFCs, as they are generally thought to damage the protective ozone layer around the Earth.

CASE STUDY

GONE ON A BOSMAN

An example of how European law can impact upon national matters was the interesting case of Jean-Marc Bosman, a Belgian footballer. Out of contract with Belgian club RFC Liège, Bosman sought to play the game in France with Dunkerque. When Liège demanded a transfer fee and the French club declined to pay, Bosman referred the matter to the European Court of Justice.

In the Court's 1995 judgement, it was ruled that clubs could no longer request a transfer fee if a player out of contract wished to play elsewhere in Europe, as it violated the right established in the Treaty of Rome for European workers to work in any member country. Since then, players have regularly moved from one club to another in these circumstances and are commonly said to have 'gone on a Bosman' (Blanpain & Inston, 1996).

LEARNING

The legal environment is not static.

The European Union

As discussed in this chapter, the EU has provided impetus in Europe for harmonization of legal standards in many areas. Regulations and directives made by the Council of the EU are binding in all member states. They must be implemented within the domestic legal frameworks of each member state by agreed dates, or penalties may be imposed on the state.

Business Organizations and the Law

As I have explained, business is concerned with the conversion of inputs to outputs. The prevailing legal systems in a territory constitute a controlling, constraining framework within which these activities are conducted.

Business can be helped by laws (for example through assistance to collect invoice payments on time) as well as constrained by them (for example through prohibition of disposal of effluent waste into drainage systems).

Table 1.1 shows some examples of legal influences on various business activities that may be of interest to auditors.

TABLE 1.1 Legal Influences on Business Activities

BUSINESS ACTIVITY	*POSSIBLE LEGAL INFLUENCES*
Start up	Company formation and registration
	Mandatory insurance (such as employer's liability)
	Taxation
Operations	Employment
	Contracts
	Health and safety
	Environment
	Producer responsibility
	Product safety
Extending the buildings	Planning
	Construction standards
	Environment
	Fire safety
Deliveries	Consumer protection
	Road safety
	Transportation of dangerous goods
Being an auditor	As above, possibly in entirety
	Negligence
	Defamation

TIP

As an auditor, it is a very good idea to have a readily accessible source of legal advice. Organizations often have employed or retained legal practitioners. If you are self-employed, you might be able to access a similar service through your professional body or a Chamber of Commerce. Alternatively, a direct relationship with a professional legal practice may be useful.

Areas where such legal advice may be valuable include the following:

- Review of the terms and conditions of a contract
- Review of employment contracts
- Review of the wording of audit reports
- Advice on legal matters (specialist matters related to audit or on more general matters such as getting your invoice paid)

The Environment Related to Resources

Businesses exist to produce goods, services, information, and other outputs from their inputs. The critical inputs are a supply of resources, human beings to 'do the work' and 'buy the goods', and the necessary technologies and intellectual capital for the process. Earlier in this chapter, I called these (as economists do) the *factors of production*—land, labour, and capital.

TIP

In the UK, annual statistics on these and many other demographic factors are published by the Office for National Statistics and available from the Stationery Office (TSO). Refer to these (or similar sources) when detailed information is required.

Land and Natural Resources

Natural resources include land, underground or on-the-ground deposits, oceans and rivers, flora and fauna, and the weather. An important distinction is that between renewable resources and non-renewable resources. Either way, it is generally true that our resources are finite in supply (though small areas of land can be reclaimed from the sea at large cost).

Land use is renewable. We can build on land, demolish the structure, and build again. We can plant crops each year (though to do so on former industrial land probably would not be a good idea, unless an extensive 'clean-up' had been completed).

Fossil fuels are not renewable. When we extract oil, refine, and introduce it to an internal combustion engine, that fuel is gone forever. Quite how many years' supply of hydrocarbon fuels remain is the subject of debate and probably uncertain at this time with the discovery of new oil sand and shale deposits.

In recent years, 'the environment' has taken centre stage, both politically and of necessity. Predicted rates of increase in average global temperatures, and changing climates, have demanded attention, particularly in relation to reducing CO^2 emissions. At the same time, recycling and energy efficiency have become prominent issues in many developed areas of the world. Whether we are doing enough remains to be seen.

Labour (People)

Human beings are important as both producers, and as consumers of goods and services (see Figure 1.2). On 15 November 2022, the world population passed eight billion. It was about 3.5 billion just 50 years ago. Many production processes are people-intensive, and, accordingly, having a suitable supply of educated, motivated, and affordable staff is important to most businesses. Many organizations will want to develop the education and motivation of their staff throughout their working careers and reward successes in these areas with promotions and salary increments.

Of interest to organizations seeking to develop their business in a particular territory will be the following factors, helpful in identifying a workforce:

- Total population of the territory
- Age profile within the population
- The size of the working population
- Mobility of the working population
- Retirement age
- Occupational structure of the working population
- Education level of the working population
- The length of the working week
- Wage levels, minimum wage levels, and trade union involvement

CASE STUDY

WHAT COULD GO WRONG WITH 'SPONGY HANDS'?

A promotions company's business was buying and selling concert souvenirs for 'B-list' artistes and touring bands (i.e., newly rising stars, and fading has-beens).

When it identified large foam 'spongy hands' printed with the artist's name as a potentially attractive product, it purchased 1,000 at a unit cost of 20p (a fifth of 1 GBP), and sold out at £3 (GBP 3) each on the first night of a multi-night concert tour—a markup of 1,400 per cent.

In the second year as the act's memorabilia agent, the organization trebled its order and sold out on the second night of the tour. Expecting to 'make it big' in the third year, the purchase order was multiplied by ten—still at the year-one cost price. This time, the organization was not able to sell even one 'spongy hand'. Why?

The year was 1997, and the artist's name was Paul Gadd, known to his fans as 'Gary Glitter'.

LEARNING

Some external factors are way beyond the control of an organization.

TIP

As an auditor, make time each month to update your CPD records—your personal development and the audits you have conducted or participated in.

For example, I maintain my IOSH CPD on my professional body's online system. I hear that they must chase members hard to update their CPD records at the specified review points. Some end up in disciplinary hearing or even expelled from membership. Don't let it be you they're chasing or expelling!

Marketers will conduct similar analyses for organizations seeking to establish the size and distribution of markets.

Capital and Technical Factors

Flows of capital to successful businesses are undiminished, and these are providing for unprecedented changes in the input-process-output cycle. There has already been a simply massive change in technology in recent years, and, as Fareed Zakaria observed in *Newsweek* magazine, perhaps worryingly for some:

> The 21st century will be the century of change. More things will change in more places in the next ten years than in the previous 100.
>
> (Zakaria, 2006)

Technological change leads to new products and services (including profound changes to the life expectancy of human beings, as previously incurable conditions are now diagnosed and routinely treated), new markets, increased automation, and displacement of people from processes, faster exchange and storage of data and information, and greater possibilities for intrusion and loss of privacy.

The Internet continues to transform the way in which people shop, pay, access information, and communicate, both for business and socially. It is difficult to predict how the Internet will change in the future; suffice to say, it will.

> Iceland has the highest number of people online. It's 95%. The UK lags that, and the US lags quite a long way behind that at 78%.
>
> Tim Berners-Lee, inventor of the World Wide Web (in Luscombe, 2012)

As of 2023, there were 5.16 billion Internet users worldwide, which is 64.4 percent of the global population. Of this total, 4.76 billion, or 59.4 percent of the world's population, were social media users (Statista, 2023a).

A-FACTOR 5

Top management should balance the influences of the competing external and internal environments to face its target market(s) with aligned and well-communicated business objectives.

Research into new processes, materials, crops, pharmaceuticals, vehicles, and sources of energy (etc.) turns up new developments all the time. Barriers to making these technical developments include lack of skill in the workforce, or redundant skills, where technology has moved at a pace so quickly that parts of a workforce have not retrained quickly enough. Exhaustion of natural resources (as noted previously), particularly fossil fuels, could impede future technical developments.

All these three inputs to businesses—land, labour, and capital—are interconnected. For example, the productivity of human beings and the efficiency of plant and equipment will be impacted by the technology available to them at any point in time. These inputs are essential to organizations, because without them, conversion to outputs could not take place.

As an auditor, it is likely that you will be exposed to new technology, ranging from the R&D department in the auditee's organization and the new carbon fibre, lean-burn aircraft you flew to site in, through to the new audit reporting software on your iPad. You might get invited to a preview of the new *flying car* (as I was—see page 2). As part of your CPD (continuing professional development) programme, you should devise a plan to keep abreast of developments applicable to your role, specialisms, and interests.

SENIOR MANAGEMENT'S INTERPRETATION OF THEIR BUSINESS ENVIRONMENT

Senior management's role is to take account of the external and internal issues that are relevant to their organization's purpose and that affect its ability to achieve the intended outcomes of its management systems, based on their most realistic interpretation of the opportunities and/or threats faced. The most usual place for this to be written down will be in the business plan, with specific objectives and key performance indicators.

Management commonly expresses its analysis of the prevailing *Context* (aka business environment), and any subsequent developments expected, in a series of corporate documents. The first of these is often a statement of a *Vision* of how the organization will be in the future. A *Mission Statement* sets out the purpose of the organization, and a series of *Corporate Objectives* for the plan year and perhaps beyond will be established. These documents are commonly used as a means of cascading (or sharing) the essential activities (as seen by top management) in a consistent manner and language through the organization. Figure 1.5 shows how this cascade might look.

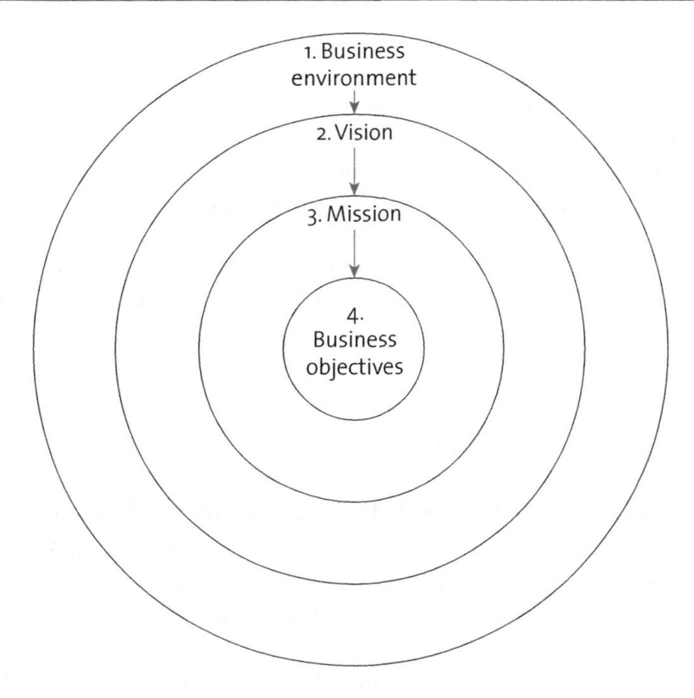

FIGURE 1.5 Connecting business environment (Context) to Vision, Mission, and Business objectives as a cascade of communications to interested parties (stakeholders).

Shareholders usually have some means of replacing the board if they don't like the direction the organization is heading. In a public limited company, directors retire in rotation and offer themselves to the AGM for re-election. In recent years, there have been some very vocal expressions of stakeholder anger inside and outside of AGM venues.

As an auditor, you'll learn in this book that your work concerns the probability of occurrence of harm and the impact of the consequences upon the achievement of the business or organization's objectives. You should also be prepared to challenge the appropriateness of such objectives in appropriate circumstances, for example if the *Risk* is too great.

WHAT IS RISK?

Ask a manager in the twenty-first century, *What is risk?*, and, as likely as not, you'll be told that it is an estimation of the *probability* and *consequence* of harm occurring. Health and safety managers have been busy in many organizations, and risk assessment documents are common in developed territories and well-run organizations. In this understanding, some managers will use words such as *frequency* or *likelihood*, and some will use words such as *impact* or *severity*. Either way, most will know that risk concerns a reasoned view of the future that can be predicted and planned for. The greater the foreseeable risk, the greater the imperative for control.

Sometimes, the level of risk is driven by the level of fear. Gardner (2009) says that fear among people can be a constructive emotion. When we worry about risk, we pay more attention to it and take more urgent action. Occasionally, fears burst into full-bore panics—racist police officers, paedophiles lurking in churches, and Internet chatrooms the latest. In the 1990s it was road rage and asbestos. A decade earlier, it was herpes. Satanic cults, mad cow disease, school shootings, and crack cocaine have all raced to the top of the public's list of concerns, often driven by reported events. Organizations must think about risks too.

CASE STUDY

POLAR EXPLORERS DICE WITH DEATH (A TRUE CASE STUDY?)

Your author Dr Stephen Asbury and a companion were exploring the North Pole when they came upon a huge polar bear. The bear growled angrily and rubbed its stomach in a hungry manner, clearly relishing the hearty meal which had just walked into its territory.

As experienced visitors to polar climes, the two intrepid adventurers were both wearing 'tennis racket'–style snowshoes, as a risk control measure to spread their load on the ice. Asbury began removing his snowshoes, clearly contemplating a dash for safety. Watching him, his companion said, 'Stephen, you can't possibly outrun a polar bear in its own terrain!'

As a risk specialist, what was Dr Asbury's response?

Stephen replied, 'I don't have to. I only have to outrun you!'

LEARNING

Know what can hurt you, and focus on the priorities, the *Big Rocks*.

A-FACTOR 6

Risk is the effect of uncertainty on objectives—or anything which may impact upon the achievement of objectives. It is generally quantified in terms of its *inherent* and *residual* probabilities and consequences. Value creation and value protection are the essence of an organization's success.

> No director can ignore the risk to the reputation of his [*sic*] company and its brand that health and safety and environmental expectations present.
>
> Sir Nigel Rudd, one of *The Times* Power 100, and holder of four directorships of FTSE companies.
>
> (Eves & Gummer, 2005)

Definitions of Risk

'Risk', then, can and should be defined as the scale of any type of potential impact on objectives, whether the source is HSEQ, financial, reputational, motivational, legal, or otherwise. The opposite face of risk is 'opportunity', achieved by addressing an opportunity to reap a reward. A good way for auditors to think about and refer to these two opposing outcomes:

- Value protection (where the harm is the potential for impact upon the achievement of the organization's objectives)
- Value creation (the reward for addressing a suitable opportunity)

Half a dozen other common definitions of risk are presented subsequently:

- The combination of the severity of harm with the likelihood of its occurrence—Health and Safety Executive (HSE, 1997: 54).
- A combination of the hazard and the loss and, in any given set of circumstances, risk takes into account the relevant aspects of both—Boyle (2002: 18).
- The chance of a particular situation or event, which will have an impact upon an individual's, organization's or society's objectives, occurring within a stated period of time—Fuller and Vassie (2004: 5).
- The effect of uncertainty on objectives—ISO 31000:2018 (ISO, 2018c).
- The potential that a chosen action or activity (including the choice of inaction) will lead to a loss (an undesirable outcome). The notion implies that a choice having an influence on the outcome exists (or existed). Potential losses themselves may also be called 'risks'. Almost any human endeavour carries some risk, but some are much more risky than others—Rodway (2023).
- [The] combination of the likelihood of occurrence of a work-related hazardous event(s) or exposure(s) and the severity of injury and/or ill health that can be caused by the event(s) or exposure(s)—ISO 45001:2018, terms and definitions, 3.21.

However we define it, an effective leader and line manager should know about the foreseeable risks their organizations may face, rank them in order of importance, and take appropriate action to control them.

Inherent and Residual Risk

Risk can be assessed or expressed in two main frames:

- Inherent risk
- Residual risk

Inherent risk is an assessment of risk exposure *prior* to when the effects of the selected business control framework and other control measures are accounted for. Some call this the *initial* or *unmitigated risk.*

Residual risk is the remaining risk exposure *after* the mitigating and compensating factors of the business control framework and other control measures are accounted for. Some call this the *mitigated risk*.

Some intentionally selected control measures have the tendency to reduce the probability of an event, such as preventative controls like a well-trained workforce or fixed guards on machines. Some other controls tend to reduce the consequences, for example, early detection, containment, mitigation, and restoration controls. Other controls can reduce both probability *and* consequence, for example, elimination and substitution controls, such as eliminating obsolete chemicals or substituting with chemicals of lower toxicity.

Whichever control measures are selected to control *inherent risks*, their adoption should follow a hierarchy known as *E-SEAP*, which you will read about in the forthcoming *Tip* on page 42.

A Brief History of Risk

Risk has a fascinating history, beautifully narrated by Peter Bernstein in his book *Against the Gods* (1996). You would not have to go back in time many years for the modern clarity of approach and measurement of risk to be lost. A well-educated individual of a thousand years ago would not recognize the number zero (0), and they probably would not pass a basic mathematics test. Five hundred years later, few would do very much better. Without some form of measurement, some numbers, risk was a matter of gut feel or superstition.

The 'power of numbers' arrived in the West in the early thirteenth century, when a book entitled *Liber Abaci* appeared in Italy—its author was the mathematician Leonardo Pisano, more commonly known as Fibonacci.

Fibonacci is best known for a sequence of numbers which provide the answer to a problem about how many pairs of rabbits will be born during the course of one year from one starting pair, assuming that every month each pair produces another pair, and that rabbits start breeding in the second month after birth. The answer is 377, and the starting number and month-end totals for the year are:1, 2, 3, 5, 8, 13, 21, 34, 55, 89, 144, 233, 377. Each successive number is the sum of the two preceding numbers, and the ratio of one number to the previous number in the sequence is approximately 1.6. In fact, as you go through the sequence calculating this ratio, it gets progressively closer to a number called the 'golden ratio', which features in nature (for example in the shapes of shell spirals, leaves, and flowers) and in architecture (for example in the United Nations headquarters in New York). Playing cards are similarly proportioned. The Fibonacci sequence also features in the book and film *The Da Vinci Code*, where a dying Jacques Saunière leaves a code for Robert Langdon, played by Tom Hanks, to decipher.

Fibonacci introduced the 'power of numbers' to the West for the first time, but using them to assess risk remained many years distant. Bernstein (1996: 1) comments on the development of human understanding of risk:

> What is it that distinguishes the thousand years of history from what we think of as modern times? The answer goes way beyond the progress of science, technology, capitalism and democracy . . . The revolutionary idea that defines the boundary between modern times and the past is the mastery of risk: the notion that the future is more than

> a whim of the gods and that men and women are not passive before nature. Until human beings discovered a way across that boundary, the future was a mirror of the past or the murky domain of oracles and soothsayers.

Bernstein gives an interesting account of our history, suggesting, 'The ability to define what may happen in the future and to choose among alternatives lies at the heart of contemporary societies.'

Hazard and Risk

A modern definition of 'hazard' is 'the potential for harm'. The word 'hazard' is said to derive from an Arabic word for dice, *al-zahr*.

We have seen a sample of definitions of risk, though there are many others. The word 'risk' is said to derive from the early Italian *risicare*, meaning 'to dare'. To dare implies the freedom to choose and, as a result, possibly to fail.

CASE STUDY

VERY LOW PROBABILITY, CATASTROPHIC CONSEQUENCE EVENTS, OR *BLACK SWANS*

When anyone asks me how I can best describe my experience of nearly forty years at sea, I merely say uneventful. Of course there have been winter gales and storms and fog and the like, but in all my experience, I have never been in an accident of any sort worth speaking about. I have seen but one vessel in distress in all my years at sea . . . I never saw a wreck and have never been wrecked, nor was I ever in any predicament that threatened to end in disaster of any sort.

From a paper presented by E.J. Smith in 1907 (quoted in Toone, 2004).

On 15 April 1912, RMS *Titanic* sank with the loss of 1500 lives. The captain went down with the ship. His name—E.J. Smith (ibid.).

LEARNING

The absence of prior loss is no guarantee that a future loss is impossible.

Dice is a game of luck—of pure chance, of potential for harm, the hazard of losing your stake. While lots of things have potential for harm (hazard), managers can choose to dare or not—they can choose how and when (and whether) to respond to hazards. This choice influences the probability of the harm occurring and the consequences should the harm be realized.

This 'daring' to participate actively in the business environment involves choices for managers (and options for auditors to recommend alternatives if necessary). The first choice is to *terminate* the activity, or not start it in the first place (which provides a case for HSEQ practitioners to review proposed projects at the concept stage) if the risk is beyond the appetite of the organization or is, in absolute terms, too great.

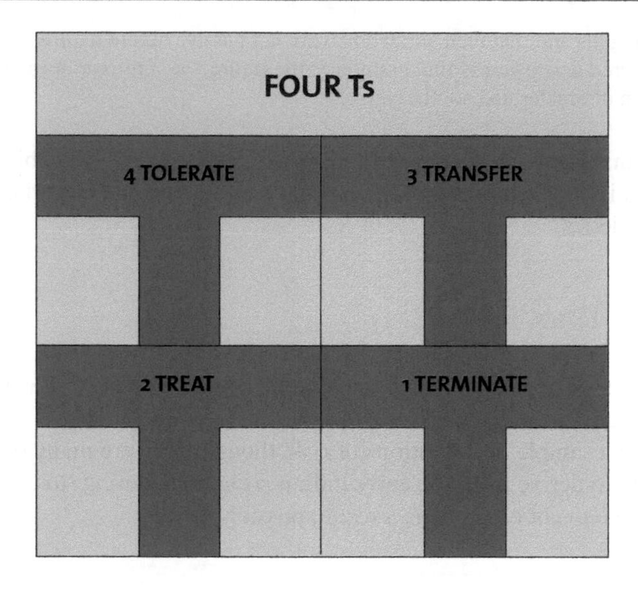

FIGURE 1.6 The Four Ts: Four choices for managing a risk.

If the risk cannot or should not be terminated, it can be *treated* or mitigated to a lower residual level by application of a structured means of control (or management system). Other choices include *transferring* the risk to someone else (for example by insuring it financially or by participating in a joint venture).

After the choice or choices have been made, the residual risk is *tolerated*. Tolerating a risk in this context means managers are 'not happy with the situation' or 'living with the risk'. Managers need to be vigilant and review how well their risk control measures continue to be effective. Tolerating risk is ultimately fatalistic and defeatist, as it implies passive acceptance.

These four options are the 'Four Ts' illustrated in Figure 1.6.

TIP

There are several hierarchies for OH&S control, but the most recent (and I think the best) is from clause 8.1.2 of ISO 45001:2018.

- **E**liminate the hazard
- **S**ubstitute with less hazardous materials, processes, operations, or equipment
- Use **E**ngineering controls and/or reorganise the work
- Use of **A**dministrative controls, including procedures and training
- Use of adequate **P**ersonal protective equipment (PPE)

Remember *E-SEAP* as your aide-memoire.

An engineer who was key to the development of NASA's Saturn V rocket, responsible for mankind's six forays to the Moon, characterized risk in this way:

> You want a valve that doesn't leak, and you try everything possible to develop one. But the real world provides you with a leaky valve. You have to determine how much leaking you can tolerate.
>
> Arthur Rudolph, quoted in his *New York Times* obituary (in Bernstein, 1996: 2)

The Origins of Insurance as a Risk Transfer

As noted, there are several ways of dealing with a risk. One of these is to insure or transfer some of it against loss. Insurance works when the premiums of the many reimburse the losses of the few.

Following the Great Fire of London in 1666, there was a demand for insurance. At that time, businesspeople in London would meet in coffee shops; one of these people turned out to have a particular role to play in the history of insurance. Edward Lloyd opened a coffee shop on Tower Street, and by 1688 it had become firmly established. It was a popular place for London's sailors, so popular that it moved to larger premises in Lombard Street in 1691. Responding to the needs for shipping information from his customers, Lloyd provided a schedule of arrivals and departures of ships from the port of London. Thus, the *Lloyd's List* was born and later used by ships' captains to consider the risks associated with various shipping routes.

A-FACTOR 7

R = P x C (Risk = Probability x Consequences)

Ship owners seeking insurance would go to a broker, who in turn would 'sell' (or transfer) portions of the total risk to a few individuals, who would confirm their agreement to cover a percentage of any loss by signing their name to a contract. Such 'writing their name under' each other to cover the full value of the ship (or other asset) became known as *underwriting*.

By 1771, 79 of these underwriters had each subscribed £100 to form the *Society of Lloyd's*—these were the original 'Lloyd's Names'. The names committed all their assets to secure their insurance promise. That commitment was the principal reason for the rapid growth and excellent reputation, held to this day, of insurance underwritten at Lloyd's. After several relocations, it moved to its current location at One Lime Street and was opened by Her Majesty the Queen in November 1986.

The Practicalities of Understanding Risk

The basic system for the transformation of inputs to outputs and back to inputs were discussed earlier in this chapter (and see Figure 1.2). Figure 1.7 shows the reality of this for any business; this is the essence of enterprise. On the left-hand side are shown the

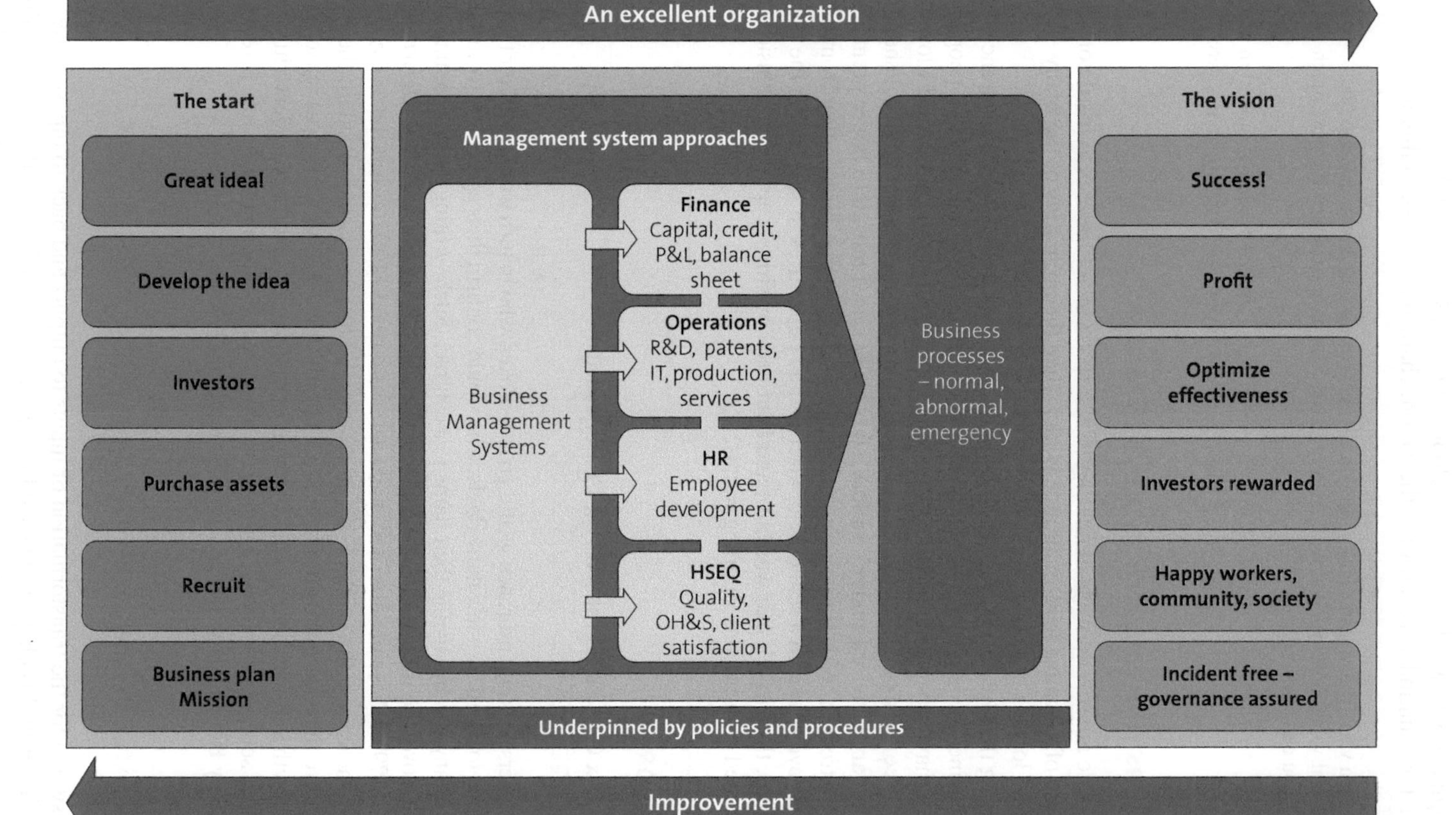

FIGURE 1.7 The essence of enterprise.

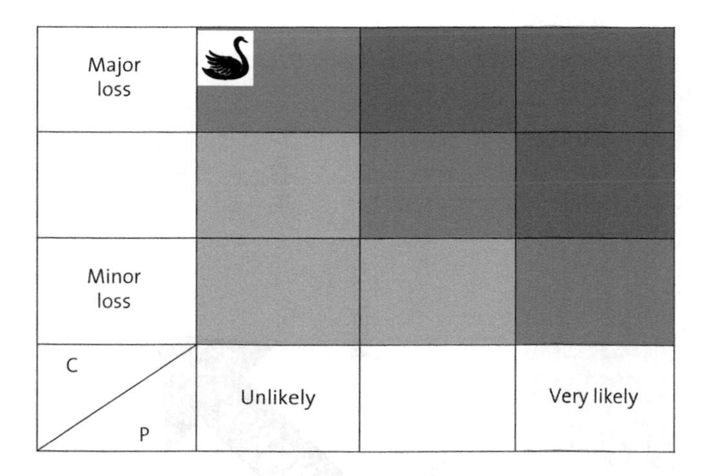

FIGURE 1.8 A simple risk-ranking matrix, showing the 'Black Swan' characteristic.

aspirations of the entrepreneurs, finding funding for the enterprise, and investment in the necessary resources. On the right-hand side is the *vision* for the achievement and overall success of the enterprise, stated in whatever terms the entrepreneurs have decided upon. Connecting the two sides is *risk management*; these are treatments delivered by the effective implementation of assured business management systems and business control frameworks (BMS/BCF).

A-FACTOR 8

Where the assessed inherent risks are judged to be potentially significant, managers and auditors alike should apply the 'Four Ts' and 'E-SEAP' to effectively mitigate these.

Of course, not all organizations are equally successful. Many examples of organizations which dared and failed are listed in the prefaces. How (even whether) any organization identifies and responds to the risks in its environment will be a significant feature in its success (or failure).

Figure 1.8 shows a simple risk-ranking matrix. You'll see the black swan presented in the top left corner. This represents the *Black Swan* category of risk which is characterized by very low probability and catastrophic consequence (Taleb, 2010). Correct use of the matrix provides for prioritization of management's (and auditor's) attention according to its relative position by using a hierarchy (E-SEAP) for selecting mitigation control(s).

Figure 1.9 shows how an organization has developed this approach to assess in greater detail a variety of potential risk areas, including risks to people, the environment, assets, and reputation and legal risks. This is often referred to as a 'five by five' matrix.

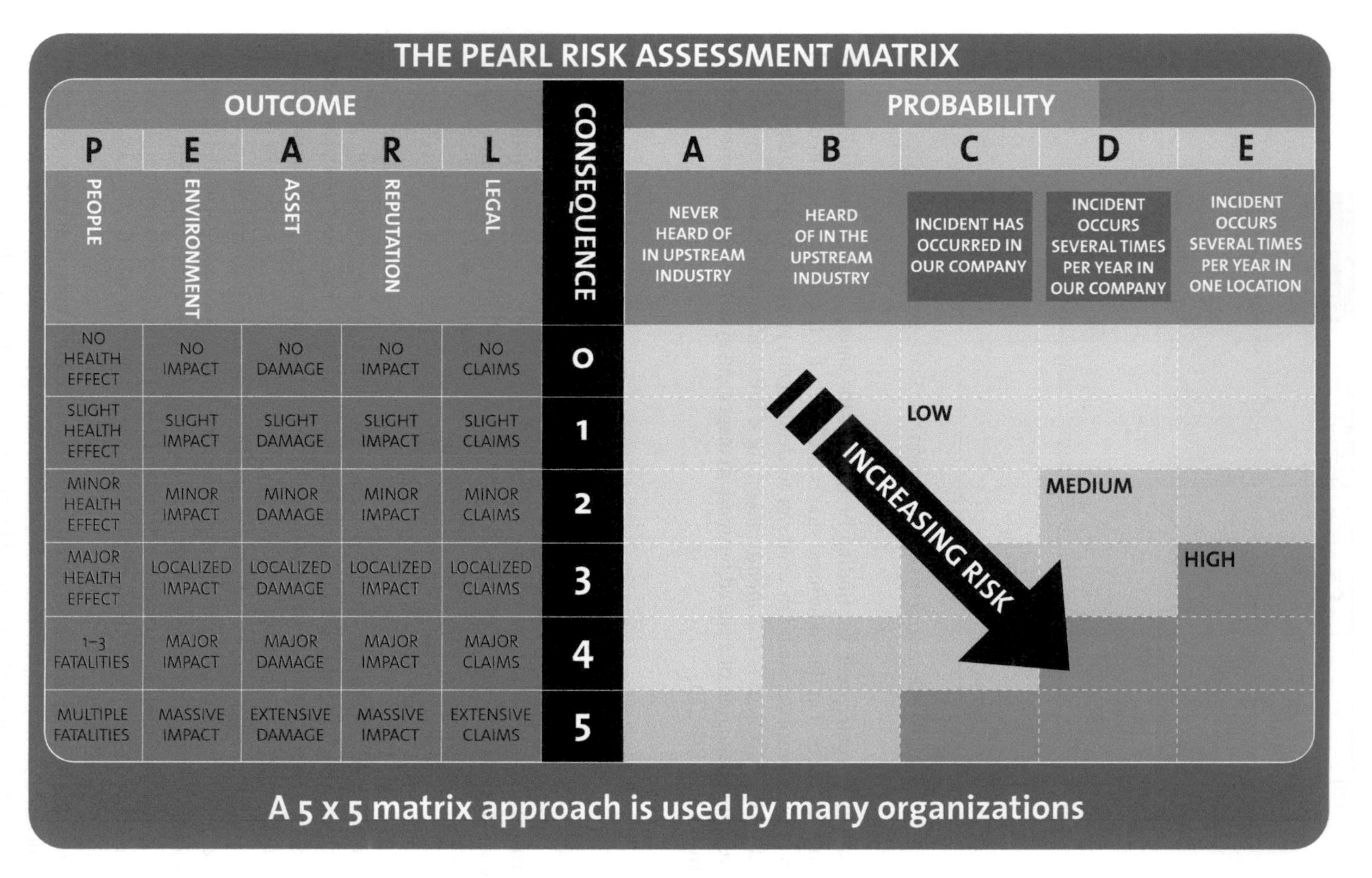

FIGURE 1.9 A more developed risk-ranking matrix, the PEARL matrix, used for assessing probability and consequence of people, environment, asset, reputation, and legal risks.

There are other numbers of graduations, and these are sometimes presented using different orientations. This matters not; it is the principle and the relative positions of the identified risks that matter, rather than any absolute 'scoring'—especially for auditors.

A-FACTOR 9

Recognize that ultimately an audit is an independent and balanced assurance to interested parties regarding an organization's ability to meet its business objectives in increasingly volatile business environments.

CCPS (2007: 214–216) reminds us of the importance of maintaining a dependable practice in hazard identification and risk analysis (HIRA), using seven steps, as follows:

1 Document the intended risk management system.
2 Integrate HIRA activities into the life cycle of projects or processes.
3 Clearly define the analytical scope of HIRAs, and assure adequate coverage.
4 Determine the physical scope of the risk system.
5 Involve competent personnel.
6 Make consistent risk judgements.
7 Verify that HIRA practices remain effective.

The greater the inherent risk, the greater the priority for response to the risk—terminate, treat, transfer, tolerate—using the hierarchy (E-SEAP) as described.

These days, some organizations (and most auditors) are rightly interested in activities with significant consequence potential, regardless of any probability. Some people (including me) call these very-high-impact losses *Black Swans*, as (like those birds) they don't come around very often. High-impact losses can end careers and organizations. Auditors should absolutely consider this combination when selecting at-risk activities for their work plan; you'll read more about this in Chapter 5.

BP CEO Tony Hayward in 2010 (he resigned later that year), commenting on BP's 'impossible' oil spill disaster in the Gulf of Mexico, which claimed eleven lives, killed thousands of animals, and spewed millions of gallons of oil into the sea (quoted in CNN, 2010).

> We're sorry for the massive disruption it's caused their lives. There's no one who wants this over more than I do. I would like my life back.

Just how low a residual risk should be depends upon the 'appetite' for risk in the management of the organization. Some readers will be familiar with terms such as ALARP/ALARA—'as low as reasonably practicable/achievable'—or 'so far as is reasonably practicable (SFRP). These concepts are summarized on page 97. The bottom line, however, is that each risk must be managed well enough to prevent significant impacts upon objectives.

The 'Four Ts' illustrated in Figure 1.6 assist a manager in maintaining or excluding activities, processes, or services. Where 'Treat' is the selected strategy, there is an established hierarchy of control, which I referred to earlier as E-SEAP, repeated for prominence:

- Eliminate
- Substitute
- Engineering controls
- Administrative controls
- PPE

Auditors and Risk

An essential first step for an auditor is to independently consider 'significant risks to objectives' in the *Context* of the business environment where the auditee's organization operates. It is unlikely that any two business environments are the same, even if they are in the same county or country. In my auditing process, time is specifically allowed to do this, and you'll read how to conduct this analysis in Chapter 6.

However, in summary, the process for estimating risks is by using a risk assessment approach to assess qualitatively the significance of each identified area of potential risk, using the independent variables of probability and consequence. While the auditee's (H&S, E, or Q) managers might have done this, you'll be repeating it INDEPENDENTLY. In effect, you as an auditor will be following in their footsteps in order to select a sample of significant risks for review and verification in your audit team's work plan.

The three risk assessment questions I proposed back in 2005 (Asbury, 2005) invariably assist auditors (and no doubt they may assist management) to decide on the significance of the risks they have identified:

1. How often will this happen (the likelihood, frequency, or probability)?
2. How big could the impact be (the severity, impact, or consequence)?
3. Who might be affected by any occurrence (which interested parties, stakeholders)?

Note that 'interested parties' (also known by the interchangeable term 'stakeholders') includes five specific groups—shareholders, employees, suppliers, customers, and society at large. You'll read more about interested parties in Chapter 2; and see Figure 2.11.

I recommend that HSEQ auditors focus upon the relative rather than absolute positions of risks identified within the auditee's activities. A useful idea is to focus on the top (say) six to twelve risk areas, ultimately selecting a sample size determined by the time and auditor resources available (aka *intensity*) to the audit team. With more time, an audit can focus on a larger sample.

There are many quantitative risk evaluation or estimation methodologies and software toolkits available within organizations and commercially. Too much focus on precise risk-scoring by auditors can easily become counterproductive; this is supported by my experience of over 1000 audits and feedback from over 15,000 training course delegates. Therefore, it is wise to avoid the 'numbers game' (Asbury, 2005). Qualitative methodologies, such as simple matrices, are generally better suited for use by HSEQ auditors, who will have less time than the full-time site managers to select and investigate by auditing a sample of risks.

In Chapter 2, I will describe the evolution, development, and rise of business management systems (BMS), and business control frameworks (BCF), aka *management system standards* (MSS), as a reliable approach to making a successful transition from business vision to business reality. This, of course, includes compliance with legal and other mandated requirements.

Reasonably, some may ask where HSE and other laws fit into all of this. Legal requirements for worker health and safety were first established in the UK in 1802 and significantly amended in 1961, 1974, and 1992. England's first environmental law came into force in 1307, when Edward I ('Longshanks') issued a proclamation prohibiting the burning of sea coal in London while Parliament was in session (Willis Corroon, 1996), while today in the UK, organizations are required to adhere to the newest legal requirements for Integrated Pollution Prevention and Control (IPPC). In the twenty-first century, organizations are increasingly likely to use systematic techniques to meet their legal requirements. BCF HSEQ MSS have provided systematic solutions to compliance, and such frameworks commonly require the identification of relevant legal and other requirements as a mandatory element of conformity (e.g., ISO 45001:2018 clauses 6.1.3 *Determination of legal requirements and other requirements*, and 9.1.2 *Evaluation of compliance*).

In Chapter 2, I will also highlight the relatively new and important theme of corporate social responsibility (CSR) within a risk-based approach, as it relates to the increasing expectation upon organizations from interested parties for transparency of operations and control and remediation of any consequent impacts. The business response to CSR has led to both effective responses with meaningful improvement targets and lots of 'greenwash' (such as pictures of waterfalls, trees, and smiling children). My risk-based approach to HSEQ auditing supports the former and meaningfully addresses the latter.

Test your knowledge with the first chapter assessment test (see Appendix 4).

Management Systems and Business Control

2

Only by using a *structured means of control* can an organization convert high-cost controls into business-assuring, profit-enhancing control.

—Dr Stephen Asbury

We have stable international HSEQ standards for the foreseeable future:

ISO 9001:2015 was reviewed and confirmed in 2021. It will not be reviewed again before 2026.

ISO 14001:2015 was reviewed and confirmed in 2021. It will not be reviewed again before 2026.

ISO 45001:2018 was reviewed and confirmed in 2022. It will not be reviewed again before 2027.

—International Organization for Standardization (ISO)

INTRODUCTION

For a summary of this chapter, you can listen to the author's Microlearning™ presentation, which you'll find on the book's Companion Website. Then, delve deeper by reading on for further details on the Companion Website. (See Appendix 4.)

Over the years, there has been an evolution of organizational controls from 'nothing' to a set of legal requirements (particularly for health and safety and, latterly, environmental protection), to engineering and organizations' standard-setting, up to the latest management thinking exemplified by Dekker (2014) in *Safety Differently*, Hollnagel (2014) in *Safety-II*, and *Adaptive Safety* as discussed by Rae and Provan (2019). Table 2.1 summarises the evolution of safety controls from the 1900s to date, including of management system standards (MSS). Each step along this journey seems to have improved control and reduced losses.

DOI: 10.1201/9781003364849-3

TABLE 2.1 Origins of the Major Schools of Safety Thinking

TIME	*MAJOR SCHOOL*	*SUMMARY*
1900s and beyond	Moral responsibility	Theories inspired by engineering, physical sciences, epidemiology, sociology, psychology, and anthropology led to a belief of moral responsibility to engineer or organise preventive measures (Asbury & Ball, 2016).
1910s and beyond	Taylorism and procedures	Scientific management and the relationship between work and rules (discussed in Chapter 1). An imprint that workers need to be told what to do, and the need for them to be supervised.
1920s and beyond	Accident proneness	Use of psychology and eugenics to explain patterns in industrial safety data; that human performance is variable in ways that could explain accidents (Sass & Crook, 1981).
1930s and beyond	Heinrich and behaviour-based safety	Heinrich (1931), reviewed by Manuele (2002, 2011), promoted the idea that accidents and injuries were preventable, using a row of dominoes to explain how distant causes can lead to injuries and how the sequence could be broken by removing the causative factor in the sequence. Krause (1990) built on Heinrich's thinking using observations to eliminate workers' unsafe acts.
1940s and beyond	Human factors and cognitive systems engineering	Human factors emerged from engineering psychology to represent an important 'hinge' between human, systems, and safety (Reason, 1990, 2013).
1950s–60s and beyond	System safety	The earliest commitments that safety should be built into the system to map and resolve conflicts between safety and other factors.
1970s and beyond	Man-made disasters	Safety taken from the engineering space following high-visibility disasters. Disasters with socio-technical systems bring accidents centre stage, setting the agenda for conversations still being conducted today.
1980s and beyond	High-reliability organizations	Approaches (Carter, 1986) that emerged, predominantly in the US, from societal preoccupation with preventing disasters—'is there a limit to the complexity we can handle?' and 'were there things we should not build or do at all?'
1990s and beyond	Safety management systems	Deming (1982) provides a framework for quality management (discussed in Chapter 2). The Swiss Cheese model (Reason, 1990) became an important icon for systematic 'barriers' or a 'defences-in-depth' approach.
2000s and beyond	Safety culture, process safety, lean safety	Encouragement to develop a safety culture (Cooper, 1998) to focus on things that can be found and fixed before they contribute to an accident. Understanding human error (Reason, 2013). Process safety becomes mainstream post-BP Texas City (CSB, 2007; Hopkins, 2009). Lean safety (Hafey, 2009).
2010s and beyond	Resilience engineering, behavioural economics, safety differently, safety-II, safety anarchy, adaptive safety	Identifying and enhancing the positive capabilities of people that allow them to adapt/understand and enhance how people build adaptive capacity to function with imperfect knowledge. Behavioural economics or nudge theory (Marsh, 2013). Safety differently (Dekker, 2014; Knutt, 2016). Safety-II (Hollnagel, 2014). Safety anarchy (Dekker, 2017). Adaptive safety (Rae & Provan, 2019; Mindfulness, behaviours (Kao et al., 2019).

For example, in the UK in 1974, there were 651 fatal injuries to workers (HSE, 2015). When comparable data is examined for 2020–1 (i.e., adjusted to mirror the reporting standards of 1974), this shows 92 fatal injuries to employees. This represents an overall decrease of 86 per cent. Likewise, over a similar period, there was a 77 per cent reduction in reported non-fatal injuries (HSE, 2016).

Answering the question 'why have these improvements occurred?' is rather more complicated. Superficially, you might say 'because we have tried harder to get it right'. Space here does not permit a full examination, but I'll point out some of the leading research and other opinions. Certainly, there are two initial competing factors.

The first of these factors is the number of people engaged in work every day. It has increased in most territories—in the UK, from 20 million people in 1974 to 32.75 million people by January 2023 (Statista, 2023b).

The second factor is the hazards that workers encounter in their employment following progressive changes to the industrial architecture. For example, in the UK, between 1978 and 2015, the number of workers engaged in manufacturing, mining, and quarrying fell from 26.4 per cent to 8.1 percent, while those in the service sector rose from 63.2 per cent to 83.1 per cent (ONS, 2016). Offices generally present fewer hazards than factories and quarries. To contextualize the situation in your own industry and/or country, I encourage you to review the data from local sources.

In some quarters, health and safety has been perhaps overplayed, with *risk assessment* used as throwaway language for almost anything related to OH&S. Ball and Ball-King (2011) say that, done well, risk assessment is highly-beneficial—it saves lives and prevents injuries, when we focus on the risk priorities. But done badly, they say, it can damage public life. Some may not properly engage common sense or line management to determine the magnitude of the 'real' risk. Between 2012 and 2022, the UK OH&S regulator established an initiative to 'myth-bust' some of the decisions alleged to be for 'health and safety reasons'. Of over 400 cases heard, over 90 per cent were decided as 'not related to health and safety'. Figure 2.1 provides a for-fun image of health and safety gone mad!

But irrespective of 'the numbers' and 'the myths', I also believe we're managing our processes better these days in many sectors, through some supply chains, and within some individual organizations. During the research for this fourth edition, I conducted a semi-structured review of organizations using certificated (and non-certificated) PDCA-framed quality, environmental, and OH&S MSS which I shall summarise. Registered ISO 9001 systems for quality, ISO 14001 for environment, and ISO 45001 for occupational health and safety management systems (OH&S-MS) (amongst others) have in some cases become preferred or even qualifiers for the submission of tenders and for the conduct of work in many sectors including aerospace, construction, oil and gas, public sector, utilities, rail, and others.

I have worked with scores of organizations using such structured means of control. As you will read in this chapter, the position on the effectiveness of OH&S-MS (Robson et al., 2007) from an academic perspective remains contested for now. However, that systematic review of the literature from 1887–2004 is now 20 years old. I have discussed its updating with its original lead researcher Dr Lynda Robson, and we have agreed that this is overdue. In my opinion, it would be desirable to update this research in the (near) future.

In the meantime, the number of registered users of MSS in the field is growing (as you may imagine, data on non-registered MSS is harder to come by). The first ten

FIGURE 2.1 Business control gone mad—for safety, please use a life jacket during water activities.

Source: Photograph by David and Rachel Brown, used with permission.

ISO 45001:2018 registrations/certifications in the world were announced and widely communicated in March 2018:

- CBRE
- Colas Rail
- EMCOR UK
- Eurovia UK
- Interserve
- Morgan Sindall
- OKI UK
- OPG
- Overbury
- Ringway Jacobs

As you may have read in the preface to this edition, the latest available data is from the 32nd annual ISO survey (ISO, 2023a) for the year to 31 December 2022. Whilst there are some issues in the data (Oxebridge, 2022) relating on one hand to 'fake' data from China and, on the other, to improved reporting by UKAS due to its new *CertCheck* database, the latest survey reports that there were 2.4 million valid ISO certifications in 12 specified *Type A* MSS covering 3.2 million permanent work or service sites, including

- 1,265,216 certificates covering 1,666,172 sites for ISO 9001:2015, which represents a 17% increase in certificates and a 15% increase in sites on the previous year (2021 = 1,077, 884 certificates and 1,447,080 sites);
- 529,853 certificates covering 744,428 sites for ISO 14001:2015, which represents a 26% increase in certificates and a 22% increase in sites on the previous year (2021 = 420,433 certificates and 610,924 sites);
- 397,339 certificates covering 512,069 sites for ISO 45001:2018, which represents a 35% increase in certificates and a 38% increase in sites on the previous year (2021 = 294,420 certificates and 369,897 sites); and
- 213,638 certificates for nine other ISO MSS (including ISO IEC 27001:2013, ISO 22000:2018, ISO 13845:2016, ISO 50001:2018, and ISO 20000-1:2018), covering 327,217 sites (ISO, 2023a).

Note that a *Type A* MSS contains requirements against which an organization can claim conformance, whereas a *Type B* MSS does not. This latter typically presents guidelines or supporting information on a *Type A* MSS.

When I worked with McDonald's (the burger and fries company) in the late 1990s, I was surprised to observe the degree of standardization across their business. And not just on the menus. There was a common standard for fire protection (Ansul). A standard design for non-slip footwear (Shoes for Crews). And the *Safety Project—Vision for the Future* standard which I co-authored for issue to almost 40,000 restaurants.

Across the millennium period, my client was Coca-Cola in Central and Eastern Europe, a business of over 80 bottling plants in 27 countries. I co-authored HSE and fleet safety standards (known as 'Loss Prevention Standards' in the company) and conducted structured and thematic audits against these to identify and initiate risk improvements.

For most of the first 20 years of the twenty-first century, I was engaged by an alliance of competency providers founded by BP and Shell, and growing to include Chevron, Repsol, Halliburton, Oxy, and Saudi Aramco, together close to 40 per cent of global energy supply. Each had implemented PDCA-framed HSE-MS in their global operations.

A semi-structured review of websites of some of the world's largest and/or well-known organizations, conducted during my preparations for this edition, adds the following A-Z of organizations which have adopted the principles of PDCA-MSS:

Airbus, Alstom, Anglian Water, Apple, Belgrade Airport, Birmingham Airport, BMW, Daimler, Emirates Flight Catering, Fiat Chrysler, First Abu Dhabi Bank, Ford, General Motors, Ghana National Gas, Fujitsu, J. C. Bamford Excavators Limited (JCB), KonicaMinolta, Korita Aviation, Lufthansa, Menzies, Qatar Airways, Nakilat, Nissan, NVIDIA, RBI Hawker, Saab, Shanghai Automotive, Sheremetyevo International Airport, Skanska, Taiwan Semiconductor, Tesla, ThayerMahan, Thyson, Toyota Motor Manufacturing, U&M Minerals and Construction, United Utilities, Volkswagen, Weber Saint Gobain, and Yorkshire Water.

Most of these are registered as compliant with their MS standard/s by recognized external certification or registration bodies. If you wished to check the HSEQ and other MSS certification status of your competitors, peers and customers, there is a useful certification verification tool provided by British Standards Institution at www.bsigroup.com/en-GB/validate-bsi-issued-certificates/.

In conclusion, the context for HSEQ-MS is becoming such that not being ISO-certified could become a barrier to doing business in some sectors. If you chose to adopt such an approach, it seems that you would be in good company. But we must review their efficacy.

RESEARCH INTO THE EFFICACY OF HSEQ MANAGEMENT SYSTEMS

> Millions of companies around the world have adopted management system standards to both convey superior operational performance and to improve their operations. Yet because these standards impose requirements on operational processes and procedures, it is largely unknown whether adopting these standards actually bears any relationship with operational performance.
>
> (Viswanathan et al., 2021: 1).

HSEQ-MS

Bandura (1997) advises that those that choose to adopt a high self-efficacy expectancy—those who want to improve—are likely to be more successful than those with low (or no) self-efficacy expectations—those with little or no desire to improve. Heras et al. (2002) agree that better-performing companies are indeed the ones that tend to seek such certification/s of their management systems. Hafey (2009) says that lean management moves the focus from compliance with legislation to a focus upon improvement initiatives. More recently, Viswanathan et al. (2021) agree with Heras et al. (2002) that OHSAS 18001—the predecessor standard to ISO 45001:2018—attracted already-safer establishments; in other words, that certification was a signal of superior *ex ante* performance.

Research from Australia, the UK, and the US, including Heras et al. (2002), Gallagher et al. (2003), Corbett et al. (2005), Bennett and Foster (2007), and BAB (2015), show that organizations with certified management systems achieved superior return on assets and/or perform better than non-certified organizations. Christman and Taylor (2002) found that being certified to ISO 9001 is rewarded by buyers and NGOs, accelerated organizational growth (Terlaak & King, 2006), led to superior financial performance and growth (Levine & Toffel, 2010; Terlaak & King, 2006; Corbett et al., 2005), and has led to improved compliance (Gray et al., 2015).

Similarly, adopting EMS ISO 14001 has been shown to reduce toxic chemical emissions and to improve environmental regulatory compliance (Dasgupta et al., 2000; Potoski & Prakash, 2005a; Potoski & Prakash, 2005b).

Saraf (2019) analysed the efficacy of QMS ISO 9001:2015 across six hypotheses. Responses from 365 participants indicated that the competitive advantage for operational performance (identified as cost effectiveness, product innovation, on-time delivery, product performance, product quality and reliability, and organization's responsiveness) from implementation was more significant when compared to organizations in the same industry than when compared to other best-in-class organizations. This seems a significant finding—who would not wish to outperform the operational performance of their own sector peers?

OH&S-MS

The value and effectiveness of adopting an OH&S-MS is a contested field, and I will summarize the literature.

Some research suggests that OH&S-MS 'will fail routinely', though it is not always clear what is what is meant by 'fail'. Gardner (2000) reports the failure rate of quality management systems as '67–93%', and Robson et al. (2007) suggest that the failure rate of OH&S-MS 'would be at least as high'. This has not been my experience, though, as a consultant in the field, I suspect that I have tended to work with more committed clients, but I acknowledge the hypothesis.

The truth is that, in the field of OH&S, supportive (or confutative) research appears limited. Reviewing the literature published between 1887 and 2004, Robson *et al.* (ibid.) identified just 23 sources meeting their relevance criteria (and of these, only 13 meet quality criteria). The overarching relevance and quality criteria of this paper is where the contestation I have mentioned is based. Later, da Silva and Amaral (2019) identified a further 21 sources published between 2007 and 2018. Research presently in progress at Harvard University (Viswanathan et al., 2021) suggests that certification has led to improvements in safety performance. All three will be reviewed in greater detail in this section.

There is research correlating low injury rates to elements of OH&S-MS, including Cohen (1977). Kolluru et al. (1996) who, along with Dalton (1998), described similar systematic approaches and commended the benefits including reduced incident rates and increased worker morale. Like Cohen (ibid.), Mearns et al. (2003) show that a more developed OH&S-MS is correlated with lower injury rates. Bottani et al. (2009) studied 116 companies, encompassing OH&S-MS adopters and non-adopters, to assess whether adopters experience significantly higher performance against the following four criteria: (i) definition of safety and security goals and their communication to employees; (ii) risk data updating and risk analysis; (iii) identification of risks and definition of corrective actions; and (iv) employee training. They reported results that show that companies adopting OH&S-MS exhibit higher performance against all four criteria. These four criteria represent parts of Redinger and Levine's (1998) 'universal model' for an OH&S-MS. They will later also correlate with parts of ISO 45001 (ISO, 2018a).

Suan (2017) reports on a survey from construction, saying that results show in mandatory and voluntary OH&S-MS in organizations that the outcomes of safety performance and productivity are positive.

Tiwari and Shukla (2018) introduced an index to objectively quantify the effective implementation of an OH&S-MS in SMEs. They say that applying their index determines the effective implementation of the OH&S-MS and that this 'helps to reduce the accident rate and incident rate'.

There has been research on (so-called) world-class companies (including Collins, 2001; Morton, 2016). When OH&S professionals talk about world class, they generally mean best of the best, best in the class, best in the world as identified by Saujani (2016), based on Hansell's (2012) five key qualities found among world-class companies:

1 Visible senior management leadership and commitment
2 Employee involvement and ownership
3 Systemic integration of OSH and business functions
4 Data-based decision-making and system-based root-cause analysis
5 Going beyond compliance

On DuPont, Stewart and Stewart (2002) identify that in the world's safest companies, 'safety has unquestioned priority and meticulous attention is given to using the best safety practices'. Lorriman and Kenjo (1994) in their study of Japanese implant organizations in the UK are starker, saying the alternative to becoming world class and competing with global peers is 'to . . . go out of business'.

Of course, there must be some relationship between a system and its implementation. Unless it is effectively led and culturally normative, its potential may remain unfulfilled as exampled by Baird (2005). Likewise, reports on the 2005 *BP Texas City* explosion (such as CSB, 2007; CCPS, 2007; Atherton & Gil, 2008) describe an unfortunate focus on lower risk personal safety over higher risk process safety which was ultimately catastrophic.

Karapetrovic and Willborn (1998) explain that the (then) current trends in management point towards comprehensive management systems that, they said, provide for competitive performance. This aligns with my position that independently led OH&S-MS audits can be used to confirm compliance with the planned arrangements.

Gallagher et al. (2003) report a 'false sense of security' arising from the presence of an OH&S-MS. This position can be contested, as what is unclear is the sense of definition of the scope of the OH&S-MS when they say that 'the definitional requirements for an OH&S-MS have been watered down making it more likely that organisations can claim to have a system, but less likely that it will be effective'. I would argue that this cannot be so if an independent audit, which is now a mandatory part of common OH&S-MS including ISO 45001:2018, is competently led, methodologically correct, and followed up by committed leadership. It is, of course, what this book is all about, though (of course) I also acknowledge that organizations can still 'say anything' if they choose to—unless the audit report is published internally and/or externally and held to scrutiny. We will consider this point later the necessity of a holistic approach to management and independent auditing.

Despite their warnings, Gallagher et al. (2003), supported by Bennett and Foster (2007), also report that 'OH&S management systems can live up to their promise' and that an OH&S management system 'has the potential to provide a useful contribution to health and safety'. Nair and Tauseef (2018: 4) explain the potential of this opportunity, advising that

> results indicate that the company with a formal management system is highly committed to focusing on measuring the inputs into the system by using leading indicators thereby lowering their losses and injury rates. Whereas management system deficient organization focus more on failures to correct and improve.

There are, however, other voices including Baird (2005) who did not report a positive outcome in utilizing OH&S management systems, explaining that senior management was not sufficiently engaged. The company reported by Baird (ibid.) did not seem to understand 'audit' either; the approach described appears to be of periodic inspections of workers' behaviours instead of independent and systematic verification of the efficacy of the OH&S-MS. Darabont et al. (2020) warn that 'failure of OHS management system can have serious consequences on the quality management system and also the environmental management system'. As Baird (2005) confirms, this too would suggest a lack of senior engagement.

In 2007, Robson et al. published the results of the first systematic literature review which considered the effects of OH&S-MS interventions as reported in eight international, bibliographic databases from their inception (from as early as 1887) until July 2004. Using a search strategy (Robson et al., 2005) and deleting duplicates, 4837 sources were identified for review. Of these, just 23 (0.47 per cent) met the study's relevance criteria—in this case, a minimalist operational definition requiring a management element and at least one other element from the Redinger and Levine (1998) universal OH&S-MS framework. Nine of these sources related to legally mandated OH&S-MS, and 14 were voluntary. Thirteen of these (0.26 per cent from a total of five countries—Australia, Canada, Norway, UK, and USA) were reported as meeting quality criteria, with just one judged to be of high methodological quality.

In one of the 13 studies identified by Robson et al. (2007), Edkins (1998) reported particularly significant positive changes in the intervention group than the comparison group, although they question whether this was due to the OH&S-MS or, instead, to the personal qualities of the new safety manager. In another, Bunn et al. (2001) reported a 24 per cent decrease in injury frequency rates and a 34 per cent decrease in lost time over three years. Yassi (1998) reported a 25 per cent reduction in insurance workers' compensation costs, and Alsop and LeCouteur (1999) reporter a 52 per cent reduction in the same. The study's results were generally positive; however, the reviewers (Robson et al., 2007) concluded that the body of evidence was insufficient to make recommendations either in favour or against OH&S-MSs.

Nine years later, in their review of global occupational safety and health practice and accidents severity, Jilcha and Kitaw (2016) conclude that 'Even though, there are quit [sic] increasing research trends in the workplace safety and health control, they lack integrated and universal management system studies'.

Picking up (roughly) where Robson et al. (2007) left off, da Silva and Amaral (2019) published the results of their systematic literature review 2007–18. It was based upon systematic review of literature using the protocol of Preferred Reporting Items for Systematic Reviews and Meta-Analyses (PRISMA). After a search in the databases

Scopus, Science Direct, and Web of Science, applying inclusion and exclusion criteria, 21 articles in English language remained for analysis. Their results identified the methodologies, tools, and indicators used by organizations in OH&S management, highlighting weakness in the use of epidemiological indicators in the proactive management of OH&S, and the predominance of focus on occupational safety led to detriment of focus on occupational health.

There are no other high-quality, systematic reviews of the effectiveness of OH&S-MS, although there are some narrative reviews, including Frick and Wren (2000), Gallagher et al. (2003), and Saksvik and Quinlan (2003).

Viswanathan et al. (2021) raise two hypotheses:

- Will safer establishments adopt OHSAS 18001 (H1—selection effect)?
- Will adopting OHSAS 18001 lead to safer establishments (H2—treatment effect)?

The researchers obtained establishment-level injury data for 1995 to 2016 from the US Bureau of Labor Statistics' Survey of Occupational Injuries and Illnesses (SOII), corrected to per 200,000-hour incident rates (representing 100 full-time employees working 40 hours per week and 50 weeks per year). The SOII data set covers 2,187,431 establishments. They also obtained the details of every organization in the US certified to OHSAS 18001 between 1995 and 2018 from ten major international certification bodies—1381 certifications—as well as certification data for ISO 9001 and ISO 14001. Dun and Bradstreet NETS Manufacturing Database for 2014 identifies 4,027,800 establishments. The researchers used fuzzy, geocode, and manual matching of the SOII, certification, and NETS data sets. Of the 1381 certified organizations, 1309 had records in NETS, and 522 with records in both SOII and NETS. The data allowed for comparison between adopters and non-adopters; their sample comprised 107,513 establishments, including 279 certified to OHSAS 18001, for a total of 461,478 establishment years. Their results evidence a positive selection effect (H1); organizations with fewer injuries and illnesses were more likely to pursue certification.

A matched sample of 274 pairs was developed for use in a differences-in-differences model to evaluate the extent that adopting the standard (treatment effect H2) has a causal effect on annual injury and illness cases. The results indicate that OHSAS 18001 adoption prompts a 20 per cent decline in all injury and illness cases, a significant treatment effect. The researchers indicate future research on the effects of use of ISO 45001:2018, which replaced OHSAS 18001. They say that the greater emphasis it provides on leadership and worker engagement might mean that their results might underestimate the benefits it might yield. I look forward to seeing their future work.

Despite my personal views, reflected from my assignments, and in the adoptions and positive results reported by others, the systematic review by Robson et al. (2007) provides strong argument that at that time there was insufficient proof of effectiveness as most of the limited previous studies had flaws. The further reviews by Jilcha and Kitaw (2016) and da Silva and Amaral (2019) add little weight to that conclusion, as the high-quality literature base remains apparently so narrow. I am interested in the ongoing work of Viswanathan et al. (2021) and intend to follow it; it certainly accords with my own field experiences of almost 40 years.

Twenty years on from that first literature review by Robson (ibid.) from 1887 up to 2004, a further systematic enquiry becomes increasingly timely and desirable. As well as the peer-reviewed evidence, it should additionally contain field evidence. I have discussed this with Dr Robson (2017, *pers. comm.*), and we have a preliminary agreement to progress.

Now that we understand the potential merits of MSS, we'll move on to review the history of their use in business and organizational control.

A BRIEF HISTORY OF BUSINESS CONTROL

I've no idea how the ancient Egyptians (or whoever) built the pyramids in c. 2500 BCE, despite watching the National Geographic Channel—massive structures, huge stone blocks from miles away, great symmetry, and cosmic alignment. The Hammurabi Codex dating to 1750 BCE provided a collection of 282 rules, established standards for commercial interactions, and set fines and punishments, but hardly a 'management system'. More a *lex talionis* principle—'an eye for an eye'. Many ideas have been suggested, but there is little or no certainty. Similarly, we know very little about the absolute origins of management systems, but many people believe that General Sun Tzu was one of the first to understand the benefits of structure in control. Certainly, his book *The Art of War* (Tzu, 2009) has been regarded as a masterpiece of strategic thinking since its first translation into a European language in 1772. Its significance was quickly recognized, and such towering figures of Western history as Napoleon and General Douglas MacArthur have claimed it as a source of inspiration.

Sun Tzu

The earliest historical record of using a systematic framework for controlling an activity and giving some assuredness of a consistency of outcome possibly dates 25 centuries to the Zhou (say 'Joe') dynasty of ancient China. My suggestion is that the origins of modern management systems go back to the activities and writings of the military general, strategist, and philosopher Sun Tzu. Traditional accounts place Tzu as a military general serving under King Helu of Wu (*c*.544–496 BCE). One story from the legend of Sun Tzu's life is as follows:

> As a test of Sun Tzu as a general, the King required him to train the royal concubines as soldiers. Tzu organized the women into two groups, making the King's two favourite concubines the commanders. He ordered the 'soldiers' to turn right . . . but they only laughed; he repeated his order, with the same result. He then ordered that the two commanders be killed. When the King objected, Sun Tzu explained that it was a general's responsibility to ensure that the soldiers under his command understood the orders given to them—but that if the soldiers understood their orders and failed to follow them, the responsibility rested with the officers. Following the execution of the two women, the concubine soldiers followed their orders perfectly. Managers today might learn much from this story.

Sun Tzu's now famous book *The Art of War* presents a philosophy for managing conflicts and winning battles, set out in 385 points, in 13 chapters. It is accepted as a masterpiece on strategy and has been frequently cited and referred to by generals and theorists since its publication, translation, and distribution the world over. However, only in recent times (since the mid-to-late 1950s) has 'strategy' been associated with 'management' per se. For two-and-a-half millennia, 'strategy' was exclusively a military term, and a term, interestingly, defined as 'the art of war'.

The full text of *The Art of War* (provided by its publisher Pax Librorum) is available for download from this book's companion website (https://routledgetextbooks.com/textbooks/_author/asbury/).

From Sun Tzu through to the twenty-first century AD, there has been an evolution of management systems events and thinking. Some of the milestone events and a sample of the key thinkers responsible for this evolution are listed in Figure 2.2.

Fast-forwarding this evolution 2500 years, I and many others feel that another significant moment in the evolution of modern business control and management systems was initiated on 14 October 1900, in Sioux City, Iowa, USA, with the birth of W. Edwards Deming.

Deming

Dr William Edwards Deming (14 October 1900–20 December 1993; pictured in the 1950s in Figure 2.3) was an American with career experience across all the fields of statistics, lecturing, writing, and consulting. He trained to be an electrical engineer and worked briefly as an engineer in Chicago before becoming a statistician, working in the US Bureau of Census. His PhD, awarded by Yale, was in mathematical physics. He is best known for his work in Japan, where he taught senior management how to improve design, service, product quality, testing, and sales through various methods, including the application of statistical methods. His continuous quest for understanding processes and deviations from the norm led him to become one of the founding fathers of the quality movement.

In the years following World War II, Deming was called to work in Japan seven times between 1947 and 1965. Working with fellow Americans Walter Shewhart and Joseph Juran, he developed production and management theories that became known in Japanese industry as the 'right first time' philosophy. Through his work, Deming made a significant contribution to Japan's subsequent reputation for innovative, high-quality products and to its economic power. He is regarded as having had more impact on Japanese manufacturing and business than any other individual not of Japanese heritage.

> Management is nothing more than motivating other people.
>
> Every business and every product has risks. You can't get around it.
>
> Lee Iacocca, President and CEO of Chrysler, 1978–92

Deming's ideas were later exported around the world in the quality revolutions of the 1980s and 1990s. Academics and industrialists have credited Deming, Shewhart, and Juran with giving birth to an industrial revolution through their development of statistical control of quality levels into a new way of managing business.

1750 BCE	544-496 BCE	13-15th Century	18th Century	19th Century	1900-1950s	1960-70s	1980s	1990s	2000s	2010 →
Hammurabi Codex	Sun Tzu, *The Art of War*	Leonardo Pisano "Fibonacci", *Liber Abaci* Luca Pacioli	Adam Smith (1723-90)	First H&S Act (UK, 1802) Henri Fayol (1841-1925) Frederick W. Taylor (1856-1915) Henry Ford (1863-1947) Toyoda family / Sakichi Toyoda (1867-1930) Alfred Sloan (1875-1966) Max Weber (1864-1920) Tomas Bata (1876-1932)	Safety First Council elected (1916), renamed RoSPA (1941) British Safety Council founded (1957) Joseph Juran (1904-2008) Peter Drucker (1909-2005) William Edwards Deming (1900-93) Jehangir Tata (1904-93)	Konosuke Matsushita ("God of Management"; 1894-1989) Akio Morita (1921-99) Charles Handy (1932-) NEBOSH founded (1979) BS 5750 (1979)	Henry Mintzberg (1939-) IISO and IMSO merge to become IOSH (1980) Tom Peters (1942-) Elizabeth Moss Kanter (1943-) Kenichi Ohmae (1943-) NEBOSH qualifications become Certificate and Diploma (1987) ISO 9001 (1987)	Michael Porter (1947-) HSG65 (1991) First OH&S degrees (1991) BS 7750 (1992) Cadbury Committee (1993) BS 8800 (1996) ISO 14001 (1996) OHSAS 18001 (1999)	ILO-OSH 2001 Enron (2001) IOSH granted Royal Charter (2003) Sarbanes Oxley Act (2004) IOSH Individual Chartered status (2005) Nassim Nicholas Taleb (1960-) ISO 31000 (2009)	IOSH becomes largest H&S organisation in the world (2010) OSHCR (2010) ISO Annex SL (2012) ISO 9001 (2015) ISO 14001 (2015) ISO 45001 (2018) ISO 19011 (2018)

FIGURE 2.2 Milestone events and a sample of key thinkers in the emergence of risk management as a theme of interest tc individuals, organizations, and regulators.

FIGURE 2.3 W. Edwards Deming in the 1950s.

Source: Reproduced by permission of the W. Edwards Deming Institute.

Deming was the author of six books including the seminal 5th book Out of the Crisis (1982) which includes Plan-Do-Check-Act (see Figure 2.4), and his 'Fourteen Points for Management' (see Box 2.2). Space here does not permit me to discuss 'Diseases and Obstacles' and his 'Lesser Categories of Obstacles', but they are worth your additional reading. His 6th book The New Economics for for Industry, Government, Education (1993) includes his 'System of Profound Knowledge' (see Box 2.1), which provides a basis for application of the Fourteen Points. According to Deming (and see Box 2.1):

> The prevailing style of management must undergo transformation. A system cannot understand itself. The transformation requires a view from outside. My aim is to provide an outside view—a lens—that I call a system of profound knowledge. It provides a map of theory by which to understand the organizations that we work in. The first step is transformation of the individual. This transformation is discontinuous. It comes from understanding of the system of profound knowledge. The individual, transformed, will perceive new meaning to his life, to events, to numbers, to interactions between people. Once the individual understands the system of profound knowledge, he will apply its principles in every kind of relationship with other people. He will have a basis for judgement of his own decisions and for transformation of the organizations that he belongs to.

Deming said that, once transformed, the individual would

- set an example,
- be a good listener but not compromise,
- continually teach other people, and
- help people to pull away from their current practices and beliefs and move towards the new philosophy without a feeling of guilt about the past.

Plan-Do-Check-Act

At the heart of Deming's legacy to the business world is his adoption of four connected process improvement steps widely known today as the 'Deming Cycle' of PDCA (Plan-Do-Check-Act) (Deming, 1982: 86–90). This was later referred to by Deming (1993: 132) as Plan-Do-Study-Act (PDSA).

In his own writing and teaching, Deming called this cycle of activities the 'Shewhart Wheel', after his friend and mentor Walter Shewhart. Whatever it is called, this cycle or wheel can be used for various purposes, such as running an experiment:

- Plan (or design) the experiment.
- Do (or implement) the experiment by following the plan.
- Check (or verify) the results by testing.
- Act (to improve) upon the results.

My simple interpretation of the PDCA cycle or the 'Deming Wheel' is shown in Figure 2.4. It provides a cyclical process for determining the next action, the 'wheel' illustrating a simple approach to continual improvement through informed decision-making.

The shorthand PDCA mnemonic has borne the test of time, despite the efforts of standards writers, certification bodies, business sectors, consultants, and academics who

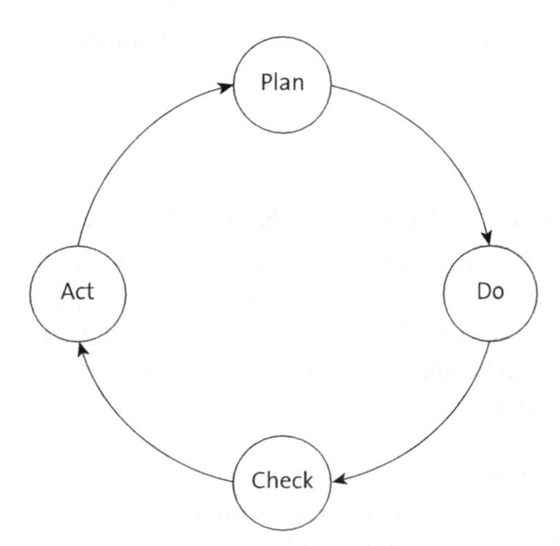

FIGURE 2.4 The PDCA cycle, commonly known as the 'Deming Wheel'.

have substituted Deming's simplicity with complexity. PDCA can also be applied as a 'wheel within a wheel' to illustrate the relationship of a single process to the corporate and strategic processes.

Deming saw that the elimination of waste could be achieved by aligning processes coherently and then carrying them out in a manner which was sufficiently close to the laid-down standards as it could be. The armaments industry was one industry which saw the potential of such an approach to manufacturing, since every time an item of munitions fails to explode upon impact, all the resources that had been consumed in the lead-up to launching the weapon had zero payback; the enemy's soldiers and equipment were not destroyed as intended, nor the war won.

Some observers feel that the outcome of the Falklands War (or, in South America, *Guerra de las Malvinas*)—an effective state of war between Argentina and the UK between 2 April and 14 June 1982 over the long-disputed territories of the Falkland Islands, South Georgia, and the South Sandwich Islands—might have had quite a different outcome if more of the bombs launched by the Argentinian air force had exploded on impact with their British targets. Would these munitions have exploded as designed if they had been manufactured and assembled by Toyota?

Management Truths

Deming set out his *System of Profound Knowledge* (Box 2.1) and his Fourteen Points of Management (Box 2.2) some forty years after his teaching had been listened to, accepted, and benefitted from by the Japanese.

Deming's ideas remain vibrant signposts for management system writers, managers, and auditors today. He saw immutable truths for systems of management; for example, a line manager must understand that all people are different and understand the interaction of psychology and statistical variation. The number of defective items that one quality inspector finds may depend upon the size of the workload presented to them, but another inspector, not wishing to penalize anybody from within their own team, may pass an item that is just outside the acceptable manufacturing tolerances.

BOX 2.1

DEMING'S SYSTEM OF PROFOUND KNOWLEDGE

Deming believed that to effect transformation of the style in which something is currently being managed there had to be an external perspective. He called this a System of Profound Knowledge; it was his approach to understanding organizations and had to be applied through the transformation of the individual, who, once transformed, would

- set a good example,
- be a good listener but will not compromise,
- continually teach other people, and
- help people to move into a new way of working.

The system can be illustrated in four parts which are all interdependent upon and interrelated to each other:

1 Appreciation for a system
2 Knowledge about variation
3 Theory of knowledge
4 Psychology

Therefore, leaders of organizations that required transformation, and the managers involved, needed to learn the psychology of individuals, the psychology of a group, the psychology of society, and the psychology of change. Some understanding of variation, including appreciation of a stable system, and some understanding of special causes and common causes of variation are essential for management of a system, including management of people.

Deming, W. Edwards, Out of the Crisis, *excerpts from the System of Profound Knowledge, © Massachusetts Institute of Technology, by permission of the MIT Press.*

BOX 2.2

DEMING'S FOURTEEN POINTS FOR MANAGEMENT

Deming's Fourteen Points for Management have been one of his abiding contributions to the transformation of organizations. He said that problem-solving, big or small, was insufficient. If management really wanted to signal that they intended to win in business and aim to protect stakeholders' interests, they had to sincerely adopt and effectively implement his points.

1 Create constancy of purpose toward improvement of product and service, with the aim to become competitive and to stay in business, and to provide jobs.
2 Adopt the new philosophy. We are in a new economic age. Western management must awaken to the challenge, must learn their responsibilities, and take on leadership for change.
3 Cease dependence on inspection to achieve quality. Eliminate the need for inspection on a mass basis by building quality into the product in the first place.
4 End the practice of awarding business on the basis of price tag. Instead, minimize total cost. Move toward a single supplier for any one item, on a long-term relationship of loyalty and trust.

5 Improve constantly and forever the system of production and service, to improve quality and productivity, and thus constantly decrease costs.
6 Institute training on the job.
7 Institute leadership. The aim of supervision should be to help people and machines and gadgets to do a better job. Supervision of management is in need of overhaul, as well as supervision of production workers.
8 Drive out fear, so that everyone may work effectively for the company.
9 Break down barriers between departments. People in research, design, sales, and production must work as a team, to foresee problems of production and in use that may be encountered with the product or service.
10 Eliminate slogans, exhortations, and targets for the workforce asking for zero defects and new levels of productivity. Such exhortations only create adversarial relationships, as the bulk of the causes of low quality and low productivity belong to the system and thus lie beyond the power of the workforce.
   - Eliminate work standards (quotas) on the factory floor. Substitute leadership.
   - Eliminate management by objective. Eliminate management by numbers, numerical goals. Substitute leadership.
11 Remove barriers that rob the hourly worker of his right to pride of workmanship. The responsibility of supervisors must be changed from sheer numbers to quality.
12 Remove barriers that rob people in management and in engineering of their right to pride of workmanship. This means, inter alia, abolishment of the annual or merit rating and of management by objective.
13 Institute a vigorous program of education and self-improvement.
14 Put everybody in the company to work to accomplish the transformation. The transformation is everybody's job.

Source: Deming, W. Edwards, Out of the Crisis, *Fourteen Points for Management,*

As mentioned, fear invites 'improved' figures, and bearers of bad news often fare badly in all but the most generative, culturally advanced organizations. And so, to keep their jobs, people present to their boss only good news; a committee appointed by the CEO of a company may report what the CEO wishes to hear. Would they dare report otherwise?

> [Shell] had been engaged in accounting manoeuvres since 1997–98, including a flawed internal audit function; Shell had engaged as [group reserves auditor] a retired Shell petroleum engineer—who worked only part time and was provided with limited resources and no staff—to audit its vast worldwide operations.
>
> Michaels et al. (2004)

> [On Enron] A eureka moment. It suddenly struck Mintz as so obvious. The executives entrusted with reviewing all of the LJM transactions . . . approached their duties casually, giving everything just the onceover. They seemed to figure that somebody else was doing the tough analysis. But no one was.
>
> Eichenwald (2005)

There are also two other 'fears' important enough to raise here—managers' or auditees' fear of finding out, or fear of being found out (FOFO or FOBFO). Finding out that something is not as it should be negates 'conscious avoidance' and the possible defence of 'plausible deniability'. The fear of being found out, or exposed, is sometimes called the 'imposter syndrome'. Emetophobia is the sick fear of secrets being revealed. We will discuss these 'fears' in more detail in Chapter 4, *Relationships with Auditees*.

Other observations by Deming are also relevant to today's corporate practices; for example, accounting-based key performance indicators potentially drive managers and employees to achieve targets for sales, revenue, and costs by manipulation of processes and the way in which data is presented.

MODERN HSEQ MANAGEMENT SYSTEMS AND PDCA

Modern management systems probably emerged from US Department of Defence (DoD) Military Quality Standard MIL-Q-9858 (Lucius, 2002). Between 1958 and 1996, MIL-Q-9858 provided guidance for a quality management system for defence work. When it was withdrawn without replacement, organizations, suppliers, and buyers moved to the then-new ISO 9000 suite of quality standards. The ISO 9000 suite had itself emerged from BS 5750, first published in 1979. A timeline of the development of ISO 9001, ISO 14001, and ISO 45001 since 1979 is provided in Figure 2.5.

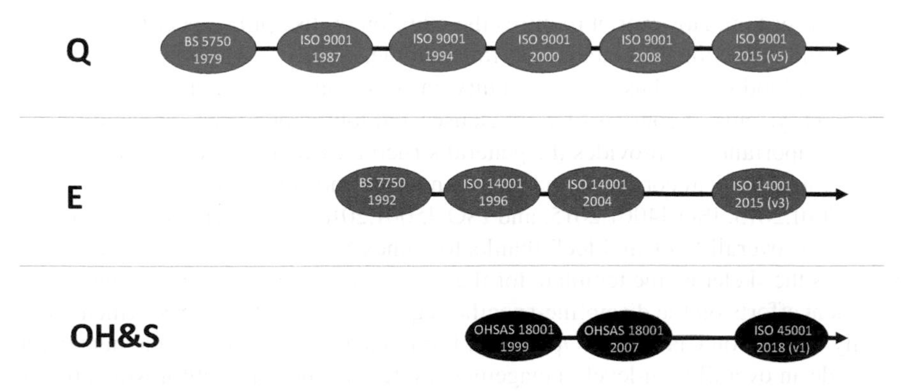

FIGURE 2.5 Timeline since 1979 of certifiable Quality (Q), Environment (E), and Occupational Health and Safety (OH&S) management system standards

As you will read, thanks to ISO Annex SL, virtually all HSEQ (and other) management systems, frameworks, or business control implementation models in use today have elements that can be mapped to the four interconnecting stages of the Deming Wheel (PDCA). Whether owned by the International Organization for Standardization (ISO), national certification bodies, sector groups, or by an organization developing its own framework, the overall approach used will generally align to PDCA.

A sample of HSEQ and other management system standards and reference frameworks is available for download from the companion website (https://routledgetextbooks.com/textbooks/_author/asbury/).

Annex SL

In April 2012, ISO published its 'Annex SL' (ISO, 2012a; previously ISO Guide 83) of the *Consolidated ISO Supplement of the ISO/IEC Directives*. This has had (and will continue to have) significant impacts on all management system standards owned by ISO for writers, implementers, and management systems auditors.

In the last ten years or so, many organizations have become interested and started to seek to implement and certify multiple management system standards in an integrated approach. I included separate illustrations of the structures of ISO 9001:2015, ISO 14001:2015, and ISO 45001:2018 in the Preface (see pages xxxiii–xxxv). And now here, I suggest a structure for an integrated ISO 9001/14001/45001 management system (on page 73). This move towards systems integration has led to a need for organizations to be able to combine these standards in an effective and efficient manner. And to audit them in an integrated manner too. While the main clauses of these MSS map across to PDCA, the reality of the detail is that they also have requirements, terms, and definitions that are subtly or substantially different. In the past, this had caused confusion and inconsistent understanding and implementation.

All ISO technical work, including the development of standards, is carried out under the overall management of the Technical Management Board (TMB). ISO/TMB produced Annex SL with the objective of delivering consistent and compatible MSS.

The Annex (ISO, 2012a) describes the high-level structure (HLS) for a generic management system. However, it requires the addition of discipline-specific requirements to provide fully functional standards for, for example, quality, environmental, service management, food safety, business continuity, information security, or energy management systems. It consists of eight clauses and four appendices. Appendix 3 is especially important as it provides the general structure, which is in three parts: high-level structure, identical core text, and common terms and core definitions.

ISO 9001:2015, ISO 14001:2015, and ISO 45001:2018 (and other ISO MSS) now have the same overall 'look and feel' thanks to Annex SL. For MSS writers, Annex SL provides the skeleton, the template for their work. They can concentrate their development efforts on the discipline-specific requirements of their MSS, which will be mainly focused on Clause 8—Operation. For management system implementers, it will provide an overall high-level management system framework within which they can select which discipline-specific standards (HSEQ? security? IT? food safety?) they need. It is very much the intention of the standard to remove the conflicts and duplication, confusion, and misunderstanding that arise from different MSS. ISO intends that all their MSS should be readily compatible.

For management system auditors, this will mean that all audits have a core set of generic requirements that need to be addressed, no matter the discipline. And for auditor training, it could drive the development of training addressing this common core set of requirements, with additional training for discipline-specific requirements.

The main clause numbers and titles of all ISO management system standards will be identical:

1 Scope
2 Normative references
3 Terms and definitions
4 Context of the organization
5 Leadership
6 Planning
7 Support
8 Operation
9 Performance evaluation
10 Improvement

The Scope, Normative references, and Terms and definitions will have content which is specific to each discipline, and each standard will have its unique bibliography.

ISO 9001, ISO 14001, and ISO 45001

Between 1958 and 1996, US Department of Defence (DoD) MIL-Q-9858 was the *standard document* of the defence contracting system. It provided guidance for quality management systems for thousands of defence contractors large and small. When it was cancelled in October 1996, without replacement, this community moved to adopt the ISO 9000 series of quality management standards in its place.

ISO 9001 was adopted by the DoD in April 2001. This resulted in a sharp upturn in the number of ISO 9001 registrations/certifications. This switch to what was effectively (then) an industrial quality standard allowed government departments to move out of the business of certifying quality in supply chains, instead leaving quality assurance to third-party auditors. This would pave the way for much of what was to come in the next twenty years and beyond.

Figure 2.6 shows the main terms used in the clauses of ISO 9001:2015, ISO 14001:2015, and ISO 45001:2018 as aligned to ISO Annex SL—Context of the organization, Determining the Scope of the MSS, the role of Leadership (and for ISO 45001, the importance of worker participation), Planning, Support and Operation, Performance evaluation and Improvement.

You will immediately see how these cyclical elements reflect Deming's recognition that senior management must *plan* what needs to be achieved, in both quantitative and qualitative risk-based terms, ensure an effective implementation in accordance with the developed plan (*do*), conduct monitoring and measurement to *check* (or study)

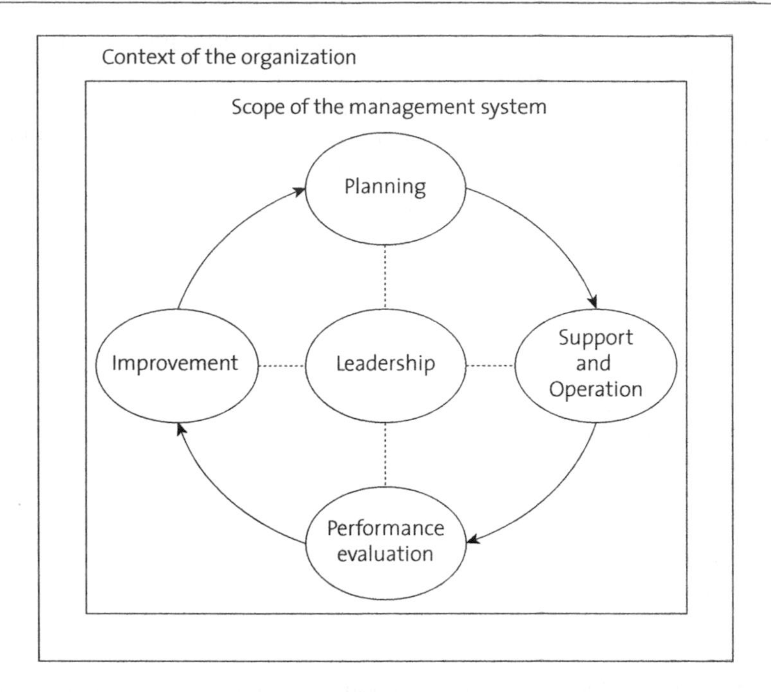

FIGURE 2.6 ISO 9001, ISO 14001, and ISO 45001—the elements of successful ISO HSEQ management.

what has been done, and *act* to reconfirm current practices, or alternatively to initiate improvement where this has been revealed to be necessary ('better next time').

Now that we have stable ISO standards for HSEQ for the foreseeable future, I have taken the opportunity for this fourth edition to present a suggestion for an integrated ISO 9001/14001/45001 MSS framework. It is shown in Figure 2.7. It combines the high-level structure from Annex SL with the clauses and numbering from this 'holy trilogy' of HSEQ standards. The elements shaded grey (8.2–6) relate only to ISO 9001:2015, and preparedness and response to emergencies (of all types) is moved to clause 8.7. Business continuity planning (BCP) is not explicitly addressed by any of these three standards but, in my opinion, could also be incorporated here. In effect, this could be to include the requirements of ISO 22301:2019 (ISO, 2019), another of ISO's externally certifiable MSS, if required.

ISO standards are available for purchase from www.iso.org/store.html. At the time of writing, ISO 9001:2015 and ISO 14001:2015 were each 145 CHF (about £130 GBP/$156 USD), and ISO 45001:2018 is 166 CHF (about £150 GBP/$180 USD). There are new ISO general guidelines for implementing ISO 45001:2018 (ISO, 2023c).

ILO-OSH 2001

Figure 2.8 illustrates the occupational safety and health (OSH) framework provided in 2001 (with a second edition following in 2009) by the International Labour Organization (ILO, 2009). The ILO aimed to create worldwide awareness of the dimensions and

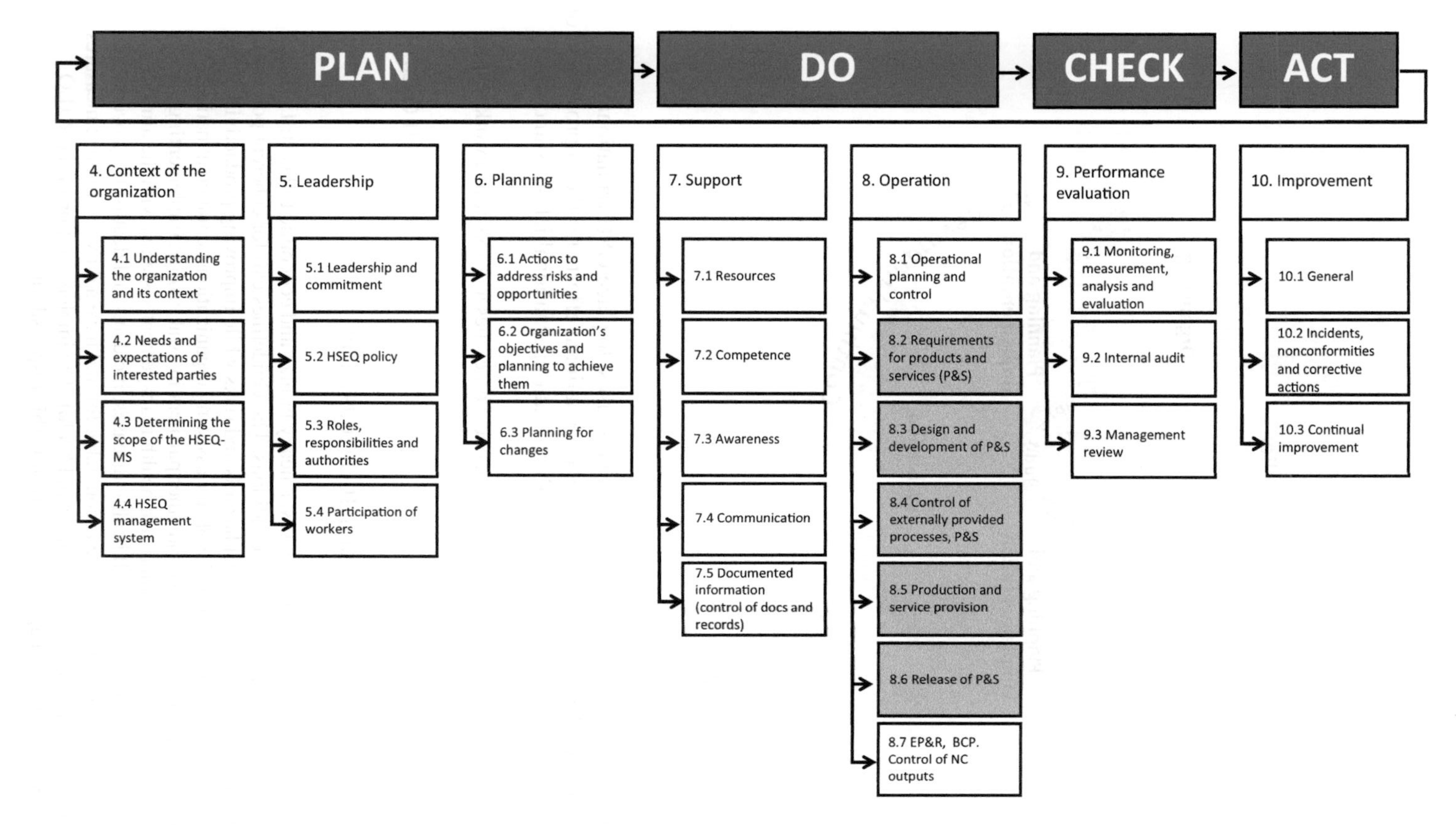

FIGURE 2.7 Possible structure for an integrated ISO 9001/ISO 14001/ISO 45001 (HSEQ management system)

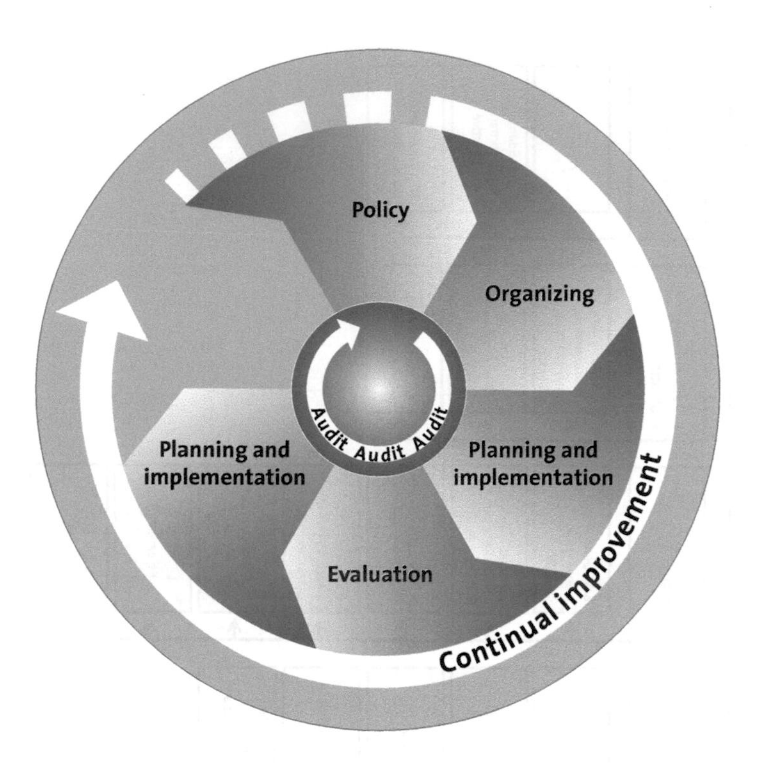

FIGURE 2.8 ILO-OSH 2001.

consequences of work-related accidents and diseases, to place OSH on national and international agendas, and to provide support to national efforts for the improvement of OSH systems and programmes in line with relevant international labour standards.

> Every 15 seconds, somewhere in the world, one worker dies and 153 have a work-related accident.
>
> ILO (2016)

Commenting on its OSH standard, ILO (2009) said:

> At the onset of the twenty-first century, a heavy human and economic toll is still exacted by unsafe and unhealthy working conditions. The Guidelines call for coherent policies to protect workers from occupational hazards and risks while improving productivity. They present practical approaches and tools for assisting organizations, competent national institutions, employers, workers and other partners in establishing, implementing and improving occupational safety and health management systems, with the aim of reducing work-related injuries, ill health, diseases, incidents and deaths. Employers and competent national institutions are accountable for and have a duty to organize measures designed to ensure occupational safety and health. The implementation of these ILO Guidelines is one useful approach to fulfilling this responsibility.

The ILO guidelines may be applied on two levels—national and organizational. At the national level, they provide for the establishment of a national framework for occupational safety and health management systems, preferably supported by national laws and regulations. They also provide precise information on developing voluntary arrangements to strengthen compliance with regulations and standards, which, in turn, lead to continual improvement of OH&S performance.

At the organizational level, the guidelines encourage the integration of OH&S management system elements as an important component of overall policy and management arrangements. Organizations, employers, owners, managerial staff, workers, and their representatives are motivated in applying appropriate OH&S management principles and methods to improve performance.

IOGP Guidelines

Figure 2.9 illustrates a now-withdrawn example of an HSE-MS which was developed by a sector for its members' use. These guidelines (OGP, 1994; now IOGP—International Association of Oil and Gas Producers) were prepared for the oil and gas sector's E&P (Exploration and Production) Forum by the Safety, Health and Personnel Competence (SHAPC) and Environmental Quality Committees (EQC), through their Health, Safety and Environmental Management Systems (HSE-MS) Task Force (OGP, 1994).

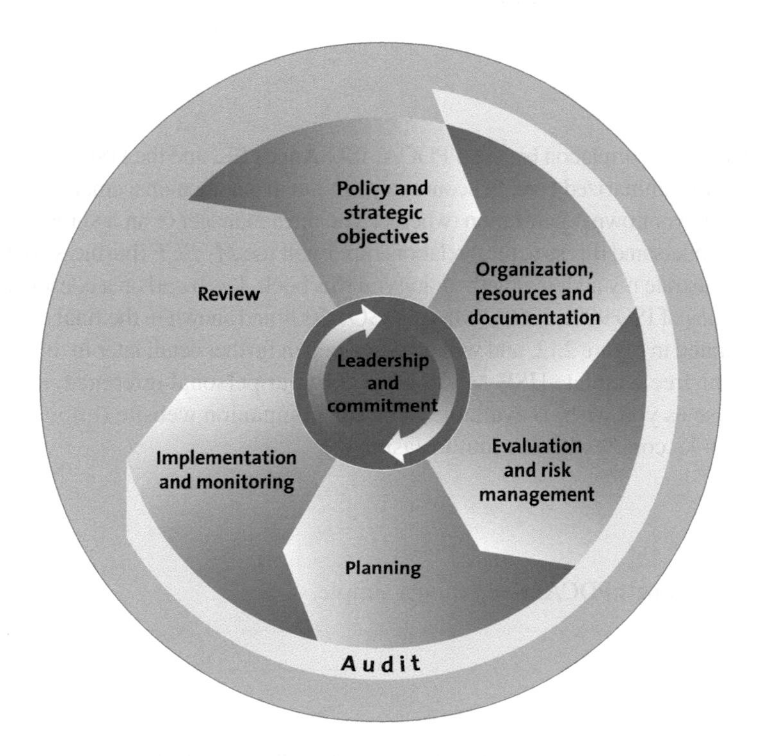

FIGURE 2.9 Example of a sector's own HSE-MS.

The IOGP HSE-MS guidelines were prepared to assist in the development and application of HSE in oil and gas exploration and production operations. The standards were particularly applicable to the sector because Forum members participated in their development, and accordingly, the guidelines had gained wide acceptance.

IOGP replaced its HSE-MS with Report 510 Operating Management System Framework in 2014 (IOGP, 2014). The Framework is designed to help companies define and achieve performance goals and stakeholder benefits, while managing the broad and significant range of risks inherent in the oil and gas industry. 'Operating' applies to every type of upstream or downstream company activity, from construction to decommissioning, throughout the entire value chain and life cycle of the business and its products. The Framework offers an integrated approach and the flexibility to address some or all of a wide range of risks, impacts, or threats related to occupational health and safety, environment, social responsibility, process safety, quality, and security.

Oil and gas exploration and production activities are subject to extensive legislation and regulation concerning HSE. All major organizations in the sector likely have their own HSE policies and strategies to satisfy their own operating and regulatory requirements, and management systems are a principal component of such strategies. The guidelines say that they were developed both by individual companies and by national and international bodies. There is wide recognition of the benefits of objective- or goal-setting approaches, a fundamental aspect of this type of approach which draws on the management principles of ISO MSSs.

Mapping HSEQ-MS

Table 2.2 maps out the correlation between PDCA, ISO Annex SL, and the HSEQ-MS frameworks I have summarized here. I encourage you to map out the management system framework used in your own organization (whether you are a manager or an auditor) so that you completely understand this structural relationship. I will use *My BCF* (business control framework) to illustrate my auditing methodology in this book. It is based on a consolidation of the current editions of ISO 9001, ISO 14001, and ISO 45001 and shown in the final column of Table 2.2, illustrated in Figure 2.12, and will be described in further detail later in this chapter.

A copyright-free template HSE-MS manual for your personal reference, to amend and adapt for use as you wish, is available from the companion website (https://routledgetextbooks.com/textbooks/_author/asbury/).

A-FACTOR 10

Remember Deming's PDCA. Keep things simple.

A-FACTOR 11

Do not permit the terminology or detail used to describe any MSS or business control framework to distract you from the structured simplicity of Plan-Do-Check-Act.

TABLE 2.2 Correlation between PDCA, ISO Annex SL, ISO 9001, 14001, and 45001, the former OGP HSE-MS, and My BCF

PDCA ELEMENT	*ANNEX SL*	*ISO 9001* *ISO 14001* *ISO 45001*	*ILO-OSH 2001*	*OGP*	*My BCF*
Plan	4. Context of the Organization 5. Leadership 6. Planning	4. Context of the organization 5.1 Leadership and commitment 5.2 Policy 5.3 Roles, responsibilities, and authorities 6. Planning (risks, opportunities, and objectives)	Policy Organizing Planning	Policy and strategic objectives Organization, resources, and documentation Evaluation and risk management Planning	Context Leadership Planning
Do	7. Support 8. Operation	7. Resources, Competence, Communication, Documented information 8. Operational planning, Management of change, Outsourcing/contractors, Emergency preparedness	Implementation	Implementation	Support and Operation
Check	9. Performance evaluation	9.1 Monitoring, measurement, Evaluation of compliance 9.2 Internal audit 9.3 Management review	Evaluation Audit	Monitoring Audit	Performance evaluation
Act	10. Improvement	10. Improvement, Non-conformity, and corrective action	Action for Improvement	Review	Improvement

Corporate Social Responsibility

As an element of overall corporate responsibility, corporate *social* responsibility (CSR) includes several HSEQ-related themes and will be of interest to some auditors. CSR is essentially about organizations deciding positively to progress beyond the baseline of legal compliance to integrate socially responsible behaviours as core values in recognition of the sound business and wider benefits of doing so.

Since organizations and the challenges they face differ widely, government interventions need to be carefully considered, well designed, and targeted to achieve their objective. The UK government's approach is to encourage and incentivize the adoption and reporting of CSR through best-practice guidance and, where appropriate, intelligent regulation and fiscal incentives.

According to Asbury and Ball (2016), the six key themes of CSR include the following:

- Legal requirements—satisfying corporate legal responsibilities
- Health and safety—protection of workers and others from the harms caused by work hazards
- Environmental sustainability—emissions to air, discharges to land and water, end-of-life disposal, biodiversity, climate change
- Ethical trading (value chain)—product safety, product quality, responsible marketing, ethical supply chain relationships
- Workforce rights—human rights, equal opportunities, fair pay, working time, employee engagement
- Community effects—community relationships, community development, emergency relief

CASE STUDY

CSR AND LABOUR RELATIONS

Hon Hai Precision Industry Co. Ltd., trading as Foxconn, is a Taiwanese multinational electronics manufacturing company. It is the world's largest maker of electronic components. Its most notable products are Apple's iPad, iPod, and iPhone.

Foxconn has been involved in several controversies, largely relating to how it manages its employees in China where it is the largest private-sector employer. In 2012, Apple hired the Fair Labor Association to conduct an audit of working conditions at Foxconn.

Suicides among Foxconn workers have attracted media attention. One was the high-profile death of a worker after the loss of a prototype phone, and there were several reports of suicides linked to low pay in 2010. Suicides of Foxconn workers continued into 2012, with one in June of that year.

On Friday, 21 September 2012, a new iPhone hit the world's shops, selling over five million units in its first weekend. On Monday, 24 September 2012, Foxconn had to halt production at a plant in northern China after a fight broke out, apparently in a workers' dormitory. Foxconn confirmed that a 'personal dispute' had escalated into an incident involving about 2000 workers, resulting in injuries to 40 of them. The BBC reported that 5000 (yes, five thousand) police had attended the scene. Later, comments on a Chinese blogging site suggested the fight broke out after security guards beat a worker.

LEARNING

Manage labour relations well for good employee relations and to mitigate negative media around key launch events.

Asbury and Ball (2016) provide a useful CSR-MS model, which is reproduced as Figure 2.10. Their work (ibid.) provides additional reading, including a self-guided workbook, for those requiring further and deeper understanding of CSR.

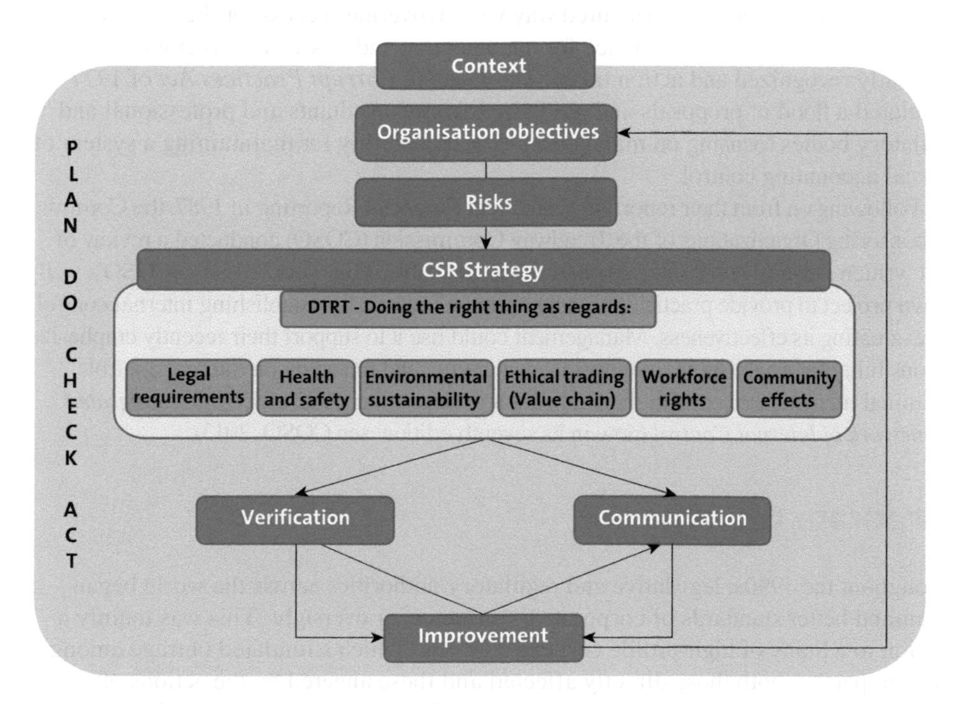

FIGURE 2.10 The Asbury and Ball CSR Management System model (2016).

INFORMATION FOR MANAGEMENT ABOUT CONTROL

It soon became apparent, when teaching people around the world how to audit management systems, that a critical prerequisite for such audits was often missing. There was extensive internal control guidance available for auditors, produced over many years by the professional auditing bodies, but virtually nothing had been specifically written for the auditee (*auditee* being the term used for the senior management representative within the scope of an audit).

The auditee's bookshelf/hard drive/email inbox is often overflowing with manuals, policies, governance guides, values and ethics statements, principles of business, vision and mission statements, laws and regulations, approved codes of practice, client and company rule books, mandates, organization charts, job descriptions, reporting relationships, accountabilities, roles and responsibilities, competence standards, process maps, training matrices, minutes of meetings, action plans, insurers' reports, plans, standards, strategic and tactical reviews, manuals of authority, audit reports, procedures, risk registers, and a whole lot more. But there were very few high-level overview documents written from management's perspective that described how these discrete internal controls could and should be implemented in a coordinated and complementary manner that would tie management's activities in a structured way with delivering success for their organization.

In the US, this lack of guidance for management and boards of directors was eventually recognized and action taken. *The Foreign Corrupt Practices Act* of 1977 stimulated a flood of proposals and guidelines from consultants and professional and regulatory bodies focusing on management's responsibility for maintaining a system of internal accounting control.

Following on from their report on Fraudulent Financial Reporting in 1987, the Committee of Sponsoring Organizations of the Treadway Commission (COSO) conducted a review of what written material about internal control was available. This work led on to COSO's well-known project to provide practical, broadly accepted criteria for establishing internal control and evaluating its effectiveness. Management could use it to support their recently emphasized responsibility for establishing, monitoring, evaluating, and reporting on internal control. A seminal moment arrived with the publication in September 1992 of COSO's *Integrated Framework of Internal Control* (now in its seventh edition; see COSO, 2013).

Corporate Governance

Throughout the 1990s, legislative and regulatory authorities across the world began to demand better standards of corporate governance, or oversight. This was mainly a reaction to a litany of high-profile corporate failures which stimulated outrage among innocent parties, both those directly affected and those angered by the actions of companies operating in their countries, cities, and towns. These outraged citizens were voters, and legislators took note that the majority wanted those responsible for running organizations to be held more accountable for their actions than had been the case in the past. Many professional accountancy bodies across the world had long accepted the need for global standards. The International Federation of Accountants now ensures that all accountants and auditors worldwide subscribe to a global code of ethics. And there has been growing support for international standard-setters to develop and promote international standards of accounting and auditing. You will read about professional standards for HSEQ and other auditors at the end of Chapter 3.

The interdependence of countries' economies requires high standards which are globally accepted, applied, and enforced as the most effective solution—balancing the needs of regulatory authorities with the needs of commercial and other organizations. It is widely accepted that such standards are what give investors confidence in the companies they invest in, and other interested parties (aka stakeholders) the confidence to buy from, work for, supply to, and live next door to.

Such standards require that organizations and their senior management, throughout the world, in both private and public sectors, must demonstrate

- accountability (of managers to their stakeholders),
- integrity (to attract financial and social support), and
- transparency (of their operations and financial position, as reflected in their statutory and voluntary reports to stakeholders).

The COSO Framework (COSO, 2013) became an accepted reference on internal control in the USA and around the world. Its implications for corporate governance led other countries to follow with their own expectations: the Cadbury Committee reported in the UK in 1992, followed by the Greenbury Committee in 1995, then the Hampel Committee in January 1998, and the Turnbull Committee in September 1999. The Criteria of Control Board (CoCo) of the Canadian Institute of Chartered Accountants reported in Canada; Marc Vienot first reported in France in 1995; the Peters Commission reported in the Netherlands in June 1997; and Kon TraG was published in Germany in March 1998. In the last 20 years or so, most developed and developing countries have issued guidance regarding the corporate governance of major companies registered in their jurisdictions.

Essentially, all these corporate governance standards have the same message: an organization's senior management (in particular, the directors of a public limited liability company) must take responsibility for two things:

- Really understanding what the risks and opportunities of the organization are, and what it does to enhance its performance based on this knowledge
- Informing interested parties (stakeholders) about what the company has been doing in a transparent and trustworthy manner

Developed during three of the most turbulent years in the USA's corporate history, COSO's Enterprise Risk Management: Integrated Framework was originally published in September 2004. It was intended to enable organizations to meet corporate governance expectations, setting out principles and concepts which could become a common language and giving clear direction and guidance on enterprise risk management. Updated in 2017, the revision is titled Enterprise Risk Management: Integrating with Strategy and Performance (COSO, 2017) and addresses the evolution of enterprise risk management and the needs of organizations to improve their approach to managing risk to meet the demands of an evolving business environment—considering risks in both the strategy-setting process and in driving performance.

Internal Control Reference Frameworks and Structured Means of Control

COSO's Integrated Framework of Internal Control (COSO, 2013) has stood the test of time and now in its seventh edition, it remains a broadly accepted standard for satisfying an organization's reporting requirements. In addition, the Enterprise Risk Management: Integrating with Strategy and Performance (COSO, 2017) now provides a more robust and extensive focus on the broader subject of organizational risk management. It was not intended to and has not replaced the Framework—rather it incorporates the internal control framework, allowing organizations to move toward a fuller risk management process.

A copy of the COSO Framework and the ERM summary documents are available for download from the book's companion website (https://routledgetextbooks.com/textbooks/_author/asbury/).

A-FACTOR 12

To carry out a successful and effective management system audit, an auditor needs a relevant internal control reference framework against which the auditee's performance can be assessed.

The obvious question, then, is: 'What constitutes a suitable internal control reference framework, or "structured means of control"?' In my opinion the best answer is: 'The structured means of control or reference framework currently being used by the organization. And if the organization doesn't have a reference framework, now might be the time to select, develop or purchase a suitable one!'

> Present circumstances don't determine where you can go; they only determine where you start.
>
> Dr Stephen Asbury

You will occasionally find an auditee that is not using a particular framework, or group of frameworks, because their organization does not have a corporate-wide internal control or risk management approach or framework. A problem that does arise for managers in trying to select an appropriate structure for control is the multiplicity of control frameworks they're asked to comply with these days. Happily, ISO Annex SL has brought some relief here, and sometimes the type of audit will naturally lead the auditor towards a particular reference framework. For example, ISO 9001 is the natural framework to select for quality management assurance, unless there is a more specific sector standard. As I have said, it's often true these days that ISO 9001 is a 'qualifier' for even an invitation to pre-tender with some organizations. Likewise, ISO 14001 is the natural choice for environmental management, and ISO 45001 for occupational health and safety management.

CASE STUDY

WE WANT GOOD H&S STANDARDS, BUT OUR STAKEHOLDERS ARE NOT INTERESTED IN EXTERNAL CERTIFICATIONS, GONGS, OR WHISTLES

This is what an SME hi-tech manufacturing organization told its OH&S adviser. There were discussions about the most suitable reference frameworks, including ILO-OSH 2001 and ISO 45001. In the end, the organization selected the UK regulator's guidance framework known as HSG65 (HSE, 2013), as it was simple, recognized, and free of charge and copyright in use.

LEARNING

Sometimes good things are free and are all you need.

The five parties (stakeholders) interested in the performance of any organization are shown in Figure 2.11. It is immediately apparent that different stakeholder groups have different requirements and aspirations. For example:

- Investors may desire higher sales prices, while customers obviously prefer discounts and lower prices.
- Employees prefer higher wages, while investors may prefer to pay lower wages.
- Organizations may prefer to keep employed numbers low, while society might want more permanent jobs to be available.

Management's role includes always balancing these competing requirements and aspirations.

A-FACTOR 13

Only by using a 'structured means of control' can a manager convert high-cost *controls* into business-assuring, profit-enhancing *control*. This approach provides a higher level of assurance, as *control* is systematic, replicable, and repeatable.

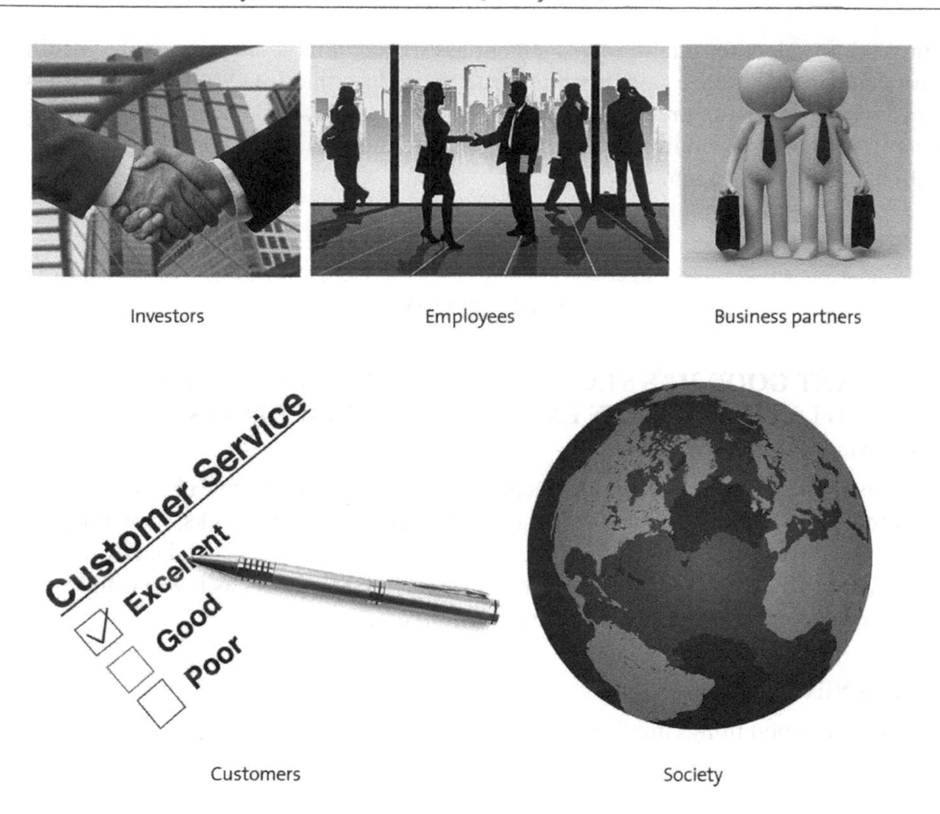

FIGURE 2.11 The five groups of interested parties or 'stakeholders'.

Source: Shutterstock

A-FACTOR 14

Whatever the auditee's reference framework, an auditor needs to have their own 'structured means of control' in mind, which they can use to simplify the complexity of an auditee's framework. This can also be useful if there is a vacuum.

In the remainder of this chapter and beyond, I will describe and refer to a simply structured management system standard (a business control framework/BCF) based on ISO 9001:2015, ISO 14001:2015, and ISO 45001:2018 which you can adopt and use as a management system and for MS auditing, unless your organization or auditee has adopted an alternative framework. It provides for a robust analysis of the context, considers the business objectives, and gives an overview of assessing risk. This reference framework is, as you might expect, aligned to the Deming Cycle/PDCA and ISO Annex SL. Whenever necessary, it will allow managers/auditees and auditors alike to map out—and thus understand—any of an auditee's internal controls.

CASE STUDY

SPACE SHUTTLE *COLUMBIA*, 1ST FEBRUARY 2003

On 1 February 2003, NASA Space Shuttle *Columbia* was destroyed during re-entry into the Earth's atmosphere at the end of a 16-day science mission. All seven of the crew were killed. Within two hours, an independent investigation board was established, in accordance with procedures put in place following the 1986 *Challenger* Shuttle disaster (Atherton & Gil, 2008: 211–9).

The investigation reported that, 81.9 seconds into its flight, *Columbia* had been stuck by a large piece of insulation foam which had detached from the external fuel tank. Impact tests showed that the foam damage could be sufficient to create a 16-inch hole in the protective heat-resistant tiles on the leading edge of the wing. It was also found that there had been some foam loss on previous space shuttle flights. Foam loss became an accepted part of every mission, and with each successful landing, safety concerns seem to have faded away.

Many of the issues identified by the *Columbia* investigation had been identified in the report on *Challenger* seventeen years earlier—hazard evaluation, perceptions of NASA's invincibility, management of change, engineering authority, orbiter integrity, and pressure on programme schedules.

LEARNING

We all have a role in ensuring that lessons are learned and remembered over the longer term. In this case, 17 years represents over a third of a typical career span, and this probably means that for many the people working on *Columbia, Challenger* was something they had only read about rather than experienced first-hand.

TIP

On the companion website (https://routledgetextbooks.com/textbooks/_author/asbury/), you are able, as a purchaser of this book, to download for your personal use a template HSE management system manual, which is aligned to ISO 14001:2015 and ISO 45001:2018. This will provide you with a flying start should HSE-MS be your current interest.

MY BUSINESS CONTROL FRAMEWORK (BCF)

The strength of my structured means of control illustrated in Figure 2.12 lies both in its simplicity, and in its alignment to PDCA, ISO Annex SL, and ISO 9001/14001/45001.

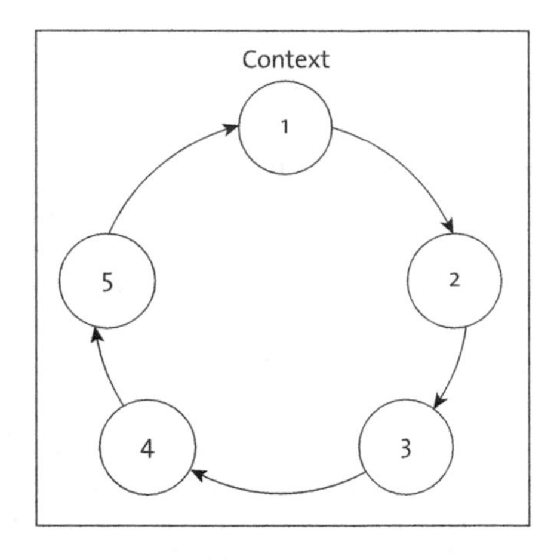

FIGURE 2.12 My Business Control Framework (BCF).

It is readily adaptable for all types of organizations, and it reflects all the features expected of modern risk-based HSEQ management systems for worker protection and operations integrity. Throughout the rest of this book, reference to *My BCF* is to *this* business control framework.

Business controls for all processes in the framework can be typified across five 'control mechanisms'—leadership; planning; support and operation; performance evaluation; and improvement. A small number of characteristics and examples of each mechanism follow, and a fuller list is shown in Table 2.3.

Summarizing Table 2.3, the characteristics and contents are as follows:

- Leadership: developing policy, setting organizational roles, responsibilities, and authorities, determining the scope of the management system
- Planning: identifying risks and opportunities, determining legal and other requirements, setting objectives and planning to achieve them
- Support and Operation: resources, competence, awareness, creating, updating, and controlling documentation, operational planning and control, managing change, procurement, contractors, emergency preparedness and response (including control of nonconforming products)
- Performance evaluation: active and reactive monitoring, measuring and analysis, evaluation of compliance, internal audit, management review
- Improvement: recording non-conformities, corrective actions, continual improvement

All BCFs comprise various categories of control, aligned to PDCA, and which are generally rooted in good management practice. These categories can therefore be considered both components of a business control system and as essential criteria for an effective management system: 'wheels within wheels', as Deming might have said.

TABLE 2.3 A Guide for Mapping Typical Controls with the Five HSEQ-MS Elements

	LEADERSHIP	*PLANNING*	*SUPPORT AND OPERATION*	*PERFORMANCE EVALUATION*	*IMPROVEMENT*
Characteristics	Taking overall responsibility and accountability	Hazard identification	Resources (finance, assets, people, for monitoring and measuring); an environment for the operation of processes	Processes for monitoring, measurement and testing of premises, plant, procedures, and people ('the 4Ps')	Determination of opportunities for improvement
	Ensuring policies and related objectives are established and compatible with direction	Assessment of risks and opportunities to identify priorities	Competence	Environmental monitoring	Implementing necessary actions to achieve the intended outcomes of the HSEQ-MS
	Ensuring integration into other business processes	Identifying legal and other compliance requirements	Awareness	Customer satisfaction	Process for reporting, investigating, and taking action on incidents, non-conformities, and complaints
	Provides resources	Establishing objectives at relevant functions and levels	Communication (internal, amongst contractors and visitors, to customers and other interested parties)	Evaluation of legal and other compliance	Tracking improvements from records of actions and corrective actions
	Confirming the importance of HSEQ management and conforming to MS requirements	Planning action(s)	Documented information (documents and records); documents available, records retained	Internal audit—plan, conduct, reporting, and close out	Enhancing HSEQ-MS performance

(*Continued*)

TABLE 2.3 (Continued)

	LEADERSHIP	*PLANNING*	*SUPPORT AND OPERATION*	*PERFORMANCE EVALUATION*	*IMPROVEMENT*
	Directing and supporting persons to contribute		Operational planning	Management review	Promoting a culture that supports the management system(s)
	Promoting improvement		Determining requirements for products and services		
	Developing and promoting MS culture		Eliminating hazards and reducing risks (OH&S hierarchy 'E-SEAP')		
	Ensuring consultation and participation of workers		Management of change/s (to products, services, and processes; to legal or other requirements; changes in knowledge; and developments in technology)		
	Protecting workers from reprisals when reporting incidents		Outsourcing—control and traceability of externally-provided processes, products, and services		
	Focus on customers		Procurement		
			Contractors		
			Emergency preparedness and response		
Typical content(s)	Mission statement (why we're here)	Hazard registers (e.g., HAZID, HAZOP)	Budgets and staffing plans	Monitoring plans	Non-conformity reports

Vision statement (where we're going)	Registers of environmental aspects and impacts	Training plans and CPD	Maintenance schedules	Corrective actions
Values (not influenced by short-term priorities)	Register of legal and other requirements	Control, cooperation, communication competence ('the 4Cs')	Customer feedback, customer surveys, user groups	Incident, accident, near-miss reports
Purpose	Risk matrices	Meetings, working parties, and committees	Calibration records	Investigation reports
Business principles	Risk ranking	Designs (plans, inputs, controls, outputs, changes)	Inspections and checks/checklists	Improvement records
Ethics	Inherent (unmitigated) risk	Operating manuals	Statutory inspections	Confirmation of improvements
Governance and assurance	Residual (mitigated) risk	Work instructions/safe systems of work	Supervision	Internal audit committee
Management integrity	ALARP	Procedures	Safety tours	External reporting (e.g., GRI—Global Reporting Initiative)
The 'tone at the top'	Control/s	Method statements	OH monitoring	Certifications and awards
Organization charts, structure, roles, and responsibilities	Safety data sheets (SDS)	Batch numbers, catalogue numbers, manufacture/sell by/use by dates	IH monitoring	
Manual of authorities	Hierarchy for control (E-SEAP)	Permits to work/Lock-out Tag-out (LOTO)	Behavioural safety	
Reporting relationships	Process maps	Management of change (MOC)	Tests and pilots	

(Continued)

TABLE 2.3 (Continued)

	LEADERSHIP	*PLANNING*	*SUPPORT AND OPERATION*	*PERFORMANCE EVALUATION*	*IMPROVEMENT*
	Understanding laws and regulations	Risk universe	Contracts and SLAs	Non-destructive testing (NDT)	
	Golden Rules	Objectives	Engineering of systems, alarms, and warning indicators	Benchmarking	
	Policies (H&S, Q, E, right to work, bribery and corruption, other)	Targets	Emergency plans	Surveys	
	Project management	KPIs	Fire drills and emergency scenario exercises	Peer reviews	
		Goals	Business continuity planning (BCP)	Performance monitoring	
		Expectations	Waste sorting, segregation, re-cycling	Statistical process control (SPC) and process capability	
		Budgets	Decommissioning	Epidemiological analysis	
		Priorities	Security	Internal audit plan/s	
		Critical success factors	Housekeeping	Internal audit reports	
				Management reviews/ minutes	

Control—Barriers to Loss

Another way of thinking about the deployment of 'control' is to consider how it acts as one or more barriers which prevent the realization of the risk. The greater the risk (especially the possible severity of impact/remember *Black Swans*), the greater the number and reliability of the protective 'barriers' which should be in place.

Figure 2.13 shows how three example hazard barriers—equipment, process, and people—can be intentionally combined to mitigate risks.

- Equipment barriers include design, installation, statutory inspection, servicing, and maintenance.
- Process barriers include risk assessment, work management, and emergency preparedness and response.
- People barriers include worker selection, training, supervision, occupational health monitoring, and personal safety measures.

James Reason (1990) models a similar and well-known approach, calling it the *Swiss Cheese* model. I suggest that you familiarise yourself with this model if you are unaware of it. However, to summarize, Reason hypothesizes that most incidents can be traced to one or more of four levels of failure:

- Organizational influences
- Unsafe supervision
- Preconditions for unsafe acts
- Unsafe acts themselves

In the *Swiss Cheese* model, an organization's defences against failure are modelled as a series of barriers, represented as slices of Swiss cheese. The holes in the cheese slices represent weaknesses in individual parts of the system and continually vary in size and position in the slices. The system produces failures when all the holes in each of the slices momentarily align, permitting 'a trajectory of accident opportunity', so that a

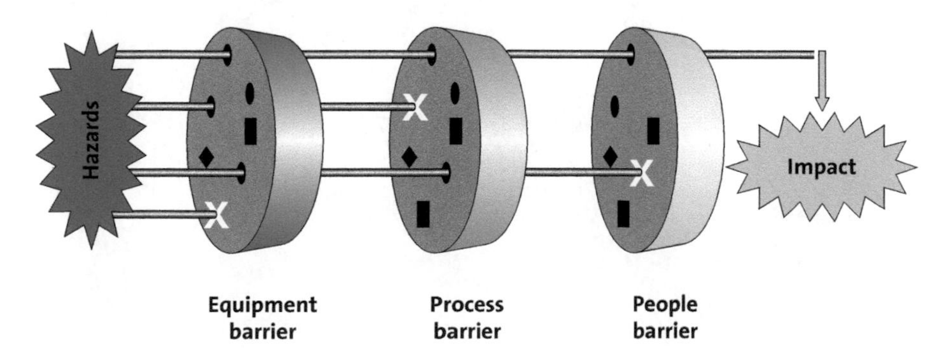

FIGURE 2.13 Layers of control provide risk-reducing barriers.

hazard passes through all the holes in all the defences and leads to loss. So no holes in the MS implementation, or only the smallest and infrequently occurring ones, are best.

My BCF—Context of the Organization

ISO Annex SL, clause 4 requires organizations with management systems developed using the Annex to understand the context of the organization prior to establishing a structured means of control. Readers should review Table 2.2 on page 77 if they wish to be reminded how Annex SL maps to PDCA. In summary, organizations should

- understand the organization and its context (clause 4.1),
- understand the needs and expectations of interested parties/stakeholders (clause 4.2),
- determine the scope of the management systems (clause 4.3), and
- understand the management systems for the specific discipline, such as environment (clause 4.4).

Figure 2.14 shows the flight deck of a NASA Space Shuttle, with its mass of instrumentation for measuring the significant parameters affecting the mission. Superimposed on the picture are examples of some of the environmental factors affecting an organization's mission. Analysis of the business environment and some useful tools

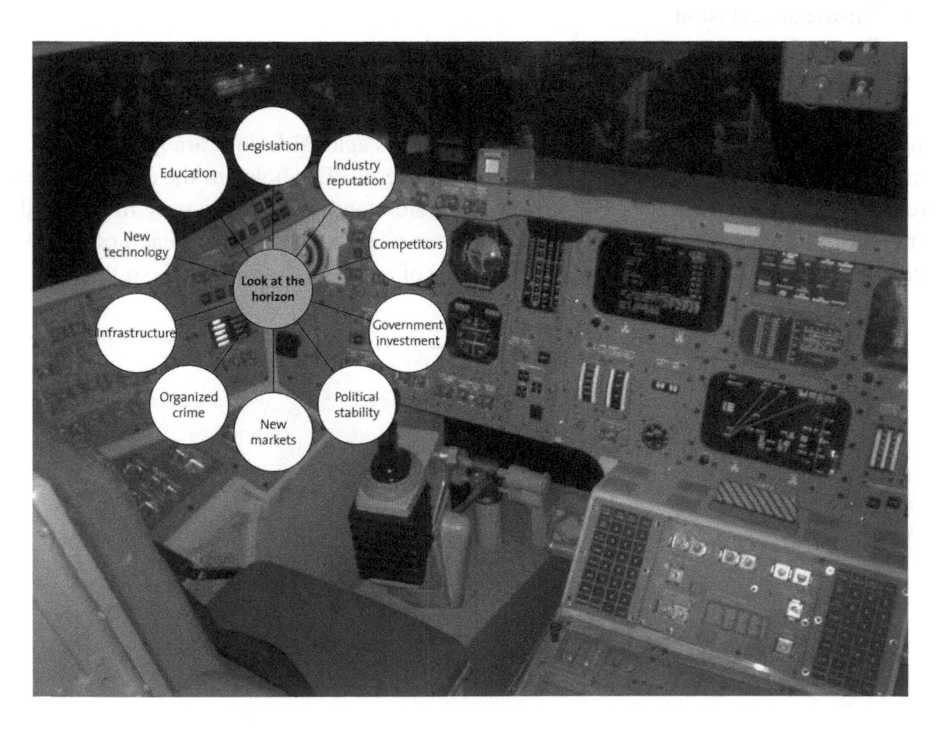

FIGURE 2.14 An organization's environmental factors.

for considering and understanding it (PEST, SWOT, and understanding the requirements of interested parties) were discussed in Chapter 1.

Business Objectives

Every organization and the people within it need to understand 'why we're here'; this is sometimes called the *Mission* or *Purpose*. It also needs to know 'where we're going'. This is sometimes called the *Vision*—a bright image of itself in the future. With such a *Mission* and with a clear *Vision* in mind, management can create strategies and specify organizational or business objectives to enable to organization and its people to deliver the *Mission* and achieve the *Vision*. *Values* are a set of core beliefs held by an organization. They act as guiding principles that provide an organization with purpose and direction; they set the tone for its interactions with its customers, employees, and other stakeholders.

The objectives should take full account of the opportunities and constraints inherent in the prevailing *Context* (business environment), without compromising the *Values*. The objectives should be SMART (Specific, Measurable, Achievable, Right, Timely). This is discussed from an auditor's perspective in Chapter 10.

Such *Mission*, *Purpose*, *Vision*, and *Values* are often set out in a statement to aid their communication.

Boxes 2.3–2.5 show samples of published *Mission*, *Purpose*, *Vision*, and *Values statements* from three leading global organizations: the Coca-Cola Company, JCB, and McDonald's.

BOX 2.3

Purpose and Vision statement: The Coca-Cola Company

Our purpose: Refresh the world. Make a difference.

Our vision is to craft the brands and choice of drinks that people love, to refresh them in body & spirit. And done in ways that create a more sustainable business and better shared future that makes a difference in people's lives, communities and our planet.

Source: Coca-Cola Company [The] (2023)

BOX 2.4

Mission, Vision, Values statement: J.C. Bamford (Excavators) Limited

MISSION

Everything we do is focused on helping our customers grow sustainable businesses to support the UK economy in the long term.

We BELIEVE in doing this fairly, without fuss & by helping our customers affordably operate JCB and complimentary plant, machinery and vehicles.

We have practical knowledge of the plant, machines & vehicles our customers use and we're ready & enthusiastic to help with the transition to zero-carbon alternatives.

VISION

It's our aim to be a trusted partner within our customers' businesses through good times and bad.

VALUES

We live by core values that place the customer at the heart of the decisions we make. We make sure we understand what our customers do and how they do it, within industries we have supported since 1970.

Source: JCB (2023)

BOX 2.5

Mission and Values statement: McDonald's

OUR MISSION IS TO MAKE DELICIOUS FEEL-GOOD MOMENTS EASY FOR EVERYONE

This is how we uniquely feed and foster communities. We serve delicious food people feel good about eating, with convenient locations and hours and affordable prices, and by working hard to offer the speed, choice, and personalization our customers expect. At our best, we don't just serve food, we serve moments of feel-good, all with the light-hearted, unpretentious, welcoming, dependable personality consumers know and love.

OUR VALUES

The backbone of our Brand is, and always has been, a commitment to a set of core values that define who we are and how we run our business and restaurants.

When we live our values every day and use them to make decisions—big and small—we define McDonald's as a brand our people, and the people we serve, can trust.

Source: McDonald's (2023)

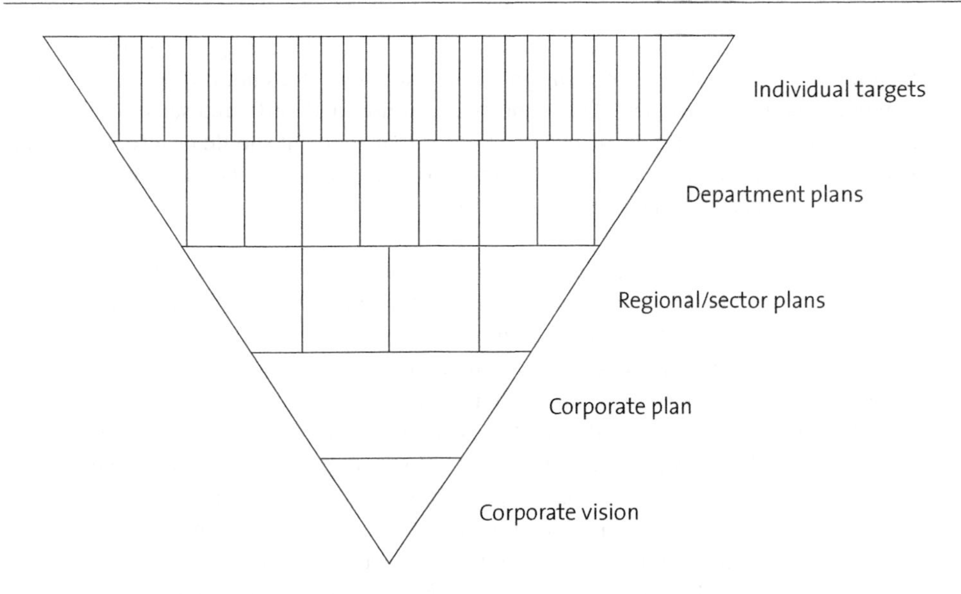

FIGURE 2.15 Achieving success by aligning objectives at all levels in the organization.

I had intended also to include the vision for the London 2012 Olympic Games, which was staged so brilliantly by my country over a decade ago. However, it proved to be too large and too detailed a document for inclusion, and so instead I have provided the full text of the British presentation to the International Olympic Committee in Singapore on 6 July 2005—which set out the vision for the London Games and which led to the success of the bid—on the book's companion website (https://routledgetextbooks.com/textbooks/ author/asbury/). It can be downloaded and used to judge for yourself how well the vision was delivered.

Even though they are a means of realizing an organization's *Vision*, business objectives are not themselves controls; rather they are the necessary start and end points for an integrated business control framework.

Business objectives should:

- guide the business processes of the organization;
- apply to, and at each level, of management;
- give explicit time frames for achievement of measurable results (i.e., SMART);
- be the result of wide participation in their development;
- be communicated to, and understood by, all employees and others whose performance can affect their achievement;
- accord with any published codes of conduct, values, or ethics; and
- form a coherent whole and be internally consistent—as illustrated in Figure 2.15.

Business Risks

Risk, defined in A-Factor 6, is the effect of uncertainty on objectives; or anything that may impact upon the achievement of the organization's objectives.

Risk assessment is a vital management activity. Discussed in Chapter 1, it allows for prioritization of significant risks over insignificant ones. This does not imply that all risks can, or indeed should, be avoided. The inability or failure to identify and seize business opportunities may itself be a significant risk. Senior management should ensure that a risk assessment process is embedded in their organization's strategy and its implementation.

An effective risk assessment process requires:

- senior management with an intimate knowledge of the political/legal, economic/financial, social/demographic, and technical/infrastructure (PEST) features of the market or environment in which their organization operates;
- creation of strategic and operational objectives which are well known and clearly understood (SWOT) throughout the organization;
- addressing risks methodically in all the organization's (business) activities;
- a structured description of the factors critical to the organization's success and the opportunities and threats that may help or hinder achievement of its set objectives;
- estimation of the organization's exposure to the factors, opportunities, and threats in quantitative or qualitative terms of probability of occurrence and the consequences of their possible impact (A-Factor 7; $R = P \times C$); and
- collating these exposures in the format of a risk profile or risk matrix (see the examples in Figures 1.8 and 1.9) which enable management to prioritize areas for risk response.

As discussed in Chapter 1 (see Figure 1.6), the strategies for risk response are choices within the 'Four Ts':

- Terminate: avoid or cease the activity or situation.
- Treat: reduce the probability and/or potential consequences of the risk by applying an MSS/BCF (structured means of control).
- Transfer: transfer some of the risk to, or share it with, another party, such as an insurer or joint venture party.
- Tolerate: knowingly accept the residual risk, recognizing that you are 'not happy with the situation' or 'living with the risk' until acceptable alternatives become available.

As discussed in Chapter 1, all organizations exist within their *Context*, their business environment, which is subject to constant and increasingly rapid changes, and which is likely to affect management's vision of risk and opportunity. Business objectives are the starting point for risk management, as these (should) guide all the processes of the organization. Risk management is a vital activity that identifies, assesses, and prioritizes risks and opportunities. Business process analysis includes the identification of critical success factors and risks, and therefore identifies:

- Which controls are needed or desirable (remember E-SEAP);
- how the organization can be configured more effectively, in line with its business processes; and
- an effective business control framework which can trigger timely reaction to changes in risks and opportunities in the business environment or in operations.

Reasonable Control—ALARP, ALARA, and SFRP

Controls should be applied to control hazards/aspects/threats in operations in a reasonable, effective, and efficient manner—not too few, and not too many. This carefully considered balance is commonly known as ALARP—'as low as reasonably practicable'. The US Nuclear Regulatory Commission uses the term ALARA—'as low as reasonably achievable'. It means making every reasonable effort to maintain exposures . . . as low as practical. In the UK, this balance is sometimes called SFRP—*so far as is reasonably practicable*—a phrase from the primary OH&S legislation in the country. It is acceptable to regard these three terms as interchangeable.

The key word present in each is 'reasonable'. It involves weighing a risk against the sacrifice needed to control it, as established in *Edwards v National Coal Board [1949] All ER 743 (CA)*, where *reasonably practicable* was said to be a narrower term than 'physically possible'. It implies that a computation must be made in which the quantum of risk is placed in one scale, and the sacrifice involved in the measures necessary for averting the risk is placed in the other. If there is a great disproportion between them, the person (employer) upon whom the obligation is imposed discharges the onus upon him. Controls selected are thus said to be *ALARP* if the cost (time, trouble, or money) of additional control measures is grossly disproportionate to the level of improvement (reduction) of the risk.

Table 2.4 shows some examples of risks in an organization's *risk universe*, based on the PEARL (People, Environment, Assets, Reputation, Legal) model first presented in Figure 1.9.

Understanding Business Processes

Business processes are logically linked groups of activities needed to fulfil the objectives of the enterprise, converting inputs to outputs as described in Chapter 1. A business process flow chart is a picture you might bring to mind. It typically comprises:

- core processes—processes which directly deliver the required product or service to the organization's customers; and
- service and control processes—processes which provide and facilitate the corporate infrastructure to deliver the core processes.

TABLE 2.4 Risk Universe—Example Categories of Risk

PEOPLE	*ENVIRONMENT*	*ASSETS*	*REPUTATION/LEGAL*
Loss of key personnel Health and safety Right to work Human rights Discrimination Morale	Emissions (air) Discharges (land, water) Complaints Sustainability Biodiversity	Security Fire and explosion Life cycle—design, commission, operate, maintain, decommission Management of change (MOC)	Prosecution Litigation Complaints Media portrayal (local/global) Strikes and disputes

Effective overall business control will result if an organization is managed as a series of core business processes and service processes, each with its own BCF/MSS, which includes a process to deliver continual improvement.

The Human Factor

Understanding how people behave in an organization is key to the success of any control framework. Critical success factors in this regard are:

- the tone from the top, set by the highest management level regarding ethical values, standards, and behaviours/actions of everyone associated with the organization;
- the quality of all levels of staff and their understanding, support, and compliance with the business controls in their area (directors must direct, managers must manage, supervisors must supervise, and workers must work in accordance with the organization's requirements);
- the provision of adequate time and resources for proper operation, maintenance, and review of the business controls;
- good communication between individuals and between groups of people; and
- reliable, timely, and useful information enabling staff to discharge their responsibilities efficiently and to measure their achievement of specified objectives.

The impact of culture and human factors was powerfully described by the late Carolyn W. Merritt (then-chairman and CEO of the US Chemical Safety and Hazard Investigation Board) in her statement to the BP Independent Safety Review Panel in Houston, Texas, on 10 November 2005:

> One of my aspirations is that all industrial managers treat safety and major accident prevention with the same degree of seriousness and rigor that is brought to financial transactions. Few people would operate a major corporation today without a strict system of financial controls and auditing, where everyone within the corporation recognizes the severe consequences for noncompliance.
>
> Merritt (2005: 4)

That same standard of diligence is not always applied to risk management and safety. If a worker gets away with a flawed safety decision one day or repeatedly, far from facing a penalty, they may actually end up rewarded, perhaps for boosting production, or being able to go home early. Like using a cell phone while driving, or taking a shortcut, one may come to believe that what was thought to be unsafe is actually safe, based on your experience. It is a phenomenon that is sometimes called *normalization of abnormalities*.

Even simple ways in which people behave influence the outcome of a day's sales, as illustrated by the following short case study.

CASE STUDY

I DON'T KNOW. I ONLY WORK HERE.

I bought a home in North Wales in 2014. Wanting to decorate it with local art, I visited a local gallery where I found a lovely Snowdonia landscape scene that I thought would look amazing in my new home. I asked the gallery assistant whether the artist was local. Her reply was 'I don't know. I only work here'.

LEARNING

Every employee of an organization must understand where they fit into the 'big picture'. This failure of management cost the gallery £1,000 of revenue that day.

Chapter 2 concludes with two case studies—the first highlighting the power of HSE-MS and HSE audit over time, and the second highlighting what can go wrong when a BCF fails.

CASE STUDY

A MEASURE OF PROGRESS OVER TIME

An audit team conducted a liability survey (an HSE audit commissioned by an organization's insurers) at a new semiconductor plant in Scotland in 1998. Good standards of worker and environmental protection were noted throughout, and these had been implemented in a highly systematic manner. Despite providing a high level of assurance to site management as well as the liability insurer, there were a small number of improvement recommendations made within a loss prevention plan.

The same audit team conducted follow-up audits again in 2005 and in 2011. On both occasions, it was immediately apparent that the auditee had continued to improve its HSE management and performance. It had set new objectives, developed new implementation plans, provided refresher training to all worker groups, and investigated and initiated a new occupational health program.

Since the first visit in 1998, a new plant director had been appointed, and he was able to provide a renewed and re-energized management commitment to continually improve performance. Over twelve years, the organization has become one of the best HSE performers in the world; it has won several sectoral, national, and international HSE awards.

LEARNING

An independent audit process can be hugely supportive of an organization's own improvement programs and can serve to provide a measure of the progress made over time.

CASE STUDY

HOT UNDER THE COLLAR

An English local authority fitted gas-fired central heating into its community housing stock. After a few days, it started to receive numerous telephone calls from its residential tenants, suggesting that homes were too cold or too hot. A council maintenance officer visited and showed residents how to adjust their thermostatic controller, advising that, of course, temperature change would not be immediate. The telephone complaints, however, continued and then increased.

After additional visits to investigate, it was found that the installing contractors had wall-mounted the thermostat controller panels but had not connected them to the boiler due to a significant cost overrun.

It seems that neither business control framework—that of the contractor nor that of the local authority–had worked as intended!

LEARNING

Sometimes control diminishes when funding diminishes.

> Good management is the art of making problems so interesting and their solutions so constructive that everyone wants to get to work and deal with them.
>
> Paul Hawken; author, environmentalist, entrepreneur (1946–)

For more on this topic, watch the one-hour webinar in which the author talks about the evolution of business management systems for continual improvement in organizational performance. (See Appendix 4.)

Test your knowledge with the next chapter assessment test (see Appendix 4).

Auditing, ISO 19011, and Initiating Audit Culture

3

An auditor is not bound to be a detective, or as was said, to approach his work with suspicion or with a forgone conclusion that there is something wrong. He is a watchdog, not a bloodhound. He is justified in believing tried servants of the company in whom confidence is placed by the company. He is entitled to assume that they are honest and rely upon their representations, provided he takes reasonable care.

—Lord Justice Lopez in *Kingston Cotton Mill Company No.2* (Law Times, 1896)

The New York Stock Exchange requires publicly traded companies to maintain an internal audit function to provide management and the audit committee with ongoing assessments of the company's risk management processes and system of internal controls. Stock exchanges throughout the world have their own norms governing such companies, and some have implemented requirements similar to those of the NYSE.

—IIA (2023a)

Market demand for management system standards has led to a huge increase in the number of subject and sector-specific standards. So, the auditing of these management systems needs to reflect the variety and number of standards being developed.

—Dr Stephen Asbury

INTRODUCTION

For a summary of this chapter, you can listen to the author's Microlearning™ presentation, which you'll find on the book's Companion Website. Then, delve deeper by reading on for further details on the Companion Website. (See Appendix 4.)

When HSEQ and operations integrity (OI) losses occur (some examples were suggested in the Introduction), it begs three questions:

DOI: 10.1201/9781003364849-4

- What was senior and line management doing?
- Where were the alerts from the auditors?
- Are our current approaches to auditing thus fundamentally flawed?

Now is the time for better management and for better auditing. The debate that has been going on for years concerning whether an auditor/audit team is a 'watchdog' or a 'bloodhound', a 'partner' or a 'policeman' is over. The job of the audit team is to act independently and with courage to report the truth, and for management to act in a timely manner upon that truth. There is a real and important difference between an audit opinion provided by an audit team to an auditee, and a performance appraisal handed down to a manager by their boss—though it is easy to see how they can be confused, and easy to understand why some managers find it hard to accept anything other than a 'good' audit opinion.

> Perhaps the perfect auditee is new to their role, and just keen to find out about the effectiveness of their management systems?
>
> Dr Stephen Asbury

FIGURE 3.1 Audit is a mirror; it reflects what is there. Do we see it as it is, or as we wish it to be?

Source: Image from iStock, used with permission.

CASE STUDY

RIGHT FIRST TIME, BEST PRACTICE OH&S

A London-based, Russian-owned, new-start high-end automotive-sector engineering company commissioned an OH&S audit at its facility. From the start, the auditee confirmed that it was committed to establishing the new facility as 'right-first-time' and wanted certainty that it could meet 'best practice' H&S standards as well as local regulations. It responded extremely positively to the audit recommendations and, at a re-audit six months later, had closed out 100 per cent of the original advice. A visit by the UK regulator referred to a status it called 'compliance plus' in post-inspection correspondence with the auditee.

LEARNING

Right first time is the vision of many well-managed, excellence-focused organizations.

A SHORT HISTORY OF AUDITING

In *The Audit Explosion*, Power (1994) confirms the origins of independent auditing in financial assurance. The 'explosion' of auditing to other disciplines since, including auditing of management systems, is reported by Power (ibid.), and supported by Humphrey (1997). The requirement for auditing an OH&S-MS originated within the *POPIMAR* structure provided by HSG65 (HSE, 1991). ISO Annex SL (ISO, 2012a) made auditing a mandatory requirement of ISO's HSEQ-MS.

The value-add from integration of discipline or theme audits is discussed by Karapetrovic and Willborn (1998). The growth in auditing has led some (Power, 1994) to suggest that we are living in an 'Audit Society'; Power's terms 'audit society' and 'audit explosion' have gained wide currency within the social sciences (Maltby, 2007).

The role and function of audit has changed over the last 130 years. In 1896, in respect of the notable case of *Kingston Cotton Mills*, Lord Justice Lopez said that the auditor 'is a watchdog, not a bloodhound' (The Law Times, 1896)—the implication being that a watchdog barks when it sees something suspicious, whereas a bloodhound actually searches for something suspicious.

Subsequent case law has extended the auditor's duty such that it is not sufficient for an auditor to rest upon the honesty and accuracy of others. Auditors must go further and satisfy themselves that evidence upon which they have relied has been taken based on sound auditing principles. In *Fomento (Sterling Area) Ltd. v Selsdon Fountain Pen Co. Ltd.* [1958, 1 All ER 11], Lord Denning put it this way:

> To perform his task properly he must come to it with an enquiring mind—not suspicious of dishonesty—but suspecting that someone may have made a mistake somewhere and that a check must be made to ensure that there has been none.

Amendments to the UK Banking Act 2009 raised the standard further—from the duty of having an 'enquiring mind' to that of having a 'suspicious mind'. The role of the modern auditor of banks and financial institutions was thereby effectively transformed from that of 'watchdog' to 'bloodhound' 113 years after *Kingston Cotton Mills*.

The development of certifiable quality management systems from 1979 (BS 5750) and 1987 (ISO 9001) led to a requirement for competent management system auditing. This was reflected in 1990 by the publication of ISO 10011–1 *Guidelines for Auditing Quality Systems* (ISO, 1990). This was replaced in 2002 by the publication by ISO of ISO 19011 *Guidelines for Auditing Management Systems* (ISO, 2002). ISO 19011 was revised in 2011, with a second revision in 2018 (ISO, 2018b).

According to Broberg (2013: 102), the traditional role of audit is to provide 'a sort of assurance for the owners and stakeholders of a company who need to be assured that the information presented about the company is true and fair'. Her work reviews this role and provides an updated view of auditing and the function of auditors. These reported movements in the role of an auditor and the audit function are reflected in changes to the auditor's duty of care. In the spread of auditing to other disciplines as reported by Power (1994; Humphrey, 1997; Karapetrovic & Willborn, 1998), it is right that I summarize the literature as it relates to 'independence' in the context of an auditor's duty of care. Certainly, ISO 19011:2018 (ISO, 2018b) advises auditors on the importance of *independence* as one of seven valued principles for MS auditing. Let's review the background to this.

In some early professional auditing literature, use of the term 'independence' is reported as ambiguous (Antle, 1981). Antle contrasts 'independence' (any situation which alters incentives to ignore, conceal, or misrepresent his [*sic*] findings) with 'conflict of interests' (trade-off between benefits and costs of truthful auditing). Simunic (1984) advises that 'any situation which increases the probability that an auditor will not truthfully report the results of his [sic] audit investigation can be viewed as a threat to independence', identifying simultaneous supply of management advisory services as an example of this.

Salehi et al. (2009: 10) concur, pointing out that 'an independent auditor is essential because of the separation of ownership from the management'. They point out, in my view rightly, that independence depend on the profession's strength and stature and is fundamental to the reliability of audit reports. Like Simunic (1984), they identify economic dependence of the auditor on the client amongst the possible causes for reducing independence (e.g., from the provision of non-audit services), and thus not producing a fair and truthful report. They claim that if auditors act independently, this can reduce the *expectation gap* (which I will discuss later in this chapter).

There are other reasons that auditors may misrepresent the facts. Bhattacharjee and Moreno (2013) advise that auditors may experience emotional reactions during the audit process. These may include, they say, moods, anxiety about the task, and like or dislike of the auditee's personnel. Their research suggests that these emotions can influence their decision-making and audit opinion(s).

There is a body of literature surrounding audit value and quality including Taylor (2004); Robson et al. 2012); Knechel et al. (2013); and Brivot et al. (2018). Taylor reported that we do not know the optimal level of audit quality and therefore whether we have 'too little' or 'too much' auditing, concurring with Simunic (1984) and Salehi et al. (2009) that audit quality will always be somewhat suspect if other services are provided which may compromise the auditor's objectivity.

Robson et al. (2012) reported the results of a study of seventeen auditing methodologies used in the public sector in Canada, focusing on aspects related to reliability and validity. The study reported wide variations in auditing methods, as well as discrepancies between actual auditing practices and ISO standards for MS auditing. It recommended research to determine the impacts of these variations.

Knechel et al. (2013) reviewed definitions of audit quality and the available frameworks [methods] for establishing the same. They summarize the research on quality indicators, including inputs, processes, and outcomes.

Brivot et al. (2018) provides two contrasting norms of audit quality. The first, which most commonly arises in Big 4 firms, is called 'the model audit', which says that audit quality arises from a technically flawless audit in which professional judgement is highly formalised and documented in a perfectly documented audit file which passes regulatory inspections. The second, which arises in firms of all sizes, is called the 'value-added audit' which considers that audit quality results from tailoring the audit to meet the client's unique needs. As a result, professional judgement can be unrestrained, and audit quality is attested by the client. Unsurprisingly, the authors report tension between these two norms, which I recognize.

My personal view is that an effective OH&S auditor must be both a watchdog *and* a bloodhound—they must have an enquiring as well as a suspicious mind (but must control this latter when preparing their audit opinion!). An influential and well-known CEO I worked with between 1991–5 clearly understood this changing role and function for audit. I recall that in his view the weight ratio of 'enquiring: suspicious' should be about 90:10. That still seems about right to me. A work plan comprising an independently selected *risk-based* sample of significant HSEQ risks and *Black Swans* from the *Context* scope is the essential starting point to *The Audit Adventure*.

CASE STUDY

IT'S ALL OK

An audit at an electrical components manufacturing company in 2016 revealed specific shortfalls in high voltage electricity (HV) safety, and an audit report prioritized this for action ('Priority 1'). At the audit closing meeting, the managing director of the company said that he felt everything was OK but that he would attend to the recommendation.

A repeat audit one year later revealed the same HV safety shortfalls. At this closing meeting, the MD repeated virtually the same statement.

Just two months later, there was serious electric shock incident, following which the company was prosecuted and fined heavily.

LEARNING

A competent manager reviews the evidence and findings reported by an independent audit team and attends to the recommendations. There can be a significant difference between what one believes to be true and the actual situation on the ground.

It is thus important for, and incumbent upon, all organizations to look at themselves afresh with integrity and honesty, and with their responsibilities to all the stakeholders in mind, at the reflection in the mirror presented to them by an audit of their HSEQ and OI management systems, and ask, 'Do we really have reasonable control of our operations?'

INITIATING REASONABLE CONTROL AND ASSURANCE

Chapter 2 provides a history of business control, and how the simplicity of the Deming Cycle has been supplemented by (some say) the complexity of subsequent control frameworks, including the HSEQ trilogy of ISO 9001:2015, ISO 14001:2015, and ISO 45001:2018.

Legal and stakeholders' expectations for statutory and corporate governance have led to a variety of codes and standards aimed at providing greater levels of assurance. Typically, the approach within an organization is hierarchical—depending, of course, upon its size. This hierarchy begins at the highest level, with specific responsibilities for the corporate body and its executive directors, extending to the appointment of an internal audit committee, which gains its independence by representation among its membership of external or non-executive directors, and the appointment of an internal audit manager. The internal audit manager is responsible for identifying and specifying a balanced mix of audits or other assurance products in a rolling audit plan and providing a flow of the resulting reports and information to the audit committee for their consideration and action.

Reasonable control of HSEQ and OI, in the context of this book, covers all aspects of the organization's business, including quality, which can impact safety, health, and environmental performance. The powerful auditing methodology described shows audit practitioners and those seeking to develop a career in auditing, in step-by-step stages aligned to ISO 19011:2018, how to conduct a risk-based management system audit that will provide (or be unable to provide) reasonable assurance that an organization's objectives will be achieved.

Performed correctly, I believe that *audit* concerns organizational improvement. Accordingly, I will be addressing both the 'down' side of risk—protection of that which is important to us—and the 'up' side of risk—opportunities to create value from our existing and future activities.

But we are getting ahead of ourselves for now.

A-FACTOR 15

An audit should provide a reflection, as if in a mirror, of the effectiveness of the auditee's business control framework.

ISO 19011

The original guidance standard for auditing quality systems was published in 1990 as ISO 10011–1 (ISO, 1990). It was withdrawn in 2002 on the publication of ISO 19011:2002. Since 1990, many new management systems standards (MSS) have been published. This resulted in the need to consider the broader scope of management system auditing, as well as provide guidance that is more generic. This is reflected in the title of the current version ISO 19011:2018, Guidelines for Auditing Management Systems (ISO, 2018b), as well as in its content.

Its relevance for internal audit managers, lead auditors and auditors, and auditees is as follows:

- ISO 19011:2018 provides guidance on auditing management systems, including the principles of auditing, managing an audit program, and conducting management system audits, as well as guidance on the evaluation of competence of individuals involved in the audit process, including the person managing the audit programme, auditors, and audit teams.
- ISO 19011:2018 is applicable to all organizations that need to conduct internal or external audits of management systems or manage an audit programme.

The application of ISO 19011:2018 to other types of audits is possible, provided that special consideration is given to the specific competence needed.

ISO 19011 and ISO/IEC 17021

ISO 19011:2018 (ISO, 2018b) is the current version of this guidance standard. At the time of writing, it is in the final stages of a planned review commenced in 2022—it was at international harmonised stage code 90.60 *Completion of Main Action/Close of Review* (ISO, 2023b). Nil (or only minor) changes are anticipated.

With the publication of the third edition of ISO/IEC 17021, titled Conformity Assessment: Requirements for Bodies Providing Audit and Certification of Management Systems (ISO, 2017a), we now have two auditing-related ISO standards.

ISO/IEC 17021–1:2015 was reviewed and confirmed in 2020 and, at the time of writing, remains the current version. It will not be reviewed again before 2025. It contains principles and requirements for the competence, consistency, and impartiality of bodies providing audit and certification of all types of management systems. Certification bodies operating to ISO/IEC 17021–1:2015 do not need to offer all types of management system certification.

Certification of management systems is a third-party conformity assessment activity, and bodies performing this activity are therefore third-party conformity assessment bodies.

Figure 3.2 shows the relationship between these two standards.

ISO 19011:2018		
Internal auditing	**External auditing**	
	Supplier auditing	**Third-party auditing** (e.g. legal audit, certification)
'First-party' auditing	'Second-party' auditing	ISO/IEC 17021:2015
		Conformity assessment (requirements for bodies providing audit and certification of management systems)

FIGURE 3.2 Relationship between ISO 19011:2018 and ISO/IEC 17021:2015.

ISO 19011—Contents

ISO 19011: 2018 has seven clauses, plus Annex A. Sections 1–3 cover scope, normative references, and terms and definitions. These do not require additional interpretation here. Clause 4 addresses Principles of auditing, Clause 5 (5.1–7) is Managing an audit programme, Clause 6 (6.1–7) is Performing an audit, and Clause 7 (7.1–6) is Competence and evaluation of auditors. The Annex (A1–18) provide additional guidance for auditors. It is available for purchase from ISO, priced at 166 CHF (about £150 GBP or $180 USD at the time of writing).

Core to the changes in this third edition is the principle of a risk-based approach to auditing—a new, seventh, principle of auditing is in clause 4. This means that internal audit managers/committees must consider the risks when developing the audit program, and auditors and audit teams must consider risks when devising their work plan, as well as when collecting evidence during an audit. Auditors should be asking themselves if significant risks have been identified, and review and verify that they are being effectively managed within the scope of every audit they complete.

Clauses 5, 6, and 7 of the standard have undergone a complete update and reorganization to reflect the Annex SL structure and the updating of ISO 9001.

ISO 19011—Clause 4, Principles of Auditing

Auditing is characterised by reliance upon several principles. Adherence to these principles is a prerequisite for providing audit conclusions, for auditors working independently from one another, to reach similar conclusions in similar circumstances. ISO 19011:2018 provides seven principles for auditing management systems.

Principles (a) to (d) relate to auditors and the audit programme manager, and principles (e) to (g) relate to the conduct of the audit as follows:

a Integrity—the foundation of professionalism based on the principle of ethical conduct, undertaking audit activities only if competent to do so, and performing their work in an impartial manner.
b Fair presentation—the obligation to report truthfully and accurately. Obstacles and unresolved diverging opinions between the audit team and the auditee should be reported.
c Due professional care—the application of diligence and judgement in auditing. An important factor in carrying out [an audit] with due professional care is having the ability to make reasoned judgement in all audit situations.
d Confidentiality—security of information. This principle addresses the need for auditors to exercise discretion in the use and protection of information acquired in the course of their duties. It also refers to inappropriate use of such information for personal gain or in a manner detrimental to the legitimate interests of the auditee.
e Independence—the basis for the impartiality of the audit and objectivity of the audit conclusions. This principle provides specific guidance on the extent of independence that needs to be achieved, while recognizing that in smaller organizations it may be difficult for internal auditors to be fully independent. It refers to internal auditors being free from bias and conflict of interest, and independent from the operating managers of the function being audited. It reflects the interpretation of independence that certification bodies generally apply.
f Evidence-based approach—the rational method for reaching reliable and reproducible audit conclusions in a systematic way. Audit evidence should be verifiable. It should be based on samples of information available, since this is closely related to the confidence that can be placed in the audit conclusions.
g Risk-based approach—an audit approach that considers risks and opportunities. The approach should substantially influence audit planning, conduct, and reporting to ensure that audits are focused on matters which are significant to the auditee and their organization.

ISO 19011—Clause 5, Managing an Audit Program

Section 5 provides guidance on how to establish and manage an audit programme, as follows:

5.1 General
5.2 Establishing the audit program objectives
5.3 Determining and evaluating audit program risks and opportunities
5.4 Establishing the audit program
5.5 Implementing audit program
5.6 Monitoring audit program
5.7 Reviewing and improving audit program

Clause 5.1 General

This clause recognizes that an organization may utilize several management system standards. They may be audited separately or in combination (a combined audit).

Audit programs should take account of the *Context* (business environment) and organizational objectives. Audit resources should be allocated to those matters of significance, that is, higher inherent risk and lower levels of performance. This risk-based approach auditing means focusing on the *Big Rocks* that can significantly damage organizations.

Clause 5.2 Establishing Audit Program Objectives

This section includes a list of considerations to take into account when establishing an audit programme, including

- identified risk and opportunities,
- auditee's level of performance,
- the results of previous audits, and
- the maturity of the management system being audited.

Clause 5.3 Determining and Evaluating Audit Program Risks and Opportunities

The individual managing the audit program should identify the risk and opportunities considered when developing the audit program and present these to the internal audit committee (or client).

For example, there may be risks associated with

- planning—failing to determine the extent, number, duration, locations, and schedule of the audit program;
- resources—provision of insufficient time, and/or training for developing the audit program, or for conducting audits;
- team selection—insufficient competence in audit team/s to conduct effective audits; and
- availability of and cooperation of auditees across the organization, and of evidence to be sampled.

Opportunities for improving an audit program include

- allowing multiple audits to be conducted in a single audit and
- planning audit dates to coincide with the availability of key auditee staff.

Clause 5.4 Establishing the Audit Program

This clause provides guidance on roles and responsibilities and the competence of the person managing the audit program. This individual should determine the extent of the audit program and, as a result, determine the program resources including

- availability of auditors;
- financial and time resources (including travel time and time zones) to implement and manage the audit program;
- requirements related to audit locations, for example, via requirements, security clearances, PPE requirements; and,
- in my experience, an example of a barrier to meeting these requirements is ineffective communication of the audit programme. Another significant risk is the absence of a positive culture in which audit is welcomed and seen as useful by plant management.

CASE STUDY

METHYL ISOCYANATE (MIC) RELEASE, BHOPAL, INDIA, 2–3 DECEMBER 1984

Around midnight on the night of 2–3 December 1984 in Bhopal, India, a pesticide plant constructed and owned by Union Carbide India Limited accidentally released around 40 tonnes of methyl isocyanate (MIC) gas into the atmosphere. In the short term the leak resulted in an estimated 2000 fatalities, 100,000 injuries, and significant damage to livestock and crops. The long-term effects are difficult to evaluate, but estimates suggest more than 50,000 people have been partially or totally disabled (Atherton & Gil, 2008: 25–9).

The Immediate cause of the release was an exothermic reaction which resulted in 2000 litres of water entering one of two 57,000-litre MIC storage tanks. The plant was designed for smaller reactions, with a refrigeration system provided. However, at the time of the incident the refrigeration system had been shut down for six months to save money, and the additional layers of protection provided by a caustic soda scrubber and flaring had been removed because these systems had been shut down for maintenance (though, if operating, these systems would, in any case, have had insufficient capacity to deal with the vapour that night). There has been controversy over the source of the water, but the current theory is that it was introduced by sabotage by a disgruntled employee.

Among several process safety lessons to be drawn from the incident, there were lessons about auditing. Between 1979 and 1982, Union Carbide had conducted three safety audits. The 1982 'Operational Safety Survey' found major safety concerns in the MIC production unit that could have led to serious personnel exposures, though Union Carbide asserts that none of the recommendations would have had an impact on the 1984 release. It was reported in the post-incident investigation and subsequent trial that it was not clear whether the local company or the US parent had the responsibility for implementation of audit recommendations.

LEARNING

Audit culture needs to be encouraged, developed, and embraced.

Clause 5.5 Implementing Audit Program

This clause describes what the person managing the audit programme should do to implement it. This includes defining the scope, criteria, and auditor resources for each individual audit and ensuring each audit has a clear objective (this clearly addresses the concern expressed earlier about audits scheduled and carried out with no clearly defined purpose or objective).

The individual managing the audit program should assign the responsibility for conducting the audit to an audit team leader, with sufficient time to ensure effective pre-audit planning.

The individual managing the audit program should evaluate each audit and determine necessity for any follow-up audits. They should review, approve, and distribute the audit report to relevant parties.

Clauses 5.6 and 5.7 Monitoring Audit Program and Reviewing and Improving Audit Program

These two sections consider the need to

- evaluate whether schedules are being met and audit objectives met,
- evaluate the performance of lead auditor and the audit team members,
- review feedback from auditees,
- identify areas and opportunities for improvement, and
- review the continuing professional development of the auditors.

ISO 19011—Clause 6, Performing an Audit

Clause 6 provides a clear interpretation of the activities associated with a competent audit. The section is structured to follow the audit process flow, which is aligned to the approach in this book. Section 6 covers the following:

6.1 General
6.2 Initiating the audit
6.3 Preparing audit activities
6.4 Conducting the audit activities
6.5 Preparing and distributing the audit report
6.6 Completing the audit
6.7 Conducting audit follow-up

Clause 6.1 General

The clause highlights that clause 6 provides guidance for the conduct of a specific audit. You will read about the necessary steps (from clauses 6.2 to 6.7) as you progress through *The Audit Adventure* ™ in subsequent Chapters 4–10.

Clause 6.2 Initiating the Audit

Clause 6.2 says that the responsibility for conducting the audit should remain with the lead auditor until the assignment is complete. The lead auditor is responsible for all contact with the auditee. It focuses on *Establishing initial contact with the auditee* and *Determining the feasibility of the audit*. You'll read how to do this later in this chapter, and about factors associated with developing a suitable and workable relationship with the auditee in Chapter 4.

Clause 6.3 Preparing Audit Activities

Information should be gathered in advance of the audit to understand the auditee's operations. It should include management system documents as well as previous audit reports. The clause describes

- performing a review of documents in advance of/in preparation for the audit,
- preparing a risk-based audit work plan,
- assigning work to the audit team, and
- preparing work documents; in *The Audit Adventure* ™, I call these documents *Audit Finding Working Papers* (AFWP).

You'll read about how to prepare for an audit in Chapter 6, and about the nature of teamwork and your relationship with your audit team in Chapter 8.

Clause 6.4 Conducting Audit Activities

Audit activities are generally conducted in a defined sequence, reflected by this clause. The sequence may be varied to suit the circumstances of a particular audit. The clause describes

- assigning roles and responsibilities of observers, when approved by the lead auditor;
- conducting the opening meeting;
- communicating during the audit;
- access to audit information;
- reviewing documents and records (documented information) while conducting the audit;
- collecting and verifying information;
- generating audit findings;
- determining audit conclusion; and
- preparation for, and conduct of, the closing meeting.

You'll read about the relationship with auditees in Chapter 4, and about conducting the audit—reviewing, verifying, and testing the effectiveness of HSEQ and OI management systems—in Chapter 7.

Clause 6.5 Preparing and Distributing the Audit Report

The audit report should provide a complete, accurate, concise, and clear record of the audit. The clause describes

- preparing the audit report (structure, draft report, final report) and
- submitting the final audit report.

You'll read how to construct, write, and submit great audit reports in Chapter 10.

Clauses 6.6 and 6.7 Completing the Audit and Conducting Audit Follow-Up

The audit is complete when all planned activities within the work plan have been carried out. Documents pertaining to the audit should be retained or disposed of in accordance with the requirements of the audit program.

The outcome of the audit can indicate the need for corrective actions or opportunities for improvement. The clause clarifies that the responsibility for keeping the individual managing the audit program informed as to the status of these actions is on the auditee.

The completeness and effectiveness of corrective actions should be verified. This might form part of a subsequent audit.

You'll read how to successfully conclude an audit in Chapter 9.

ISO 19011—Clause 7, Competence and Evaluation of Auditors

Clause 7 is the final clause in the standard. It relates to the competence and evaluation of auditors. I'll discuss selection of the audit team, and the competency and personal behaviours of its members, later in this chapter. First. I'll address internal audit committees, the types of audits, provide examples of recognized MSS against which audits may be conducted, and the creation of a balanced audit plan.

CASE STUDY

ESTABLISHING AUDIT ASSETS IN A GLOBAL UPSTREAM PETROCHEMICALS GROUP

A major international petrochemicals group completely reconfigured its auditing function, reporting through the internal audit manager to the internal audit committee (an executive committee representing top management). The organization recognized two categories of auditors from within group sources.

A team of full-time lead auditors and auditors were engaged in numbers sufficient to deliver approximately one-third of the annual audit plan. The other two-thirds of auditor resource was supplied and supported by line departments. This comprised a larger number of trained part-time auditors, who worked in their line roles most of the time but were released upon request for auditing assignments for approximately four to six weeks per year.

As well as providing breadth in its audit teams, the organization saw such short secondments as an excellent way of communicating best practices across the group. A line manager reported that he had felt he was losing a key member of staff for two weeks. When the secondee returned to his line role, however, he brought back three specific recommendations from elsewhere, which were incorporated into the local arrangements. 'I never expected that my department could be the beneficiary from an audit elsewhere,' reported the manager.

An additional feature of the reconfiguration was that managers en route to senior roles via a variety of 'fast tracks' were strongly encouraged to accept one-to-two-year postings as auditors within the full-time auditing resource. With training their company provided (based on this book), these managers were able to gain accredited auditor qualifications in a short space of time, which fully prepared them for audit assignments under the supervision of an experienced lead auditor.

Within the organization, being given an opportunity to experience audit from the practitioners' perspective is thought to provide a valuable developmental experience for such managers. The current chief operating officer (COO) of the organization worked for two years in group audit, prior to his current posting.

LEARNING

Much can be gained by a learning organization from a committed approach to audit programming and resourcing.

Figure 3.3 illustrates how attitudes to audit can be changed given positive leadership by top management and a committed audit committee positively steering the audit programme.

INTERNAL AUDIT COMMITTEES

Organizations of all types are increasingly likely to have appointed an *internal audit committee* (IAC). In different organizations this may be called an audit committee, business assurance committee, governance group, or something similar, depending upon the organization's size, ownership, and preferences. Such IAC have responsibility to

FIGURE 3.3 From groan to growth!

oversee all the activities within the organization which are generically called *auditing*. In addition, note the following:

> The [internal] audit committee refers to the governance body that is charged with oversight of the organization's audit and control functions. Although these fiduciary duties are often delegated to an audit committee of the board of directors, [they can also] apply to other oversight groups with equivalent authority and responsibility, such as trustees, legislative bodies, owners of an owner-managed entity, internal control committees, or full boards of directors.
>
> IIA (2004)

We should consider whether it is MANDATORY to have internal audit activity. Put simply, it depends on the regulatory requirements that govern the organization. In the United States, since 2004, the New York Stock Exchange (NYSE) requires publicly traded companies to 'maintain an internal audit function to provide management and the audit committee with ongoing assessments of the company's risk management processes and system of internal controls.' Stock exchanges throughout the world have their own norms governing such companies, and some have implemented requirements like those of the NYSE.

ISO 9001, ISO 14001, ISO 45001, and other ISO MSS make internal audits a mandatory requirement in clause 9.2.

Although private companies—that is, those not publicly listed—are *not* mandated to have internal auditing, many of them have established an internal audit activity as one of its core organizational governance elements. It is considered a best practice in unlisted, unregistered/certified organizations. A well-functioning, adequately resourced internal audit activity that works collaboratively with management and the board is a key resource in identifying risks and recommending improvements to an organization's governance, risk management, internal controls, and operations.

Internal auditors' unique perspective of independence and objectivity, knowledge of the organization, and understanding and application of sound consulting and audit principles make them ideal for this role.

The role of the IAC is to deliver a program of audits. Auditing is a process which evaluates activities, facilities, or processes against MSS or other framework requirements. The core concern of this book is to enable delivery of this evaluation in such a way that it provides independent, objective assurance; improves operations integrity; and makes systematic approaches to the achievement of business objectives for the benefit of all stakeholders a reality. It is a structured management process which oversees by sampling an organization's internal controls. It provides independent, objective assurance when the selected framework for control can be reasonably expected to support the achievement of objectives, and an alert to initiate improvement when it may not.

The Institute of Internal Auditors (IIA, 2023b) defines internal auditing as

> an independent, objective assurance and consulting activity designed to add value and improve an organization's operations. It helps an organization accomplish its objectives by bringing a systematic, disciplined approach to evaluate and improve the effectiveness of risk management, control, and governance processes.

A-FACTOR 16

The primary reasons for auditing are business improvement and providing assurance to stakeholders.

Some years ago, experienced lead auditor Andrew Burns-Warren (2006, pers. comm.) shared the following thoughts:

> I think the thing to remember about auditing is that anyone can be a compliance auditor, can check a list and put a tick in the box. The real benefit of auditing comes from adding value, being constructive, and helping companies to improve over a period of time, however long that is.

Whatever the definition of internal auditing, it is the role of the IAC to deliver it. There are several different types and characteristics of audits, and these are discussed later in this chapter.

The Responsibility of the Internal Audit Committee

The principal corporate responsibility of the IAC is to review and endorse the effectiveness of the organization's internal control frameworks. This responsibility is fulfilled through effective audit planning, and thereafter ensuring the timely delivery of a balanced programme of audits, as established by the plan, which identify the 'up' and 'down' sides of risk and how these, individually and together, affect the organization and its stakeholders. The overall opinion is derived from a rolling audit (or assurance) plan, often endorsed, and agreed on an annual cycle, typically for a three- to five-year window.

Information is gathered throughout the life of the audit plan, reviewed during and at the end of each audit cycle, and guides the overall opinion of the audit committee in the reports it provides to senior management.

Done effectively, this periodic (commonly annual) report to senior management provides confirmation that the audit committee and line management are listening to the auditors—their reports, recommendations, and assessments—and are ensuring that the potential for improvement across the organization is realized and capitalized upon. This lateral learning will be effective if the audit committee ensures that line management in similar parts of the organization check their control frameworks to apply specific audit findings elsewhere across the whole. Line management needs to understand the reasons for the potential and/or necessary protection and improvements and how it can achieve them—by taking appropriate, timely action on audit findings.

Corporate governance requires that significant risks are brought to the attention of shareholders and other interested parties. Increasingly, organizations are providing assurance information to other stakeholders, such as the media and the public, as a part of public accountability and transparency initiatives. You can read more about this in the section on corporate social responsibility (CSR) in Chapter 2.

The Role of the Internal Audit Committee

The role of the IAC is to

- approve the audit plan for the plan year and to endorse it for subsequent years (as noted earlier, typically for three to five years);
- endorse draft audit terms of reference (ToR) for audits in the plan year;
- facilitate access to auditors (who will often be on part-time secondment from line management for the duration of each audit);
- receive summaries of all audit reports from the internal audit manager to check and challenge as necessary;
- review follow-up of actions and recommendations;
- promote lateral learning throughout the organization;
- assess the overall performance of the audit function; and
- provide annual assurance to the main executive board.

Key Appointments Related to Audit

There are two key appointments related to internal audit in many organizations. These are

- the head of internal audit (aka the internal audit manager) and
- line managers.

Head of Internal Audit

The main functions of the head of internal audit are to

- devise a rolling audit plan for the current year and in outline for subsequent years for approval by the audit committee;
- prepare draft terms of reference for each audit, for endorsement by the IAC;
- recruit and develop audit resources, for example lead auditors, other auditors, and their administrative support;
- maintain lists of approved lead auditors and auditors (often based in line management positions) who are available for undertaking audit assignments;
- ensure that audit teams have the right mix of competences—knowledge of the audit process, knowledge of the audit subject, skill in working with others and experience of practical auditing—in their leadership and membership;
- receive audit reports when issued;
- promote lateral learning throughout the organization; and
- keep the audit plan under review and propose revision if necessary.

Line Managers

The main functions of line managers are to

- release their staff upon request to participate as auditors in other locations or departments, as per the audit plan, and
- respond to audit recommendations relating to their own site or department.

This second point for line managers is critical. In many jurisdictions, failing to respond to recommendations raised by an audit would be considered very serious, with punitive civil and/or criminal penalties a possible consequence, particularly upon discovery following a significant loss. In this context, 'respond' implies a formal and recorded decision on the recommendation followed by corresponding and timely action. This action may or may not include implementation of the recommendation, in recognition that alternative remedies may be equally effective.

TYPES OF AUDITS

There are many different types of audits; all are generally undertaken by reference to and comparison with a reference framework of one kind or another—even if sometimes this amounts to a reference to 'best practice' as it appears in the eyes of the lead auditor and the audit team. The origins of reference frameworks were discussed in Chapter 2. ISO 19011:2018 identifies three types of audits which we will review.

The selected reference framework (or frameworks) included within the organization's overall audit plan will provide the subject matter for each audit (for example, occupational health and safety, environment, quality) and set out clearly the expected structure of business controls against which each audit will be assessed.

Some readers will remember a BBC television show popular in the UK in the 1980s called *Blankety Blank*. Whether you are familiar with this show or not, imagine now the variety of terms that could precede the word 'audit':

- Site
- Health and safety
- Fire safety
- Environmental
- Financial
- Quality
- Housekeeping
- Contractor or supply chain
- Pre-acquisition/due diligence
- Pre-flight
- Annual

Some of these types of audits—and you may have thought of others—probably exist in all organizations. Some are very conceptual in nature, perhaps taking a consultancy-type approach—in order, for example, to answer the question, 'Could we increase the warranty period from one year to two?'—while others are more of a transactional or compliance check in nature—'We have to be sure that this is right in every detail before we switch it on'.

CASE STUDY

SPEND TO SAVE

A UK-based manufacturer of carrier bags was the subject of an environmental audit that highlighted possible cost savings from replacing power correction units. These were subsequently installed at a cost of £13,000 (GBP) and produced annual savings of £4,000 (GBP).

LEARNING

Sometimes the improvement activity is largely about an organization's green blood (money).

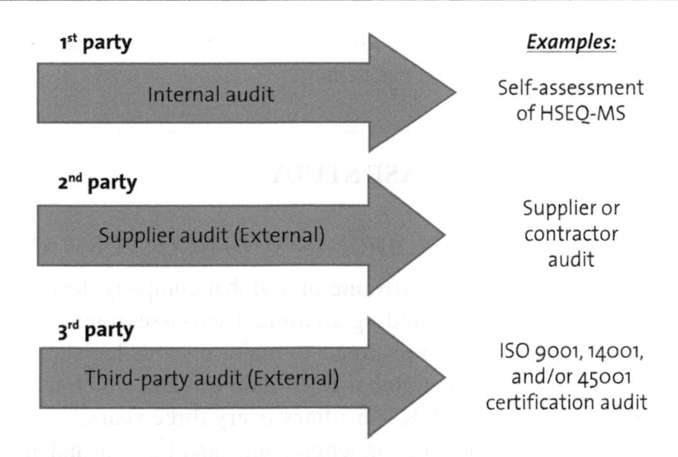

FIGURE 3.4 The three types of audit.

Audits—of whatever type—can generally be conducted in three different ways; there are three types of audits (see Figure 3.4):

- First-party audit
- Second-party audit
- Third-party audit

First-party Audits

First-party audits are internal audits and absolutely the subject matter of this book. Second- and third-party audits are external audits. I believe that those auditors can also use *The Audit Adventure* ™ to confirm that controls are effective for significant risks, *Big Rocks*, and *Black Swans*.

Organizations conduct first-party audits to audit their own activities, processes, and services. They are used to confirm or improve the effectiveness of their management systems. They can also be used to declare that their or another organization complies with an ISO or other standard (this is sometimes called a *self-declaration*).

Of course, such a declaration is credible only if first-party auditors are genuinely independent and free of bias. When you use first-party auditors for an opinion or to make a self-declaration of compliance, make sure that they aren't auditing their own work. I shall come back to the principle of *independence* later in this chapter.

Second-party Audits

Second-party audits are external audits, but within a supply chain. They're usually done by a customer, a supplier, or by others on their behalf. However, they can also be done by any party that has a formal interest in an organization. The results may comprise a client's approval to 'stay one of our suppliers' for some specified period.

Another example of a second-party audit would be an audit of how a contractor is performing against a set specification or contract.

CASE STUDY

INTEGRATING THE AUDIT MODEL AS AUDIT CULTURE ADVANCES

A UK-based oil and gas production affiliate of a global company developed a strong internal HSE-MS model, including an annual self-assessment audit plan covering the whole organization, in response to major hazards legislation. The US-based parent had a pre-existing global HSE audit process which mandated an external 'compliance audit' of its UK subsidiary every three years.

Initially, the two systems operated independently, and the external audits were viewed by the UK organization as time-consuming and resulting in very few improvement ideas. It was suggested that an experienced external auditor should join one internal audit, and this was agreed by all involved to give significant benefits, with the internal and external auditors complementing each other, providing value-adding findings, and identifying several opportunities for good-practice transfer.

LEARNING

Developing the audit model, and thus advancing the audit culture, can provide valuable learning points for an organization.

Third-party Audits

A third-party audit is an audit leading towards an external registration, certification, or accreditation. It may also be a statutory, regulatory, or similar audit.

A representation of the use of these three types of assurance activities in typical organizations is shown in Figure 3.5. The figure suggests that well over 90 per cent of all assurance activities are conducted by line management as internal audits, while independent and external management system audits likely account for considerably less than 10 per cent. This powerfully illustrates why second- and third-party processes must be substantially different to 'internal audits' if *they* are to add the desired value.

CASE STUDY

ESSO LONGFORD GAS PLANT EXPLOSION, VICTORIA, AUSTRALIA, 25 SEPTEMBER 1998

On 25 September 1998, a major fire and explosion occurred at the Esso Longford crude oil and gas processing site in Victoria, Australia, killing two workers and injuring eight others. Sales of gas supplies in the area were reduced to 5 per cent of usual volumes, resulting in 250,000 workers being sent home across the state as businesses were forced to close.

A plant supervisor was checking on a hydrocarbon release that had been leaking for about four hours, when a huge blast sent a gas and oil cloud over the area, drenching workers in fuel. The cloud ignited, resulting in a massive explosion which flashed back to envelop a reboiler heat exchanger number GP905.

The investigation reported that the immediate cause was the loss of lean oil flow, resulting in structural embrittling of the steel shell of reboiler GP905, which was followed by the introduction of hot lean oil in an attempt to stop the leak (Atherton & Gil, 2008: 170). It reported that, throughout the sequence of events, operators, and supervisors had not understood the consequences of their actions.

An external assessment of Esso's operations integrity management system (OIMS) had been carried out six months prior to the loss. The assessment report concluded that the organization had successfully implemented OIMS at the plant. However, the investigation found that the observations made by the assessment team appeared inconsistent with its own findings—particularly in relation to risk identification, analysis and management, training, operating procedures, documentation, data, and communications.

Esso was fined US$1 million for breaches of occupational health and safety regulations.

LEARNING

Audits need to be carried out by knowledgeable and experienced personnel who can independently review and verify the management systems relating to critical risks and the means by which they are implemented.

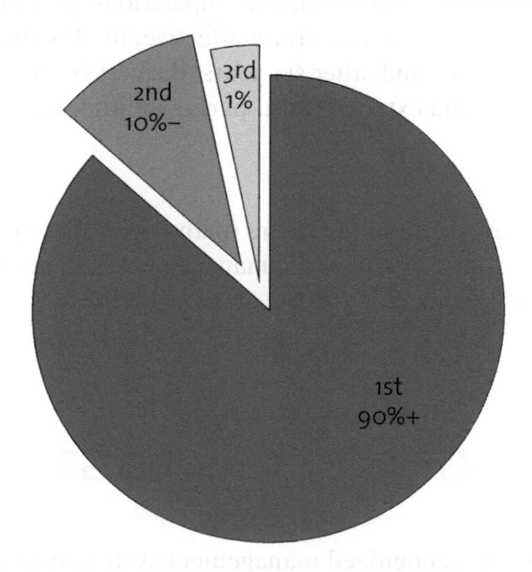

FIGURE 3.5 The deployment of assurance activities in typical organizations.

CASE STUDY

HOW ISO 9001-CERTIFIED ORGANIZATIONS SEE ISO 9001

At a seminar in Scotland attended by organizations involved with ISO 9001 in a range of business sectors, 160 attendees voted in a survey as follows:

- A quality management system based on ISO 9001 is good for business, but it can be adequately maintained by the business alone—44 per cent.
- Following ISO 9001 is a complete waste of time. Our money and efforts would be better utilized in other ways—24 per cent.
- As earlier, but to be effective it needs to be monitored by a competent third-party audit—17 per cent.
- Accredited certification to ISO 9001 provides clear benefits for any business—15 per cent.

LEARNING

Understanding the real purpose of external certification is important. If it adds value, retain it. If not, now may be the time to invest in other means of control.

CASE STUDY

THE EVOLUTION OF A QUALITY MANAGEMENT SYSTEM

A UK gas terminal obtained ISO 9001 certification for supply of products. As the globally mandated corporate internal 'flawless operations' integrated management systems culture became engrained and fully effective, the ISO 9001 systems were seen to have no added value, and, after six years, the certificate was handed back to the certification body, and external (third-party) auditing was discontinued.

LEARNING

Management systems evolve from immature (pathological) to mature (generative). Think carefully about the evolution of the management systems in your own organization or of the organization you are about to audit.

EXAMPLES OF RECOGNIZED STANDARDS

Examples of internationally recognized management system standards, for some of which registrations or certifications (or sometimes 'verifications') can be awarded or granted after third-party audits, are listed subsequently. This list has been thoroughly updated as of the date of publication. Not all are directly related to HSEQ, though some (or all) of these may be familiar to readers. Accordingly, and for brevity, we have decided not to include descriptions of each standard, and refer interested readers instead to the standards documents, details for which are provided as References. Suffice to say that most follow the PDCA structure of the Deming Cycle and ISO Annex SL discussed in Chapter 2.

Occupational Health and Safety

- ISO 39001:2012—Road traffic safety management systems (ISO, 2012c)
- ISO 45001:2018—Occupational Health and Safety Management Systems (ISO, 2018a)
- ILO-OSH 2001*—International Labour Organization, Guidelines on Occupational Safety and Health Management Systems (ILO, 2009)
- ANSI/ASSP Z10.0–2019*—American National Standard for Occupational Health and Safety Management Systems (ANSI, 2019)
- HSG65*—Managing for Health and Safety

Food Safety

- ISO 22000:2018—Food safety management systems (ISO, 2018d)
- HACCP*—Hazard Analysis and Critical Control Point; this is a system that enables food businesses to manage food safety; in the UK, the Food Standards Agency has developed six food safety management packs for different sectors of the food industry (HACCP, 2017)
- Food Standards Agency (UK), Food Hygiene Rating scheme (FSA, 2023)—see Figure 3.6.

Environment and Energy

- ISO 14001:2015—Environmental Management Systems (ISO, 2015b)
- ISO 20121:2012—Event Sustainability Management Systems (ISO, 2012d)
- ISO 50001: 2018—Energy Management Systems (ISO, 2018e)
- EMAS—Eco-Management and Audit Scheme (EMAS, 2023)

Quality

- ISO 9001:2015—Quality Management Systems (ISO, 2015a)
- AS/EN 9100/9110/9120—Quality management systems—requirements or aviation, space, and defence organizations (IAQG, 2016)
- ISO 13485:2016—Medical Devices—Quality Management Systems (ISO, 2016a)

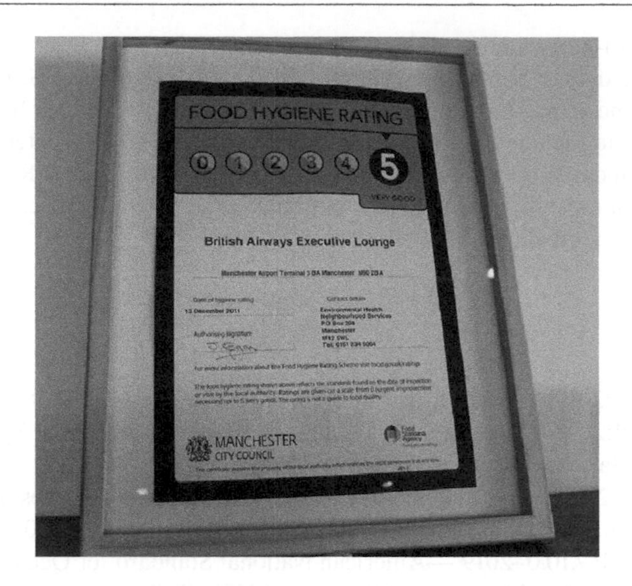

FIGURE 3.6 A food hygiene rating certificate (following an apparently successful third-party audit).

- IATF 16949:2016—International Standard for Automotive Quality Management Systems (IATF, 2016)
- ISO 29001:2020—Petroleum, petrochemical, and natural gas industries—Sector-specific quality management systems (ISO, 2020)
- The EFQM Model (EFQM, 2023)

Information Technology and Security

- ISO/IEC 20000–1:2018 Information Technology—Service management system requirements (ISO, 2018f)
- ISO 22301:2019 Security and Resilience—Business Continuity Management Systems (ISO, 2019)
- ISO/IEC 27001:2022 Information Security, Cybersecurity, and Privacy Protection—Information Security Management Systems (ISO, 2022b)
- ISO 28000:2022 Security and Resilience—Security Management Systems (ISO, 2022c)

Bribery

- ISO 37001:2016 Anti-bribery management systems (ISO, 2016b)

Integrated Standards

- ISO 31000:2018*—Risk Management—Principles and Guidelines (ISO, 2018c)
- ISO 44001:2017—Collaborative Business Relationship Management Systems (ISO, 2017b)
- Operating Management System Framework for Controlling Risk and Delivering High Performance in the Oil and Gas Industry (IOGP, 2014)

* = As a *Type B* MSS, this does not usually lead to external registration, certification, or a claim of conformance.

For this book, I have focused on my recommended audit methodology upon first-party audits, though it can certainly be used for second- and third-party audits. It can also be used for non-HSEQ audits.

Using This Auditing Methodology for Non-HSEQ Audits

While this book focuses on HSEQ and operations integrity auditing, I believe passionately that the auditing methodology it describes can be successfully and powerfully applied to reference frameworks for other topics and specialisms, including those noted earlier. Indeed, I have precisely applied this methodology to contractor, security, food hygiene, motor fleet, fire safety, and asset protection audits during the periods of preparation and writing of this book and its subsequent editions.

A BALANCED AUDIT PLAN

An important task of the head of internal audit is to balance the mix of types of audits across the countries/sites/divisions/departments of their organization, while considering the most appropriate intervals and intensities* at which to conduct these audits.

*The audit *intensity* for any single audit is the total number of auditor days (i.e., number of auditors multiplied by the duration in days of the audit). It will depend on the judgement of the head of internal audit as a result of making a risk assessment as follows.

A-FACTOR 17

A rolling, balanced audit plan is a foundational and essential preparatory component for providing assurance to stakeholders.

A-FACTOR 18

The number of auditor days (the 'audit intensity') provided to deliver any single audit from the audit plan should depend upon the depth and quality of assurance the management of the organization requires.

TIP

On my travels as an auditor and as an auditor trainer, I regularly come across organizations that have identified the types of audits as described here (first, second, and third party). I have also seen organizations that have reversed the order (third, second, and first), named (rather than numbered) the types, or have more or fewer types. This matters not—the principle remains the same.

Audit planning will typically include a risk assessment to evaluating such matters as

- complexity of scope operations,
- previous losses at the entity,
- significance of the impact of a future possible loss, and
- quality assurance of current control (that is, the audit opinion last time).

The coverage in an audit plan can be represented like a jigsaw puzzle, as shown in Figure 3.7. Note there are no overlaps, and no areas are missed. We do take, however, special note of the *interfaces*.

You should note that *interfaces*—the boundaries where one metaphorical jigsaw piece joins another—are always considered carefully. Many past losses have occurred on interfaces—such as where a permit to work is handed over from one shift to another. In totality, an audit or assurance plan will show how each part of the organization (each location, process, or combination of these) is covered, and at what frequency.

Also consider a three-dimensional jigsaw puzzle, which would show the different types of audits: under each jigsaw piece might be several layers, each for a different theme, at a different frequency (for example, annual quality audit, biannual safety audit).

Doing all audits annually, or at some other predefined frequency, overlooks the reality of varying levels of risk. Unless there are specific reasons for these (such as legal requirements), arbitrary intervals generally do not really make sense and are rightly difficult to justify to line managers on the receiving end. My own view is that an organization can gain more useful and more cost-effective audit performance at the level of expenditure it has agreed to allocate to assurance (whatever level that may be) by considering its overall governance needs. The head of internal audit and the internal

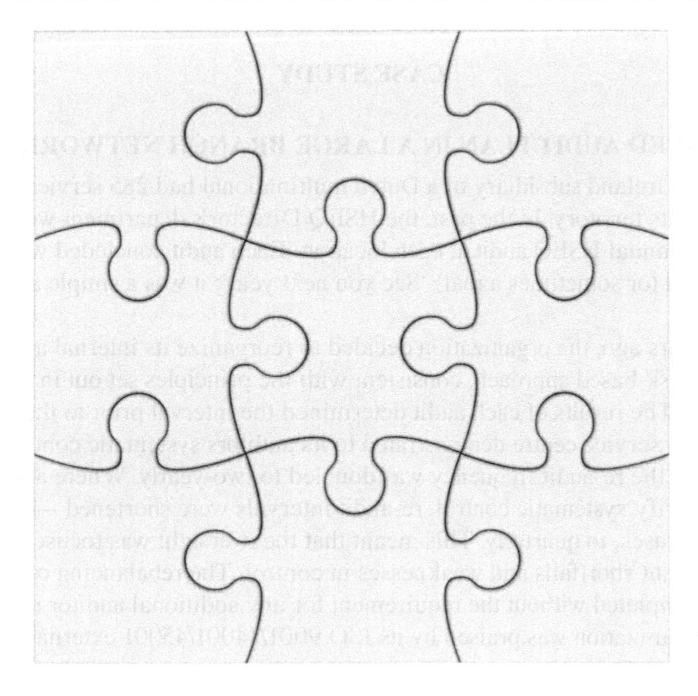

FIGURE 3.7 A representation of an organization's audit plan, in which each jigsaw piece represents a single audit. Nothing is missed, and there are no overlaps; interfaces are clearly defined.

Source: Shutterstock.

audit committee prove their value to the organization by looking at the needs of the whole and contrasting these with the needs of (and constraints on) the parts.

For example, it may be that the 'central distribution depot' is mission-critical to the organization, while the 'southern parts store' holds rarely used or obsolete stock. Accordingly, there should likely be greater intensity in the audit plan for broader, deeper audits in the centre than in the south.

A well-thought-through audit plan aligns audit frequency at each type (first, second, and third party) to the governance needs of the organization at that time. The most appropriate 'audit mix' is likely to give the highest level of assurance at a cost acceptable to the organization. This builds the culture for auditing in an organization. Keeping the audit plan under regular review remains a role of the audit committee, as described previously.

A-FACTOR 19

The internal audit committee is responsible for keeping the rolling audit plan, including 'the audit mix', under regular review.

CASE STUDY

A BALANCED AUDIT PLAN IN A LARGE BRANCH NETWORK

The UK and Ireland subsidiary of a Dutch multinational had 285 service centre locations in its territory. In the past, the HSEQ Director's department would conduct an annual HSEQ audit at each location. Each audit concluded with a metaphorical (or sometimes a real) 'See you next year'; it was a simple annual audit plan.

Two years ago, the organization decided to reorganize its internal audits to reflect a risk-based approach, consistent with the principles set out in ISO 19011:2018. The results of each audit determined the interval prior to the next one.

Where a service centre demonstrated to its auditors systematic control of its HSEQ risks, the re-audit frequency was doubled to two-yearly. Where auditors could not verify systematic control, re-audit intervals were shortened—in the most compelling cases, to quarterly. This meant that the oversight was focused upon the most important shortfalls and weaknesses in control. The rebalancing of the audit plan was completed without the requirement for any additional auditor resources.

The organization was praised by its ISO 9001/14001/45001 external auditors for these improvements.

AUDIT TERMS OF REFERENCE

Draft terms of reference (ToR) for each audit will usually be prepared by the head of internal audit some months in advance of a planned audit. Once appointed, the lead auditor will liaise with the head for clarification on the details of the particular assignment.

Each ToR document will relate to a discrete part of the overall audit plan (in our metaphor, one jigsaw piece). The terms of reference document should generally cover at least these three main areas:

- Objectives for the audit
- Reference framework(s) to be used (e.g., ISO 9001:2015, ISO 14001:2015)
- Scope of coverage

Objectives for the Audit

The objectives of an audit can vary. The IAC may have adopted house language such as *Six Sigma*, *Kaizen*, or *5S*. Such terms should remind us that continuous improvement is a major aim of a management system, including its independent audit.

The audit objectives are to provide a level of *Assurance* (high or low) to management that the reference framework/s is implemented and effective. If the level of assurance is *Low*, the auditors will likely *Alert* the auditee to the issues found; and (if they can—in terms of their time and their competence) provide *Advice* or assistance regarding possible corrective action.

A-FACTOR 20

Audit objectives can be referred to as the *3As* as an aide-memoire: Assurance, Alert, Advise.

Reference Framework(s) to be Used

Audits may include one or more reference (or business control, or MSS) frameworks. Each reference framework tells the auditors which structure of controls is to be used as reference for their audit work.

For ISO 45001:2018, they would consider this framework of management controls:

- Context of the organization
- Leadership and worker participation
- Planning
- Support
- Operation
- Performance evaluation
- Improvement

As discussed in Chapter 2, ISO Annex SL (ISO, 2012a) has harmonized and standardized all ISO-owned MSS frameworks; ISO 9001:2015, and ISO 14001:2015 have similar construction. Other reference frameworks *may* follow Annex SL, or other structures. You could refer to Table 2.2 on page 77 to assist your understanding of the applicable reference framework.

Scope

The audit scope is a statement setting out where the auditing work is to take place—the location and/or processes included. It confirms *what's included* and *what's excluded* in this audit assignment. The answer to the question *Does our audit include third-party deliveries?* should be found within the scope of the ToR.

At the start of any auditing assignment, the ToR will often have 'draft' status. It will later be discussed and formally agreed with the subject organization's management representative(s), at the opening meeting. How to do this effectively is described in Chapter 4, along with how to 'sell' your audit.

A-FACTOR 21

The terms of reference (ToR) are essentially the contract for the audit; the agreement between an organization and its auditors about *what* will be delivered by the end of the audit. No audit should commence without agreed terms of reference.

A diagram showing these main three components of the ToR appears in Chapter 6, Figure 6.2. This can be used for your own future audits as a starting point or for comparison.

TIP

When asked to join an audit team, or participate in an audit, an ideal first question to ask is *Please, can I see a copy of the terms of reference?* The ToR should confirm to you *why* you have been asked to participate in this audit—the reference framework and/or parts of the scope should align with your own competencies.

SELECTION OF THE AUDIT TEAM

The selection of an audit team usually commences with the appointment of the lead auditor. The lead auditor should be competent to lead their team to complete the job. Their typical characteristics:

- An experienced auditor, able to keep a team motivated and progress the assignment, and to bring it to consensus in reaching a conclusion and an opinion of the status of control in the auditee's organization
- Formally trained in auditing techniques provided by a recognized auditor training organization
- Participation in numerous audits as a team member before progressing to lead audit teams
- Possibly formally certificated as a lead or principal auditor by one or more of the recognized auditor registration bodies (more details of these are provided later in this chapter)

The Role of the Lead Auditor

The role of the lead auditor is a critical one, and their first involvement is likely to be up to three months before the start of the audit work on-site. Early activity for the lead auditor may be liaising with the head of internal audit in the scheduling of the audit and sometimes the production of the draft ToR. As mentioned, the final ToR will usually be finalized with the auditee just before or at the opening meeting.

The lead auditor may be involved in the selection of some or all the other members of the audit team. The level of assurance required, along with the size and complexity of the organization to be audited, will ultimately determine how many other auditors will be necessary, and the duration of the project (that is, how long the audit will last). The convention is for the head of internal audit (who may also be advised by others) to allocate the total number of audit days and then divide this by the target audit duration, to determine the number of auditors for the team (or vice versa). In practice, I have noted that there is often a compromise between the auditor days the lead auditor would like allocated to the audit team to do a thorough job and the wishes of the site, especially if—as is often the case—the site is paying the charges and expenses for each of these days.

A-FACTOR 22

For internal management system audits, the team should comprise a minimum of two members (a lead auditor plus one other auditor), with access to support for peer review.

Methods of Audit Team Selection

From here, things can vary from organization to organization and from audit to audit. Audit teams ideally comprise a mix of internal and external human resources to balance internal detailed knowledge with external breadth of experience. The three main ways that audit teams can come together are

- selection by the head of internal audit,
- selection by the lead auditor, and
- provision of auditors by the unit to be audited

Team Selection by the Head of Internal Audit

The head of internal audit may specify the team by selecting suitable individuals from the organization's list of internal auditors (full-time and part-time) and give it to the lead auditor. This can bring an audit team together reasonably quickly and can certainly prevent any suggestions that the team has any bias in terms of its formation.

Each team member will need to understand auditing, and usually each member of the team will have been formally trained by a recognized auditor training organization, though sometimes this training is 'on the job' (the forthcoming audit may indeed be a part of their training and personal development).

A possible downside to this approach is that line management for part-time auditors may resist releasing them, because, for example, 'We are busy that week'. When the selection is made very close to the planned start date, it can be very disruptive to their own department.

Team Selection by the Lead Auditor

The lead auditor may be given freedom to select the audit team. With their experience, the lead auditor will know to select people with whom they have good working relationships, who are confident auditors, and able team players.

Provision of Auditors by the Unit to Be Audited

Sometimes the organization to be audited may wish to provide one or more team members—for example, as a part of an internal training and development plan. This can be helpful to the audit team, as these individuals are likely to be better placed to know who to talk to about any given theme, where to find documents on-site, and so on. They must, however, be able to act with independence during their secondment to the team.

Balance in Team Selection

However the audit team is assembled, great care should be taken by the lead auditor to balance the professional skills and experience available—for example, too many medical doctors without audit experience could be counterproductive! It is, however, highly desirable if one or more members of the team have detailed knowledge of the activities of the unit to be audited—a medical doctor is invariably useful if the audit is at a hospital.

TIP

Base the team selection (if you can) on the requirements of the ToR. The guidance in this book which is based upon ISO 19011 can also be helpful. An audit is a project, and team selection is critical to a meaningful outcome.

ISO 19011 and Auditor Selection

Clause 7 (7.1–6) of ISO 19011:2018 (ISO, 2018b) is concerned with determining auditor competence to fulfil the needs of the audit programme, including factors to consider when deciding on the knowledge and skills required to competently audit the selected management system in the auditee's organization. Its provisions are described in the sections that follow.

Personal Behaviours

Behaviours auditors should display during the performance of audit activities include being observant, perceptive, imaginative, open to improvement, culturally sensitive, and collaborative. Keeping a cool head when others may be losing theirs is very important.

Knowledge and Skills

Generic Knowledge and Skills of Management System Auditors

Auditors require the knowledge and skills needed to audit single or multiple MSS or other control frameworks specific to different disciplines. They must understand the types of risk associated with auditing, have knowledge of organizational types, and understand general business and management concepts, processes, and related terminology.

Auditors need to be able to position discipline and sector requirements and audit findings in the wider context of an organization's business activities, governing agencies, business environment (aka context), legal and contractual requirements, and management's policies and intentions for the organization.

Discipline and Sector-Specific Knowledge of the Management System

Auditors should have the sector-specific knowledge and skills needed to be apply to the MSS. For example, this may include knowledge of

- legal requirements relevant to the specific discipline;
- the fundamentals of the discipline and the application of business and technical discipline-specific methods, techniques, processes, and practices, sufficient to enable the auditor to examine the management system and generate appropriate audit findings and conclusions; and
- risk management principles, methods, and techniques relevant to the discipline and sector, to enable the auditor to evaluate and control the risks associated with the audit programme.

Generic Knowledge and Skills of an Audit Team Leader

A lead auditor needs the following knowledge and skills:

- Balance the strengths and weaknesses of the individual audit team members.
- Develop a harmonious working relationship among the audit team members.
- Manage the uncertainty of achieving audit objectives.

Knowledge and Skills for Auditing Management Systems Addressing Multiple Disciplines

Auditors need the knowledge and skill requirements for understanding the interaction and synergy between different management systems. They can learn about this by reading this book.

ACHIEVING AUDITOR COMPETENCE

While an early predecessor to ISO 19011:2018 (i.e., ISO 19011:2002) gave quite prescriptive guidance, such as 'five years' work experience and twenty days of audit experience', the current standard acknowledges that auditor knowledge and skills can be acquired through a combination of the following, without detailing specific guidance:

- Education
- Auditor training programs
- Experience in relevant technical, managerial, or professional positions
- Audit experience

Accordingly, the head of internal audit, assisted by their lead auditors, will be very interested in evaluating the performance and capabilities of individual auditors.

Auditor Evaluation

The head of internal audit (i.e., the person managing the audit program) should establish suitable mechanisms for the continual evaluation of the performance of the lead auditors and their auditors. I suggest the following process to be considered:

- Establishing evaluation criteria; these should be both qualitative—for example having demonstrated audit skills; and quantitative—for example, the number of audits conducted.
- Selecting an appropriate evaluation method—for example, review of records, feedback, and interviews.
- Conducting evaluation—information collected about each auditor should be compared against the criteria set. When these criteria are not met, additional training, work, or audit experience and subsequent re-evaluation should follow.
- Maintaining and improving auditor competence—auditors and lead auditors should continually improve their competence through participation in management system audits and formalized continual professional development (CPD).

CASE STUDY

AUDITING EXCELLENCE WAS NOTICED BY THE CLIENT

An organization with internal management systems based on the *EFQM Model* (EFQM, 2023) requested bids for external certification to ISO 9001 (quality assurance). During bid evaluation, it met the proposed auditors and awarded the contract to an organization with limited experience in their business sector, but with an auditor clearly committed to continual improvement.

When they later required certification to ISO 14001, they rejected a bid from their current certifier, as the proposed environment auditor was judged to have a 'tick-box' approach, in contrast to their quality auditor.

LEARNING

Individual auditors' excellence will be recognized!

'Softer' Auditing Skills

This section has covered some of the harder technical and experiential attributes of HSEQ MSS auditors. Of course, professional auditors need 'softer' skills too—an ability to interact with other human beings in commercial and other environments. You can read more about these skills in Chapter 4 with respect to clients/auditees, and Chapter 8 re the audit team.

TIP

Once you have rapport, they tend to like you more, or put differently, if they like you, they'll let you live(!)

INDEPENDENCE

Many quite experienced individuals in all sorts of business and other organizations believe that independence comes *wholly and only* from outside their own organization. Some readers may have observed how 'external' advice is treated unquestioningly in some organizations, compared to advice given internally—it is often (sometimes?) more likely to be accepted and treated with unbelievable levels of reverence!

In my experience, independence is not binary. It is not something that you either have or do not have. Consider the following scenarios:

- I am auditing an organization managed by an individual who I would like to be my next boss.
- I am auditing an organization managed by an individual I rely on for our next consultancy order.
- I am auditing an organization managed by an individual whose beautiful daughter/handsome son I would like to marry next year.

How many auditors can say that they would not be influenced *at all* by these factors, irrespective of their (internal or external) employer?

So I believe that independence comes in degrees. In many ways, it is a state of mind; I can intentionally choose to act in an independent manner. The level of independence in any audit team is a factor that should be considered by the head of internal audit and, in turn, by the audit committee at the time any audit plan is approved.

In practice, maintenance of the independence of the audit will be monitored in real time (and actioned as necessary) by the lead auditor throughout its conduct. The lead auditor remains ultimately responsible for the audit and the audit opinion. This responsibility is discussed in detail in Chapter 10.

TIP

The lead auditor should resist strongly inappropriate selection of members for the audit team, for example someone who may have a vested interest in the outcome, such as a line manager from the department to be audited.

A-FACTOR 23

Recognize the importance to the overall audit opinion of an objective view from an independent audit team.

CASE STUDY

INDEPENDENCE SHOULD BE HIGHLY PRIZED

After an incident investigation, a food industry manufacturer with a strong Hazard Analysis and Critical Control Point (HACCP) process and a good overall H&S record was cited by the UK regulator for inconsistent risk assessment.

The company strongly disagreed with this finding and sought advice from an experienced health and safety management consultant, who, after further investigation, concurred with the regulator.

The HACCP process in use had identified some, but not all, food safety hazards, and the company eventually recognized this. Steps were taken to remedy the defect, and to ensure the resulting additional controls were regularly monitored and reviewed, these minor additions were readily added to the existing HACCP process.

LEARNING

Independent opinion is to be highly prized and learned from, even though the first instinct may be to reject it when it is contrary to a long-held view.

AUDITOR REGISTRATION ORGANIZATIONS

Since the mid-1990s, as use of management systems and their consequent auditing has grown and become considerably more widespread, the number of auditor registration organizations also expanded. The field has become so crowded that there are now mergers, as dominance in the marketplace is sought.

Likewise, individual auditor registrations with these organizations have grown and then levelled. Figure 3.8 shows individual registrations with IRCA (the International Register of Certificated Auditors) between 1984 and the end of 2022.

Increasingly (and rightly), auditor registration organizations require their professional members to sign up to a code of ethics or professional practice standards and to participate in programs of CPD which include both learning elements and evidence of leading and/or participating in audits. In my opinion, this mandatory approach to CPD by some of the registration organizations is commendable. I encourage all current and aspiring auditors to join such organizations, as appropriate and applicable to their sphere(s) of activity, at the appropriate grade. They typically offer such benefits as

- initial, refresher, and top-up training;
- professional recognition (often with a scale of professional and/or career grades);
- peer and client approval;
- networking opportunities;
- CPD opportunities; and
- access to preferential insurance terms.

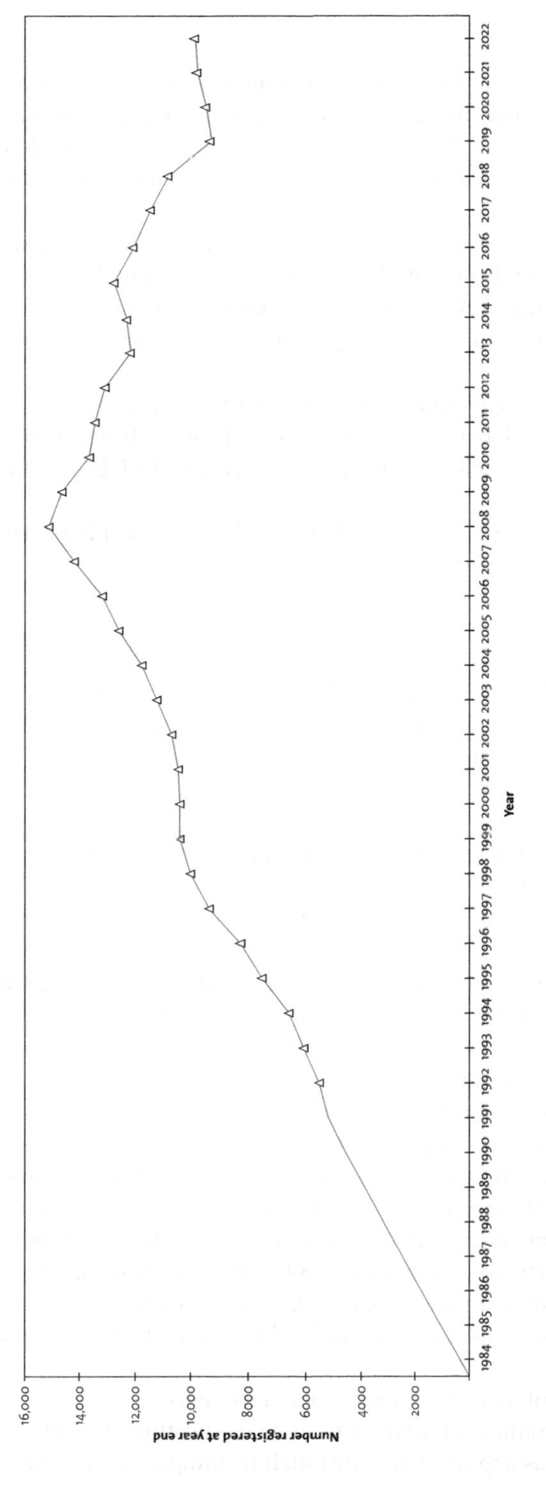

FIGURE 3.8 Graph showing the numbers of IRCA certificated auditors, 1984–2022.

International Auditor Registration Bodies

The following list represents a sample of seven auditor registration organizations, each with a short summary. The list includes a web address for each organization, to enable readers to assess it relative to their needs. There are many that claim to be the largest, the premier, and so on. There are other registration bodies; this list is by no means exhaustive:

- International Register of Certificated Auditors (IRCA)—www.quality.org
- Exemplar Global (formerly RABQSA)—https://exemplarglobal.org
- Institute of Internal Auditors (IIA)—www.theiia.org
- Institute of Environmental Management and Assessment (IEMA)—www.iema.net
- American Society for Quality (ASQ)—https://asq.org
- The Board for Global EHS Credentialing (BGC)—https://gobgc.org
- Professional Evaluation and Certification Board (PECB)—www.pecb.com

I will also summarize the history and role of the International Personal Certification Association (IPC)—www.ipcaweb.org

International Register of Certificated Auditors (IRCA)

Formed in 1984, IRCA was a part of the UK government's enterprise initiative. It was designed to make industry and business more competitive by implementing quality principles and practices. This enterprise structure included

- IRCA itself;
- an accreditation body now known internationally as UKAS;
- a national standards body, BSI Standards; and
- several commercial certification bodies.

The original reference framework/quality management standard used was the British Standard BS 5750, which evolved to become the ISO standard for quality management, ISO 9001 (see Figure 2.5 on page 69; its evolution is shown in the blue ovals). IRCA is now a part of the Chartered Quality Institute (CQI). Awarded a Royal Charter in 2006, and with 18,000 members in 100 countries, it says (CQI, 2023) that it is the partner of choice for quality management practitioners and, through its International Register of Certificated Auditors (IRCA) certification, for systems audit professionals.

IRCA is the world's original and largest international certification body for auditors of HSEQ management systems. It is based in London, UK. Over a million people have attended an IRCA certified course since 1984, and auditors from 130 countries are represented on the IRCA register (IRCA, 2023, pers. comm.).

IRCA provides auditors, business, and industry with two main services:

- Certification of auditors of management systems
- Approval of training organizations and certification of auditor training courses—it has approved over 100 such training organizations

As a registered charity, CQI and IRCA's mission (CQI, 2023) is to champion quality management for the benefit of society. Ever since it was founded in 1919 as a technical inspection association, it says it has stayed true to the same fundamental mission: to improve the ability of organizations to deliver the very best for consumers and stakeholders:

> We're committed to developing and championing ways of improving the value of products, projects, and services. This value helps organisations to be more competitive and achieve better results. It also contributes to a better quality of life for society.
>
> —CQI and IRCA

When asked about the consolidation and levelling out in auditor numbers since 2008, Ian Howe, IRCA Head of Membership, commented (IRCA, 2023, pers. comm.):

> The challenge of transitioning auditors to the 2015 iterations of ISO 9001 and ISO 14001 while maintaining high standards within the profession influenced the membership of IRCA. At that time, provision of training was revitalized through improved arrangements with a global network of training providers. Unfortunately, during the 2017–19 period, a significant number of members did not undergo the mandatory training to maintain their place on the register. From 2019 onwards however, we have seen steady growth of numbers again, and once again we are nearing the 10,000 members mark. We aim to exceed this during early 2023. The recent growth has been particularly pleasing, given the backdrop of the COVID-19 pandemic, the cost-of-living crisis, and other destabilising geopolitical situations.

Exemplar Global

Exemplar Global Inc. is a not-for-profit organization that services and supports the conformity assessment industry, including auditors, auditor trainers, and audit certification bodies. Formerly known as RABQSA, Exemplar Global has over 30 years of building certification programs. It has two principal offices, in Penrith, Australia; and Milwaukee, USA. RABQSA had previously been established from two legacy auditor

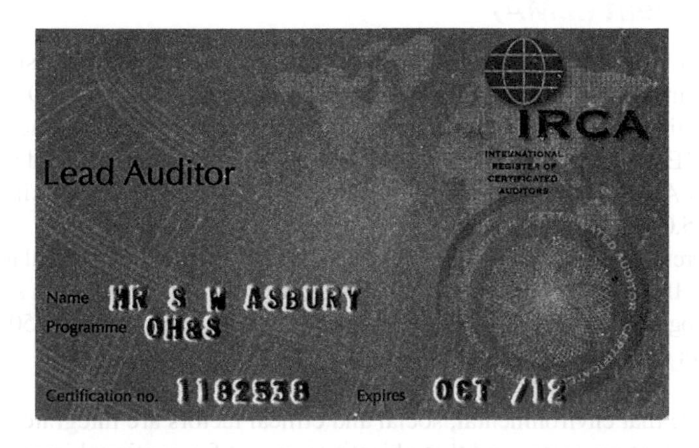

FIGURE 3.9 An IRCA OH&S Lead Auditor certification card.

registration bodies—the US Registrar Accreditation Board (RAB) and Quality Society of Australasia (QSA).

Exemplar Global's vision is to

> provide solutions that make a real difference to the lives of the conformity community. To achieve this, we provide:
>
> - Personnel certification
> - Credential management for individuals seeking further professional recognition
> - Independent certification for training providers to enhance their course quality and outcomes
> - New and innovative services that better serve all our stakeholders.

Exemplar Global certifies management system auditors, business improvement specialists, and management consultants across a range of disciplines, including quality, environmental, and occupational health and safety. Those certified are examined to ISO/IEC 17024:2012 (ISO, 2012d) and are recognized as competent, having demonstrated the required knowledge, skills, personal attributes, and additional qualifications specific to their scheme and/or scope of certification.

IIA

Established in 1941, the Institute of Internal Auditors (IIA) is an international professional association with more than 230,000 members worldwide. Its global headquarters are in Lake Mary, Florida, USA; and it has a regional network throughout the world.

The IIA says that it is 'the internal audit profession's leader in standards, certification, education, research, and technical guidance throughout the world. Generally, [it says its] members work in internal auditing, risk management, governance, internal control, information technology audit, education, and security' (IIA, 2023c).

Institute of Environmental Management and Assessment (IEMA)

IEMA is a not-for-profit organization established to promote best practice standards in environmental management, auditing, and assessment. It is certified to ISO 14001:2015.

Its origins lie in the merger in 1999 between the Institute of Environmental Management (IEM), the Institute of Environmental Assessment (IEA), and the Environmental Auditors Registration Association (EARA). It is based in Lincoln, UK, and has over 15,000 members in over 80 countries.

IEMA currently operates three specialist registers, one of which is particularly applicable to auditors. The IEMA Environmental Auditors Register has been in operation since before 1999 and is recognized internationally. It lists around 1500 auditors from over 50 countries.

IEMA says (IEMA, 2022) that its key aims are

- to ensure that environmental, social and ethical factors are integrated into our management processes and that objectives are set for continual improvement;

- to ensure that social, ethical and environmental scrutiny of products and services is integrated into our procurement processes, to achieve overall best value;
- to apply the IEMA GHG Management hierarchy, in particular to maximise energy efficiency and use of low carbon technology;
- to promote sustainable travel and actively promote and encourage the use of telephone and video conferencing where practicable;
- to maximise resource efficiency and facilitate repair, reuse and recycling;
- to ensure the wellbeing of our employees;
- to support the professional development of our employees and ensure that they all receive training appropriate to their role in implementing our sustainability policy;
- to comply with relevant legislation and regulations;
- to support our volunteers and recognize the importance and value of the work they undertake; and
- to use an accredited program to offset our residual emissions and to promote 'Net Zero' along our value chain, and beyond.

American Society for Quality (ASQ)

The American Society for Quality (ASQ) is based in Milwaukee, USA, and is an authority on quality, with members in more than 130 countries (ASQ, 2023). ASQ's vision is to empower individuals and communities of the world to achieve excellence through quality. Its vision is to be the thought leader and community of choice for individuals seeking excellence through quality.

In 2022, five of ASQ's certification programs achieved ISO/IEC 17024:2012 accreditation through the ANSI National Accreditation Board (ANAB). Globally, ASQ has issued over 400,000 personal certifications in 18 quality disciplines, including quality auditing.

ASQ maintains several auditor registers, including for Certified Quality Auditors (CQA).

The Board for Global EHS Credentialing (BGC)

BGC was established in 1960 as the American Board of Industrial Hygiene (ABIH) with one credentialing program. Now this independent non-profit organization has grown to become the Board for Global EHS Credentialing (BGC) with seven credentialing programs, including two for auditors:

- Certified Process Safety Auditor
- Certified Professional Environmental Auditor

Based in Lansing, Michigan, USA, BGC says (BGC, 2023) that it is the premier global credentialing body for professionals who protect and enhance the health, safety, and environment of people at work and in their communities.

Its mission:

> We protect people and the environment worldwide by providing precise and rigorous credentials for essential environmental, health, and safety professionals.

Its vision:

> A healthy and thriving world made better by BGC credentialed professionals.
>
> (BGC, ibid.)

Professional Evaluation and Certification Board (PECB)

PECB is a certification body based in Montreal, Canada, that provides education, certification, and certificate programs for individuals on a wide range of disciplines. It says (PECB, 2023) that it helps professionals and organizations show commitment and competence by providing them with valuable education, evaluation, certification, and certificate programs against rigorous internationally recognized standards.

PECB's mission is provide our clients with comprehensive services that inspire trust, continual improvement, demonstrate recognition, and benefit society as a whole.

PECB's principal objectives and activities are to

- establish the minimum requirements necessary to certify professionals, organizations, and products;
- review and verify the qualifications of applicants for eligibility to be considered for the certification evaluation;
- develop and maintain reliable, valid, and current certification evaluations;
- grant certificates to qualified candidates, organizations, and products, maintaining records and publishing a directory of the holders of valid certificates;
- establish requirements for the periodic renewal of certification and determining compliance with those requirements;
- ascertain that our clients meet ethical standards in their professional practice;
- represent members, where appropriate, in matters of common interest; and
- promote the benefits of certification to organizations, employers, public officials, practitioners in related fields, and the public.

The International Personnel Certification Association (IPC)

The International Personnel Certification Association (IPC) may also be of general interest to auditors, as it is a membership association for certification bodies. It is based in Chino Hills, California, with a mission 'To provide recognition to individuals who, having demonstrated competence to IPC approved schemes, can improve the performance of organisations' (IPC, 2023).

IPC replaced IATCA (the International Auditor and Training Certification Association, founded in Singapore in 1995) in 2003. IPC described the reasons for this change as follows:

> The members of IATCA recognized that there are now many sectors within business and industry and government that require and benefit from personnel certification. IATCA was established . . . expressly to address the management systems market. During [its] ten years, the requirement for personnel certification has extended into many more contexts within business, industry and government and it is now recognized that management systems form only a small part of the personnel certification market. To accommodate those changes IATCA has expanded its remit to include and contribute within those other areas, and evolved into IPC.

IPC differs from its predecessor in that its membership requirements are different. Full membership is offered only to personnel certification bodies which are accredited to the ISO standard for accreditation of personal certification bodies, ISO/IEC 17024:2012 (ISO, 2012b), which contains the principles and requirements for a body certifying persons against specific requirements and includes the development and maintenance of a certification scheme for persons.

A-FACTOR 24

First impressions count. Get the highest level of professional qualification that you can, pursue CPD, and use your (applicable) designatory letters on business cards, reports, and other stationery. Don't even think of claiming qualifications to which you are not entitled.

Trademark Infringement, Passing Off, and Misrepresentation

Please. Don't even think of claiming qualifications you have not earned, or professional memberships, registrations, or associations you do not hold. Trademarks and brands help consumers and users identify the source of products and services by identifying a word, design, or packaging with an organization. Over time, consumers become familiar with the quality and characteristics of the product or service and develop expectations. If those expectations are favourable, brand preferences can lead to repeat custom and brand growth. In legal terms, this reputation is known as goodwill.

Passing off occurs when one party deliberately or unintentionally offers its services (or goods) in a way that deceives the customer into believing they are buying the reputation or services of another party. The tort ('civil wrong') of passing off is a common law cause of action which concerns the infringement or misrepresentation of unregistered trademarks. These are also entitled to legal protection if certain conditions are met:

- The entity claiming passing off or misrepresentation must have the benefit of goodwill (i.e., it must have a reputation in the market).
- The claiming party must prove there has been a deception.
- There must be damage suffered by the claimant—courts generally presume that interference with goodwill will cause damage.

The most common remedy for trademark infringement is an injunction. Many professional organizations have taken steps to register their trademarks to strengthen the protection of their brands, and hold their members to a Code of Conduct. Disciplinary cases may be brought against those who are said to have breached the applicable Code.

As an employer and as an auditor, I have come across numerous examples of false claims on CVs and in interviews which may amount to dishonesty, misrepresentation of facts, breach of a professional code, trademark infringement, or passing off, such as in the case study which follows.

CASE STUDY

EXPELLED FROM MEMBERSHIP

An audit team in which I was the lead auditor conducted an HSE audit for a construction client in Walsall, UK. The audit work plan was based on a range of construction risks, such as work at height, entry to underground confined spaces, and possible exposure to asbestos.

A part of our expected framework for control was that the organization had ready access to competent health and safety advice. We were told that the organization had a contract for this with a locally based health and safety consultant, and that he would join the auditee's team the next day.

As the lead auditor, I was immediately uneasy about the quality of the responses given by the consultant when he was asked about his health and safety qualifications. He seemed very keen to move on to other matters, and, frankly, I suspected I wasn't hearing the full facts.

I asked the closed question 'Are you a member of IOSH?' The consultant looked me in the eye most sincerely and told me that he was a Chartered member (CMIOSH). I knew that IOSH members are issued with a membership card by their institution, so I asked to see it. I was told that he did not carry it, nor could he recall his membership number.

I asked a further closed question, 'Shall we call IOSH to confirm the details?' Knowing what I know now, I'm surprised he agreed.

When I called IOSH, I was told that the individual had been expelled from membership a few months earlier for passing off his associate membership as Chartered membership. His photograph was shown with the details of the investigation, and the decision of IOSH's disciplinary committee, on the institution's website.

IOSH continues to publish the results of its disciplinary cases (IOSH, 2023).

LEARNING

Carefully verify the facts that matter, and be disinclined to accept vague assurances, even when apparently convincing. Careful use of questioning styles can help to narrow down or precisely isolate the information you are seeking. This is a good example of the *gemba* attitude described in A-Factor 53 and elsewhere in this book.

Action speaks louder than words, but not nearly as often.

Mark Twain, 1835–1910

Test your knowledge with the next chapter assessment test (see Appendix 4).

Relationships with Auditees 4

You only get one chance to make a positive first impression. Be sure to take it!

—Dr Stephen Asbury

It's often not the message that's sent that causes offence, but the one that is received.

—Dr Stephen Asbury

INTRODUCTION

For a summary of this chapter, you can listen to the author's Microlearning™ presentation, which you'll find on the book's Companion Website. Then, delve deeper by reading on for further details on the Companion Website. (See Appendix 4.)

For an earlier edition of this book, this chapter was brought nearer to the front of the book in recognition of the huge importance of personal relationships in any auditing assignment. It remains here in the fourth edition in recognition of this importance. I put it to you that teamworking and interpersonal relationships are critical success factors for any project—including an audit.

These critical relationships include those between the lead auditor and their audit team, between the auditee and their own team, and, finally and critically, between the lead auditor and the auditee. It is this latter relationship that is explained here, as it can form (or fail) very early in the auditing process. The nature of intra-team relationships is dealt with in Chapter 8 as these generally commence after the audit has been initiated.

This chapter explains some of the most important features of the personal relationship with the auditee—from the first to the final contact. It includes ten case studies which illustrate key learning points about how to (and how not to) interact with the auditee. Together, these encourage auditors to think about the need to establish the best possible relationships with the wide variety of other individuals who could have a significant effect on the outcome of the audit. The success of such relationship building is to move the perception of the relationship from win:lose or lose:win to an expectation of win:win. In my experience, the reality is that anything other than win:win tends to be lose:lose—that is, 'nothing happens' as a result of the work done.

DOI: 10.1201/9781003364849-5

The lead auditor should start thinking about this important early aspect of their work in advance of the audit, starting with their first communication with their site contact(s). Only by thinking and acting in this way may they increase the chances of achieving the win:win outcome that the audit team, and the whole audit function, seeks. Over the longer term, win:wins tend to improve the overall perception of audit, while too many lose:loses, unsurprisingly, have the opposite effect. Therefore, on every audit day, every lead auditor, and every member of the audit team, has a role to play in enhancing the reputation of their role and profession.

WILFUL BLINDNESS AND UNDERSTANDING RESISTANCE

Auditees may have (for themselves) real reasons to resist or reject participation in an audit. As raised in Chapter 2, auditees may have wilful blindness; a fear of finding out, or fear of being found out. *Fear of Finding Out* (or FOFO) is a term used in the medical community to describe the psychological barrier that stops people from seeking medical advice for health conditions that worry them; it is probably the opposite of FOMO (fear of missing out). I believe FOFO can exist in an organizational context too. It might explain why organizations and their leaders may consciously avoid, and be reluctant to establish, what is or was occurring. *Plausible deniability* is the ability to deny any involvement in a poorly executed, illegal, or unethical activity because there is no evidence to prove involvement. This lack of evidence makes the denial credible, or plausible. It is often used in reference to situations where the senior officers of an organization deny responsibility for, or knowledge of, wrongdoing by the junior or other officers. It crops up as a defence in several investigations in the US, including the assassination of President Kennedy (Lane, 2011) and Watergate (FT, 2012). In such situations, senior officers can 'plausibly deny' an allegation even though it may be true. Perhaps in an audit context, the resistance tactic suggests forethought, such as intentionally seeking to avoid findings to plausibly avoid responsibility for one's past or current actions. I've heard auditees say, 'I'd rather not turn those stones over', 'We wouldn't have budget to deal with that', or 'I'd rather head office didn't know about this'. Other associated themes and language include

- leaving/finding a paper trail
- fear of being found out (FOBFO), of discovery, of being caught out, or caught 'red-handed'
- denial (vs confront and solve the issue)
- not having the stomach (or resources) to deal with it
- not 'rocking the boat'
- opening 'a can of worms'
- head in the sand/ostriching
- opening Pandora's box
- our secrets make us sick
- 'I was on vacation at the time'

- *emetophobia* (the fear of being overwhelmed, of secrets being revealed, anxiety pertaining to vomit)
- imposter syndrome (fear of being exposed or found out, doubting one's own abilities)
- 'Who will rid me of this turbulent (rebellious) priest?' (indirect orders said to have been issued by Henry II of England in respect of Archbishop of Canterbury, Thomas Becket), 'Someone should do something about that' (vague orders by anyone—both could lead to a plausible claim of innocence)
- Whistleblowing, and responses to whistle-blowers (see the case study which follows)

CASE STUDY

THE CONSEQUENCES OF WHISTLEBLOWING AT THE NHS

In 2023, Sky News (Sky, 2023) set up an email address for people to write in with their stories from the front line of the British National Health Service (NHS). Hundreds of people responded. But many of them said the same thing.

'The staff can see the problems in the NHS and some of them are such easy wins yet the moment anyone suggests anything, or heaven forbid, raises a complaint then their career is finished. So, we all keep quiet,' wrote one paramedic.

'Please keep me anonymous as I'm scared of repercussions—nurses are always under fear of speaking out,' said a nurse in the north-west of England.

Other organizations paint a similar picture.

Protect, a whistleblowing charity, analyzed calls to their helpline in the last year and told us that of 92 NHS workers who made contact, 77% reported retaliation or detrimental treatment as a result of speaking up. A third (33%) said they were either dismissed or felt they had to resign after raising concerns (and present this as in the other case studies):

LEARNING

Cultures of silence have long been recognized as a problem in some organizations.

Experienced auditors will recognize this paradigm. It should be met with a clear confirmation that auditors and audit teams are independent, they will select their own work plan, and will always report the facts. You'll read more about this in Chapter 6, where I will explain *the audit sample*. You'll also find an interesting *Tip* related to this on page 183.

This chapter presupposes that lead auditors, because of their prior experience, will already know many of the essential rules and techniques for establishing good interpersonal audit relationships and be capable of practising them. Another assumption is that the lead auditor is a competent auditor, and therefore that their findings and their audit conclusion will be arrived at in an ordered and defensible manner, with full documentary evidence to support their findings and sufficient logic leading to the conclusions. Aspiring and new auditors can also learn from this chapter, as well as learning from their lead auditors on their first auditing assignments. The case study which follows shows how wrong we auditors can get it.

CASE STUDY

ESCORTED OFF THE PREMISES

At an audit at a large retail goods distribution centre in North London, near the famous Wembley sports stadium, a lead auditor encountered considerable difficulty in arranging the *nemawashi* ('no surprises') briefing, as recommended in this book (See pages 235, 282, and A-Factor 74). There were several significant findings to share with the auditee, but she was either not available or did not show up as previously arranged—'Something has cropped up; we'll have to do this later.'

Instead of perhaps providing a written interim briefing note for the manager on the main issues being found during the fieldwork, the lead auditor decided to continue without.

On the final day of the audit, the auditee and her management team attended the closing meeting as planned. The lead audit team started to present the results of the audit team's work. The auditee challenged the first audit finding, saying, 'I'd have showed you where those records were if you'd asked me.' And proceeded to demand that the auditors should 'Follow me, and we'll find them now.' The closing presentation was suspended while the auditee and the auditors went to locate the critical documents. When the manager could not find them, she became extremely agitated and started talking of 'a conspiracy'. The auditee and the auditors returned to the closing meeting.

The second significant audit finding was also strongly challenged, with a further 'follow me' instruction, as the manager sought to demonstrate that the auditors had not been sufficiently thorough in their work. Once again, the essential records were not available, and the manager became even more agitated and visibly red-faced.

Over the course of the next hour, every finding was challenged in the same manner. Finally, 'blowing her top', the manager called security and said, 'Please escort these auditors off site, their work here is over.' And we were duly escorted off the premises.

A new audit was conducted a few weeks later by a different team of auditors, and the original team heard through the grapevine that they had identified similar findings and made similar recommendations. As any auditor knows, it is hard to suppress or fabricate long-tail evidence.

LEARNING

No manager likes to be surprised in front of their subordinates. Advance briefing and, if possible, agreement on some or all the major findings are an essential part of delivering an audit opinion and report successfully. Even had the audit team been unable to agree everything, they'd have known how to structure their presentation to make sure they found some early areas of agreement.

I confess. As has been widely speculated, the lead auditor escorted off-site was indeed me.

INITIAL CONTACTS WITH AUDITEES AND OTHERS

Whatever the reason for the audit, the lead auditor should always make contact, as part of their initial preparations, with the head of internal audit and, depending on the reason for the audit, possibly the chair of the audit committee. It is recommended to try to arrange a telephone conference or a meeting, as necessary, to discuss the audit generally and to get these individuals' input on any matters your early research has shown to be interesting or unusual.

The lead auditor should endeavour to find out as much as possible about the reasons for the timing of the audit. It is usually the case that internal audits are carried out because they are scheduled to take place in the organization's own audit plan. Remember—priorities in the audit plan were probably decided by the head of internal audit following on from the results of a risk assessment of the organization's business activities, and later approved by the relevant internal audit committee.

However, some audits will have been triggered for reasons other than being 'the next in line', and it is useful to have information about this. Sometimes audits will be brought forward or prioritized because an incident or loss has occurred, either in the area to be audited or in a similar facility elsewhere in the organization, or in the world. Sometimes senior management needs a level of assurance, given circumstances already existing or likely to occur in the short term in the area to be audited or in the business environment affecting that area. And sometimes an individual manager is being targeted—quite often for negative reasons!

A-FACTOR 25

Time spent on reconnaissance is seldom wasted.

A terms of reference document (which, as noted in Chapter 3, usually becomes known as the ToR) should have been prepared by or for the internal audit committee and issued to the interested parties for the audit in question. It often starts as a draft before being finalized. As a reminder, the ToR is effectively the contract for the audit, should clearly state the objectives for the audit, as well as confirming the reference framework and the scope. The objectives (the *3As*) are usually

- assure: to provide a level of assurance (high or low) to senior managers;
- alert: to identify areas needing improvement; and
- advise: to assist management with advice on how to improve.

A lead auditor's preparations normally require their review of reports of any previous audits at the scope location, including those by other internal audit or review bodies (such as a review by the health and safety committee). While these audit reports may have been accepted by management at the time they were conducted, the lead auditor may still wish to contact the people who led those audits to find out if there were any particular issues that were the focus of the auditee's or management's attention at the time they were reported. Similar enquiries may be made, at the lead auditor's discretion, with external auditors or reviewers (for example, from a statutory audit or a third-party certification or re-certification audit), or with regulators (for example following an incident investigation or prosecution).

A key characteristic of internal contacts is usually their informality, and commonly no paper trail is left. However, third-party contacts will often be more formal unless the lead auditor knows the other person.

The lead auditor needs to use their relationships with senior managers to find out how they assess the control frameworks which are about to be audited, and where they think the strengths and weaknesses may lie. If they say that they don't have a view, or don't know, that alone is useful information. The lead auditor can also ask to see a copy of the most recent management self-assessment (MSA) results, but this should only be used as a parameter to measure the audit's early findings. If there is a significant divergence between the audit's results and the MSA, then a more detailed comparison could be made later.

CASE STUDY

WORDS OR DEEDS?

This case study provides an example of how a senior manager's words can differ from their deeds.

A lead auditor conducted an OH&S-MS audit at a start-up hi-tech company in Oxfordshire, UK. All the recommendations were accepted.

After this audit, the auditor was asked to provide a retained H&S consultant advisory service by the company's managing director. Specifically, he was asked to undertake a quarterly review of the implementation of the agreed recommendations. The MD had previously said several times at company meetings that 'H&S was the highest priority' but, for twelve months after commencement, postponed each review feedback meeting at short notice.

Other company staff due to be present at these meeting increasingly expressed their concerns to the consultant that H&S frequently played second fiddle in day-to-day decision-making, and that these last-minute cancellations were just examples of that.

LEARNING

1. While being asked to assist after an audit may be flattering, an auditor must realize that their independence for future audits may be questioned; they should bear this in mind before accepting the role. (Note: 'Independence' was discussed in Chapter 3).
2. There can be a (big?) difference between what people say (or sign off on) and how those people later behave (what they do and how they act).
3. Finally, I suggest that you decide how you might handle this situation—think carefully how you could encourage the MD to reflect his words of intent in the way that he acts. The section on 'Influence' later in this chapter may help you to do so.

OBTAINING PRE-AUDIT DOCUMENTATION

The lead auditor is responsible for seeking pre-audit documentation from the site to be audited. There is no method that will absolutely ensure you obtain what you need in time to (i) sort the useful from the padding, (ii) share it with your team, and (iii) allow them time to read it before the audit. However, the tried-and-tested *Asbury* suggestion is as follows (this is covered in more detail in Chapter 6):

- Three months before the start of the audit, write to the site requesting the documents you would like. Appendix 3 provides text for a possible letter for this initial contact. Truthfully, you may not get too much of a response at this range—I'd say 20 per cent respond in a meaningful way at this time range.
- Two months out, a similar letter headed 'Second Request' is very powerful. In my experience, a further 60 to 70 per cent of auditees will send a package of information to you.
- For the remaining 10 to 20 per cent, a third letter headed 'Third and Final Request' one month out generally does the trick. This still gives you time to sort and share the information with the team.

Whenever I obtain the requested pre-audit documents, my norm is to share these with my audit team about two weeks in advance. Too much earlier and they are either mislaid or read and forgotten. Too much later and you'll hear 'I didn't have time to prepare,' 'I was travelling,' and so on.

Figure 4.1 shows me(!) hard at work a month or two before the start of an audit.

A-FACTOR 26

Request pre-audit documentation in good time. Three months before the start is an excellent start point, with a reminder at two months, and a final chase with four weeks to go.

A-FACTOR 27

Remember 'seek-sort-share'. Share the sorted pre-audit information with your audit team two weeks in advance of the audit.

FIGURE 4.1 Seek, Sort, and Share.

INFLUENCE

As an auditor, I encourage you to develop your influence. This is not the same as power, where people may do as we wish them to because of our status or title. Generally, auditors do not have power, but they can become very influential.

There are at least eleven different influencing styles, and I'll comment briefly here on each of them. It is useful to think of your own preferred style and then to consider the other ten and how they may permit you to become (even) more influential—not only as an auditor but in all your professional and life dealings. The styles:

- Logical—This means presenting your case based on logic and reasoning. The other person must have the time to sit back, think and be objective for your reasoning to be accepted.
- Educational—This concerns providing information and new ideas. People will learn from you, and you will be respected if your information or examples are seen as relevant to the identified needs.
- Selling—This means emphasizing the benefits of your suggestions. Enthusiasm will help you to sell, but few like an 'oversell'.
- Emotive—Here, you seek to appeal to emotions, feelings, and values. It can involve trying to make people feel guilty. A trusted manager can be very influential in appealing to people's emotions to get them to put in a big effort for a worthwhile cause.

- Expert—This is where you apply your superior knowledge or expertise. You need to be very credible to use this style, and you must be aware of others who may know the facts too. The expert style at meetings is to be quiet at the start, while others struggle with the facts, then to analyze the situation and suggest a course of action.
- Modelling—This means leading by example and being around long enough for people to copy you in word and deed. 'Do as I do' is much more influential than 'Do as I say.'
- Charismatic—Individuals with this influencing style rely on their charisma and strength of ego. You need a large supply of charm and humour to carry this off, but it can be successful where a straight-faced boss may fail.
- Negotiating—This concerns encouraging compromise all round, to achieve a negotiated outcome which satisfies all parties' wishes to some extent. Negotiators *never* give up—there is always a little ground to give and take.
- Joint problem-solving—This is about mutual agreement on the best decision. 'Let's work together to fix this' summarizes the style, but a high level of trust is needed. If you pull it off, commitment to the outcome is the reward. People support what they have assisted to create.
- Coercive—This is when you insist, or even threaten. If used sparingly, you become respected for being able to stand up for yourself, even in the face of strong resistance.
- Non-directive—With this style, you encourage the other person to develop their own analysis of the problem and arrive at their own solutions. Only asking questions characterizes this style.

TIP

I found Lee Bryce's book *The Influential Manager* (1991) to be an excellent read and a good source for auditors (or anyone) requiring more information on the topic of influence.

CASE STUDY

THE FACTS STACKED UP

At an audit in Nottinghamshire, UK, an audit team conducted an audit at a public-sector construction contractor. Middle managers attended the opening meeting to represent the auditee (the managing director). The lead auditor had reconfirmed the invitation for the auditee to participate, but this was declined.

The audit was conducted over several days, and very soon it became apparent that there were several significant findings—most notably that a previously active health, safety, and environmental management system implementation programme appeared to have stalled over the previous year or so. Almost all records were dated year –1, and there was very little evidence of recent activity.

The lead auditor sent a further invitation to the office of the auditee to attend the closing meeting, having briefed the attending middle managers. Once again, the auditee did not attend.

The closing meeting progressed well and with broad agreement on the findings. Two weeks after the audit, and when the final audit report had been submitted, the lead auditor was advised by his office that the managing director of the company had submitted a formal complaint about the conduct of the audit. Among several specifics, it was said that the auditor had not 'Displayed the common courtesy of inviting the managing director to either the opening or closing meetings.'

An on-site meeting was conducted at the auditee's premises a few weeks later, attended by senior management from the company, the lead auditor, and his line manager. The lead auditor was invited to lead the meeting to explain his findings. He had retained his poster-sized audit finding working papers (as you'll be encouraged to use in *The Audit Adventure* ™) and reposted them on the wall. Using these, he was able to explain risk by risk, and fact by fact, his process, his sampling, and his findings.

As the facts stacked up, the organization had no alternative but to withdraw the complaint and apologize.

LEARNING

Follow the auditing process, record the facts, and the truth is hard to challenge. And remember to archive your working papers securely.

BEHAVIOUR AND COMMUNICATION

Experienced auditors will recognize that the behaviour of the human beings implementing it will probably be the most important factor affecting whether the business control framework being audited will be robustly effective or not.

It takes a special set of personal characteristics to inspire, empower, and help lead people to work towards the achievement of specific goals. It takes a different set of characteristics to instruct them in how to do their jobs, follow up on matters of fine detail, and confirm what is expected next. No matter which set of characteristics predominates, it is possible for an individual, should they so choose, to learn from what happens and respond accordingly next time.

> Leadership is getting players to believe in you. If you tell a teammate you're ready to play as tough as you're able to, you'd better go out there and do it. Players will see right through a phony. And they can tell when you're not giving it all you've got.
>
> Larry Bird (2013), former NBA basketball player with the Boston Celtics and Olympic gold medallist, Barcelona 1992

To do their job well, auditors need a natural ability to communicate and create relationships quickly. The ideal is to develop towards is an engaging personality that enables you to develop rapport quickly, and so to make the most of the short initial period during which relationships 'form or storm'.

A key ability is putting people at ease when meeting them for the first time. It is a skill to relax and feel at ease with yourself, and at the same time to be able to put the auditee and your other contacts at ease with themselves, at all times, and in all company. Things to think about to achieve this include having a courteous, personable, and professional manner, and being able to listen carefully, speak engagingly, and explain what you want to do concisely and coherently. The A-Factors presented throughout the book and together in Appendix 1 provide readily usable and powerful ways to summarize the main points you may wish to make. Smiling also helps, especially when greeting individuals with a firm handshake.

A-FACTOR 28

A firm, but not crushing, handshake combined with a sincere smile is a powerful intimation of your professionalism as an auditor.

Relationship Auditing

The concept of auditing is well established in many fields, including finance, health and safety, environment, quality, and competency. A new area of activity for many business students is auditing business-to-business relationships.

In a commercial environment increasingly focused on personal accountability, the need for relationships which actively contribute to the success of the overall business strategy has never been greater. The desire for client/customer satisfaction has been replaced by the need for commitment and endorsement. Therefore, measuring the quality of business relationships is as important as the other traditional commercial metrics—what gets measured gets done!

If you provide any type of service to a client, you need to be sure you are delivering to expectations, if not exceeding them. And if you're purchasing services, you need to be sure you're getting the best possible performance from your supplier. Relationship auditing helps service providers and service users measure, manage, and maximize the potential of their stakeholder relationships.

CASE STUDY

HOW RELATIONSHIP AUDITING BENEFITTED A LAW FIRM

One of Europe's largest law firms decided to initiate an independent audit of one of its high-value relationships with a financial services provider. The audit uncovered that, while the client was happy with the service delivered, it was unsure whether the law firm had the resources to handle an upcoming large-scale product launch and was thus considering a panel review. Senior client contacts hadn't mentioned this concern to the firm, but a junior representative pleased to be included in the review process was eager to help.

Communicating such intelligence promptly to the right people meant the firm was able to take pre-emptive steps to reconfigure its resources and communicate that fact to the client's decision-makers. In possession of such facts, and suitably reassured, the client decided not to bother with the review and simply to assign the extra business to the law firm. The value of this single assignment paid for a three-year relationship auditing programme across the firm's top 50 clients, with a six-figure sum still 'in the bank'.

Later in the same programme, the relationship audit uncovered that an individual contact at a client company, at the time number 3 in the pecking order, was extremely unhappy with the service he was getting. The firm had previously concentrated its attention on the chief executive and the number 2, both of whom had the power of appointment. The number 3 was due to be promoted to replace the retiring number 2.

The firm was surprised, but, armed with this intelligence, it put in place a remedial strategy which, within twelve months, turned the 'renegade' into an 'apostle', and not only secured but grew the number and value of the firm's assignments.

LEARNING

Identifying and evaluating the *Big Rocks* is important, especially when you have time to take positive corrective action. All news is good news.

Identifying Individuals Whose Help You May Need

One of the first things you should do whenever commencing an audit assignment (or, frankly, any management assignment) is to make two lists. On the first list, write the names of all those people or positions you think you may need to contact during the audit. The key thing is to be imaginative and not to constrain your thinking to only those people or positions relevant to the operational activities within the auditee's area. Figure 4.2 shows just a small sample of individuals with whom relationships may need to be established before or during the various stages of the audit.

On the second list, write the names of all those people with whom you have a current relationship and who could assist you during the audit.

Audit committee: • Chair • Members	Auditee's: • Line managers • Operations staff • Support staff	Group: • Internal audit manager • Legal department
• Audit sponsor • Auditee's manager • Auditee	• Auditee's customers • Auditee's supply chain	Technical departments and SMEs: • Fire • Engineering • Maintenance • HR • IT
• Internal auditors • Previous audits • External auditors	Regulators: • Health and safety • Environment • Insurers/brokers	Personal network of contacts: • Professional body • Member groups • Social networks

FIGURE 4.2 Potentially useful contacts to be developed during the conduct of an audit.

TIP

Attending meetings of your practitioner group (for example, your IOSH branch, ASSP chapter, or RoSPA health and safety group) is a good way to make external contacts. And when you are ready, offer to present to the group on a subject of your choice. You'll be remembered in the group for a long time, especially if you deliver a memorable session.

Networking Relationships

As noted, auditors should take the opportunity to build a network of contacts, which will invariably be helpful in the future. Senior auditors will tend to find that, over time, they develop a good network of audit contacts, both inside and outside the internal audit department, with whom they can discuss issues on which they would like guidance or a second opinion. It's also true that many job vacancies are not advertised; they are recruited and filled through personal contacts.

In addition to building relationships with other auditors, an excellent and career-focused auditor will look to build relationships with senior managers in various functional and operational areas of the business in which they work. These relationships may be based on friendship, professional interest, or personal respect for audit work done in their areas. Relationships and contacts with colleagues who can give authoritative advice on legal and technical issues when requested are also useful.

Generally, the best relationships are built and rely upon the understanding by both parties that any information sought or provided is handled, unless otherwise stated, in confidence and treated as personal and non-attributable.

CASE STUDY

CONTACTS ARE ALWAYS USEFUL (AND CAN BE VALUABLE)

When he was in his mid-twenties, an HSE practitioner was teased by friends for spending too much time attending business networking meetings in the evenings. In his mid-thirties, he was the managing director of a medium-sized HSE consultancy. Many of his employees, and many of his customers, were connected with his contacts from over ten years earlier. He values those early interactions at over £5 million (GBP) to date.

LEARNING

People like to do business with people that they know and like.

TIP

Retaining the business cards of people you have met—other auditors, subject matter experts, practitioners, peers, and so on—is a good way of building your contact base. To be truly effective and valuable as members of your network, though, you need to find a good reason to contact them twice per year, at an absolute minimum if you are to maintain contact.

TIP

Join and network selectively with peers and those with common interests on LinkedIn (see links to suggested HSEQ and auditing related groups in the eBook+ and companion website, available from https://routledgetextbooks.com/textbooks/_author/asbury/), Twitter (twitter.com), Facebook (facebook.com), and similar social media platforms.

TIP

A good way to seek 'help' in an easy, non-threatening manner is to ask for 'advice'. While this seems to be a softer word, in my experience it usually delivers a powerful response. Individuals often love the opportunity to show their expertise.

BRINGING DOWN BARRIERS AND CHANGING THE PERCEPTIONS OF AUDIT

In many organizations the audit department, the auditing function, and even auditors themselves are still viewed with suspicion, even a degree of disdain, by those individuals who are likely to be audited or at least involved in audits. As a member of an auditor-training organization, I regularly hear from course delegates that this antipathy of the hunter and the hunted, the policeman and the offender, remains alive and kicking!

> Reality is what we take to be true. What we take to be true is what we believe. What we believe is based upon our perceptions. What we perceive depends upon what we look for. What we look for depends upon what we think. What we think depends upon what we perceive. What we perceive determines what we believe. What we believe determines what we take to be true. What we take to be true is our reality.
>
> Gary Zukav, spiritual teacher and author (2013)

CASE STUDY

HONE YOUR SOFT SKILLS

Food and Drug Administration (FDA) auditors in the US have been mentioned during our auditor-training class as often being very confrontational, requiring excessive detail in paper evidence, and having little concern for the efficiency of management system requirements. Most FDA audits are reported as 'competitive', with the auditees seeking to give no information other than that which is specifically requested, and the auditors trying to identify 'stones to turn over' to find hidden faults. This type of auditor-auditee relationship is not good practice and should be discouraged.

LEARNING

Hone your soft skills with auditees. Try to develop a reputation to be proud of, and don't allow a single bad experience to divert you from this aim.

The two great 'Lies of Audit' persist:

- Auditors say: 'We have come to help you.'
- Auditees say: 'We're pleased to have you with us.'

Many heads of internal audit departments, especially in larger organizations, have told me that they like to believe that this is no longer the case because of the way in which their departments have been modernized and/or have changed cultures. Many organizations have formally adopted professional auditing standards, such as those of IRCA, IIA, Exemplar Global, and others (see Chapter 3), which state that internal auditors can indeed see themselves as consultants to the business. However, each time an audit is carried out it provides a 'moment of truth' for anyone suspicious of the process.

Our challenge now, as twenty-first-century auditors, is progressively to reverse negative perceptions. Whatever the past or current perception of audit may be within an organization, it is possible to generate a receptive and creative atmosphere by focusing on building good personal relationships between *every* auditor on the team and those whose assistance will be sought in the five stages of audit specified by ISO 19011:2018—initiating, preparing for, conducting, reporting upon, and concluding.

The essential steps in building good personal relationships:

- Do not expect to be liked immediately.
- Accept that suspicion is entirely normal; do not equate it with people having something to hide; give them the benefit of the doubt.
- Ensure you have the knowledge and some examples to persuade people of the benefits of audit, and be ready to use them at every level in the organization.
- Bring as many people as possible on board at the beginning of the audit by inviting them to meet you and the audit team, hear about the audit process and how you will value their support.
- Be open about the audit process—there is nothing secret about it—and confirm that an internal audit is being done to assure management that all is well or to alert them if an intervention is required.
- Describe how that intervention might take place.
- Give pertinent examples of how audits have helped other, similar organizations.
- Explain how the auditors will focus their attention and scrutiny on 'the system' and not on individuals.
- Demonstrate the auditors' competencies for carrying out the audit in this part of the organization being audited, without saying, 'We're brilliant and we know everything!'
- Adapt your approach (for example, how you are dressed and how you speak) and interpersonal style (for example, your use of first or family names or job titles) to reflect the people with whom you are communicating.
- Start every formal interview by stating clearly why you have chosen to speak to the interviewee and how they can help you to do the audit effectively.
- Discuss the ToR and any agenda you have prepared; invite your interviewee to add other issues to the agenda and enquire/agree the order in which they would like to discuss them.
- Assure confidentiality (a principle of auditing/ISO 19011:2018), and do not breach that confidentiality either verbally or in the audit report.
- Do not judge what you find out until you are certain you have obtained the truth, and even then, ask yourself, 'So what?'
- Get back to individuals if you have promised to do so.

- Return any documents you 'borrow for reference'.
- Practise humility.
- Tell your auditee and interviewees that 'I am from Missouri,' and explain why (see the *Tip* on page 227 of Chapter 7, and the related Figure 7.5).

A-FACTOR 29

Accept that suspicion of audits and auditors is entirely normal. Do not equate it with auditees or interviewees having something to hide.

MAINTAIN GOOD RELATIONSHIPS WITH THE AUDITEE'S TEAM

The main sources of information to assist you during your forthcoming audit—in addition to the auditee and their subordinates working in the area being audited—will be the operational and support services personnel. Generally, you should find these people helpful. They will usually either give you the information needed or point you in the direction of relevant evidence to support the existence and effectiveness of parts of the business control framework.

However, we, as auditors, would not be doing our job properly if we did not check or corroborate such information by referring to an independent source, such as another person, physical evidence, or supporting documentation. The action of confirming what we have just been told can elicit emotions of frustration, even annoyance or anger, from the people who gave us the information. They may think that we do not trust them! Handling this part of any relationship, especially at the operational level, is critically important. So explain the next step of the process at the end of every interview and meeting.

Testing outside the auditee's operational area is required when auditors want to seek further confirmation of the effectiveness of particular controls from individuals or documentation in customer and supplier organizations. Auditors need to be aware that their relationships inside the auditee's department will have to be particularly well grounded to withstand the strains that can arise when the auditee's personnel are told or find out that the auditors wish to make enquiries or seek confirmation from third parties outside the immediate boundaries of the audited area.

In addition to obtaining clearance from the auditee for information checking—out of courtesy and on grounds of confidentiality—auditors must be prepared to handle such requests with sensitivity, irrespective of whether customers and suppliers are internal or external to the overall organization.

CASE STUDY

SPEAK THE LANGUAGE OF THE CUSTOMER

From 1993 to 2009, your author lived in a pleasant residential area five miles from the (then-new) Toyota motor car assembly plant in Derby, UK. The newly appointed Japanese plant director came to live in a Toyota-owned house close by.

Greatly interested in the role and origin of their new neighbour, the area's residents' association invited the director to speak at one of their regular meetings.

The talk was fascinating. The director summarized the origins of the Toyota company and its plans for developing the Derby site as a major European manufacturing facility. Towards the end of the evening, the director invited questions.

One of the most memorable questions was 'Should local people learn to speak Japanese?' By far the most memorable answer of the evening—delivered spontaneously—was, 'You should always speak the language of your customer.'

LEARNING

Always try to speak the language of your customer.

CASE STUDY

DRESS IN THE STYLE OF THE CUSTOMER

I have a colleague who conducts audits on cruise ships. He arrives at the vessel in 'whites', as he knows his first meeting will be with the captain on the bridge. Shortly after this meeting, he changes into a (intentionally slightly grubby) boiler suit, as he knows his next meeting will be in the engine room.

Likewise, no one wears a tie at Toyota. But they do in banks and most financial services companies.

LEARNING

Find out, and try to dress (where it is appropriate to do so) in the style in which the client dresses.

PREPARE THE AUDITEE AND THE ORGANIZATION FOR THE AUDIT OPINION AND ITS IMPLICATIONS

Human nature would suggest that auditors are unlikely to be welcomed with open arms when they are bearers of (what may be perceived as) 'bad news'. This bad news could take the form of serious control weaknesses, or an 'Unsatisfactory' or 'Unacceptable' audit opinion. 'If they don't like the message, they shoot the messenger,' so the saying goes. However, the extent to which this reaction will be evoked will very much depend on how the lead auditor delivers the message. Clearly, just dumping the bad news on the auditee's desk on the final day, as they head for home, is not going to help anyone!

TIP

Try to keep things on a professional level, and hard feelings to a minimum.

TIP

Provide the audit opinion at an early point, once it is known. Be prepared to explain why and how the opinion was arrived at, particularly if it may be received as 'bad news'.

CASE STUDY

STATE YOUR AUDIT OPINION EARLY

An auditor spent approximately two hours presenting the detailed findings of an audit to a management team. When the final slide announced the audit opinion as 'Unsatisfactory', the managing director calmly said, 'Please put the first slide on again . . .'

LEARNING

State your audit opinion as early as you can through meetings, or early in the closing meeting.

When applicable, contrast the audit team's early indicators of any poor control with the perhaps more optimistic expectation of senior managers (which the lead auditor may have heard about). An experienced lead auditor will 'drip-feed' any 'bad news' to the auditee through a series of *nemawashi* or 'no surprises' meetings from an early stage in the audit. *Nemawashi* is explained on page 282, Chapter 9.

In contrast, passing on a 'Good' or 'Satisfactory' audit opinion is usually rather easier and overall a more pleasant affair. This is because management will feel that they have 'passed the audit'. They are not threatened, since they see the result in the same light as having been told 'Everything in your area of responsibility is being well managed,' and they then get on with their usual job. However, such ready acceptance may not be what the lead auditor really wants, or the outcome that the audit warrants. It is probably inappropriate in most circumstances, so some auditees and their management teams need to be encouraged to rigorously challenge even seemingly good news; they should examine the methods used by the auditor to arrive at the good news, and the documentation that the audit team has prepared to support it. This process should result in auditees and management more fully understanding the strengths and weaknesses of their management control framework. Only then should they feel confident of the extent to which they can bask in the reflected glow of the audit team's assessment.

TIP

The process of 'educating' the management team is often best done through a series of 'progress' meetings held through the duration of the audit (usually after the halfway point). Such meetings can be set up as *nemawashi* ('no surprises') meetings as soon as the lead auditor has something they wish to draw to the auditee's attention.

PREPARE THE RELATIONSHIP FOR THE FINAL PRESENTATION

The lead auditor may have received or detected a warning if there is going to be any difficulty in the organization receiving the results of the audit, or resistance/'pushback' at the closing meeting. This could be because the auditee has mentioned or stated that the management team cannot (or will not) accept the audit opinion, or that he/she will not. This can be the case even though the lead auditor has initiated and participated in numerous *nemawashi* ('no surprises') meetings and responded appropriately to questions and challenges from the auditee's staff or even the auditee personally.

In this scenario, the auditee (or others) may even be prepared to make accusations and threats and/or fight every finding in the hope of unearthing some factual inaccuracy upon which the auditors have relied in arriving at their conclusions. If they are successful, they may hope to cast aspersions on the reliability of other parts of the audit teams' work.

It may be that some 'big guns'—intimidating, large, boisterous, or just very senior managers—may be assembled at the closing presentation. Auditees cannot sustain this type of behaviour on their own. They need the support of their seniors and/or subordinates, which will be provided either willingly or as result of coercion. Quite reasonably you cannot and should not expect them to come to your aid in the closing meeting, whatever they have said to you privately. A key relationship that will need to be relied upon in these circumstances will be that between the lead auditor and the auditee's line manager. In certain circumstances, others may necessarily become involved, including the head of internal audit, the audit sponsor, or even the chair of the audit committee. The case study which follows illustrates the essence of this point extremely well.

CASE STUDY

THE IMPACT OF THE 'BIG GUNS'

A highly competent technical professional had been working in the internal audit department of a major global business for a relatively short period, when she was appointed lead auditor for an audit which was to cover the engineering department of a large subsidiary.

She and her team completed the audit, and the results did not reflect well on the subsidiary's management. In fact, there were some serious issues which needed urgent management attention. This meant that the audit opinion would be 'Unsatisfactory', and possibly 'Unacceptable'.

The auditee knew this but could not bring himself to accept it, or some of the main findings, even though the supporting evidence had been fully and clearly documented and explained to him. He decided to use the closing presentation to try to cut this lead auditor down to size. Having organized a video conferencing facility, he gathered his troops in three locations. By her own admission, when the lead auditor went to the bathroom before the meeting to compose herself for what she knew was going to be a very rough ride, she was shaking so much that she could not apply her own lipstick!

The tension grew as the seats in the three conference rooms filled. Finally, the lead auditor, her audit team, and the auditee took their places. The scene was set—the auditee was going to teach this particular auditor a lesson she would not forget in a long time!

Just as the lead auditor was drawing breath to start the presentation, there was a disturbance at one of the remote locations. Everybody's eyes turned to the video screen. A late arrival had entered the conference room.

The scene had been set for a totally unnecessary demonstration of bullying tactics, which would probably have resulted in accusations, denials, and counter-accusations, perhaps many of a personal nature. The audit result would have stood, but people would have been damaged. Now the presentation went off without much more than a few clarifications; the auditee was very quiet, and the lead auditor got acceptance of the findings and a commitment to take urgent action to address the control weaknesses.

Why the turnaround?

The late arrival was the auditee's boss, who had been tipped off by a telephone call from the head of internal audit about the audit results and the planned showdown.

LEARNING

Invest the necessary time to get the essential facts relating to the audit result to a level of management that does not see it as personal criticism, and which will want to ensure the auditee listens and responds with the necessary corrective actions.

It is advisable for the lead auditor to discuss with the auditee which persons will attend the final presentation, particularly from senior management. This means that there will be enough seats in the room and enough copies of any paperwork to be distributed. You should understand that sometimes there are senior individuals who invite themselves at the very last moment. Their motives for attending may or may not be known, and it may therefore be difficult to prepare to handle any interjections from them. Their aim may be to disrupt the smooth flow of the audit team's presentation and, if the opportunity arises, to call the auditors' credibility into doubt. This can be a very difficult situation to handle, but a good way to deal with it is to raise the possibility of something like this happening with the auditee and get them to agree that they will field any such questions.

A-FACTOR 30

Don't drop your guard until the assignment is complete. It is essential to rehearse the possible nature of objections and to try to see things from the other person's perspective. Anticipate and have in mind answers to the types of questions auditees and others are likely to pose.

> Give me six hours to chop down a tree and I will spend the first four sharpening the axe.
>
> Abraham Lincoln (1809–65, 16th president of the USA)

Test your knowledge with the next chapter assessment test (see Appendix 4).

The Audit Adventure™ 5

This is your satnav for risk-based auditing. You'll always know where you are, and where you're heading. On time, to a powerful conclusion. That's what it does.

—Dr Stephen Asbury

From secret coves and great surfing beaches to World Heritage Site mining areas, we look after some of the most beautiful places in Cornwall.

—National Trust (2023)

IMAGINE AN ADVENTURE IN BEAUTIFUL CORNWALL . . .

For a summary of this chapter, you can listen to the author's Microlearning™ presentation, which you'll find on the book's Companion Website. Then, delve deeper by reading on for further details on the Companion Website. (See Appendix 4.)

My metaphor for risk-based auditing, *The Audit Adventure*™, is set at Poldhu, Cornwall, in the south-west corner of the UK. This beautiful sandy cove is shown in the photograph Figure I.2 in the Introduction, while the cartoon in Figure I.3 provides a (slightly exaggerated) representation of its topography. This powerful image of *The Audit Adventure*™ is intended to make my complete auditing approach memorable to you, lead auditors, and their audit teams. At any time, you'll know where you've been, where you are, and where you're going next.

Before your audit commences, this metaphorical topography charts the journey you and your audit team will progress to completion. At any point during the audit, it reminds of the work already done and highlights the approach and content of the work still to be done. Only when the journey is complete will you have assembled the facts—the truth—upon which to base your audit opinion.

Capturing and reporting faithfully this truth, without fear or favour, is the essence of effective auditing. I encourage you to look again at Figures I.2 and I.3 to remind yourself of the geography, prior to taking an energetic walk with me from the car park to Mrs Rimington's cafe.

The Audit Adventure™ provides a powerful and memorable mind map for the dynamics and thought processes used by effective auditors using this risk-based auditing approach.

DOI: 10.1201/9781003364849-6

FIGURE 5.1 *The Audit Adventure*™: Prepare, Conduct, Report.

It is powerful because it is aligned to ISO 19011:2018 Guidelines for auditing management systems. It is memorable because it creates a lasting image of the process. It is the approach I have used personally in over 1000 real-life audits conducted around the world.

The Audit Adventure™ mind map steers lead auditors and their audit teams through the major auditing steps described in ISO 19011:2018. The steps in *The Audit Adventure*™ comprise the following:

- **Initiate** the audit.
- **Prepare** audit activities.
- **Conduct** the audit.
- **Conclude** the audit—complete the audit and follow-up.
- **Report**—prepare and distribute draft and final audit reports.

Organizations' audit plans are (should be) generally initiated well in advance of the work at a single auditee's premises, as I described in Chapter 3. *The Audit Adventure*™ thus guides us through all the other steps shown in Figure 5.1.

A STEP-BY-STEP VISION OF THE ADVENTURE

1 An audit generally commences at a reasonably sedate pace, often with the audit team gathering a day or two before the audit starts. They may meet and have

dinner. On the first day of the audit, the team uses the background information they have obtained and studied in advance to elevate the team's collective understanding to a high vantage point, so they become familiar with the auditee's responsibility—that is, an understanding of the context (business environment), the organization's objectives, and the potentially significant (inherent) risks to those objectives. Figure 5.2 illustrates this 'high-level view from the top'.

2 After completing this independent familiarization of the context for their audit, the audit team will be 'standing in management's shoes', and in an informed position to create their risk-based work plan. This will focus on a sample of team-selected significant risks to the organization's objectives. The number of discrete risks selected for the work plan will ultimately depend on the number of auditors and the total time available to the team. This is known as the 'audit intensity' (i.e., how intense the audit will be).

3 *The Audit Adventurers* have a choice of descent routes down to the beach, using the steps, or one of the many windy paths. They will, of course, want to avoid a free fall! The descent route selected represents the auditors' selection of activities, processes, or services for their work plan for each single audit. They are guided by the route they choose to the detail necessary to review and verify the presence of the expected barriers or layers of control. The audit team knows now that the time available for preparation is behind them. As they decide on, and start down, their route to the required details, they are aware that the pace is beginning to pick up, as gravity pulls them inexorably down into the detail. As pointed out, and as shown in Figure I.4 in the Introduction, time starts to fly by.

4 The audit team could even feel out of control as they travel down the steep slope gathering ever-more-detailed evidence and data relevant to their selected

FIGURE 5.2 *The Audit Adventure*™: The high-level view from the top.

sample of risk(s). Figure 5.3 illustrates what our Audit Adventurers may find on the beach, as the tide recedes. As well as beach flora, fauna, and mineralia large and small, they may find all manner of trash deposited by the receding waters of the English Channel. Auditors will find all manner of details—sometimes even conflicting—related to the necessary elements of effective control, as their questions become ever more detailed and probing. Among all this information are the facts they need to support their overall audit opinion of assurance or alert. They just need to sort it all out.

5 After the planned amount of time spent experiencing the sights and thrills of the world-class beach, enthusiasm for refreshment compels *The Audit Adventurers* to commence the ascent back up the other side of the cove. The lead auditor must ensure that each member of the audit team has documented their detailed audit findings and initial analysis of the business control framework (BCF) before the time available for fieldwork has elapsed. Only this will give them sufficient momentum to extricate themselves from the detail and to deliver a useful, high-level audit opinion.

6 Exhausted from their fieldwork, *The Audit Adventurers* are extremely motivated to head for the best tea shop in Cornwall, owned by Mrs Rimington. This phase of *The Audit Adventure*™ replicates their climb away from the beach and towards the opposite peak. This is where management resides and where once again the auditors must speak the language of their auditee and the boardroom. It's high-level, and the view from the tea shop is quite breathtaking. The audit team relies on their momentum to push them up the final slope of *The Audit Adventure*™ and spend the last allocation of their time finding the synergies and synchronicity within their detailed findings.

7 The final push is to present meaningful, high-level, and future-focused audit conclusions that have sufficient resonance in the busy world of top management to result in action being taken to improve the business control framework, and thus performance. Cornish cream tea is served . . .

FIGURE 5.3 *The Audit Adventure*™: Know what you are looking for.

THE DYNAMICS OF *THE AUDIT ADVENTURE™*

In my experience, auditing a management system feels just like experiencing *The Audit Adventure™*. Figure 5.4 shows the two simple dynamics of *The Audit Adventure™*, 'top-down' and 'bottom-up'. Whether my audit is one or two days, or one or two weeks, I always use this satnav to know where I am with my process. It is very, very different from using a checklist, or looking around aimlessly for a few hours or days, to see what we can find.

Moving top-down enables the audit team to understand the *Context* and the *Organization's Objectives* from a high-level management perspective. They can then narrow their focus onto significant risks in sub-processes and individual activities. By fully understanding how auditee management expects their MS to operate, the team can establish priorities for their workplan before the effects of gravity drag them inevitably into the detailed and more time-consuming work of reviewing and verifying the controls around specific higher-risk activities (including *Black Swans*).

The return journey to the boardroom—where the closing presentation will likely be made—compels the bottom-up dynamic. It enables the audit team to consolidate, cluster, and build up their conclusions and their overall opinion of the adequacy of the BCF on a sound foundation of logical argument, supported by the detailed, factual evidence they have collected.

No two *Audit Adventures™* are ever quite the same, regardless of how often you visit Poldhu. The gradient of the top-down route selected will vary from adventure to adventure, and the time spent on the beach at the bottom may differ, depending on the results of the initial review of the BCF. The steepness of the bottom-up route back will depend on the volume and the significance of the evidence gathered. But with this

FIGURE 5.4 *The Audit Adventure™*: Top-down, bottom-up.

process in mind, you'll always be in control. And similar audits conducted with similar process will identify similar findings.

> **A-FACTOR 31**
>
> *The Audit Adventure*™ comprises two simple dynamics: 'Top-down' and 'bottom-up'.

AN AUDIT IS A PROJECT

An audit is a project. Like other projects, it comprises a series of interrelated activities with a start, a middle, and an end. Like every well-managed project, it should have a realistic timing plan, an agreed budget, and a clearly specified objective. Your effectiveness as a lead auditor depends on carrying out all five stages of an audit—initiating, preparing, conducting, reporting, and concluding—in a way that covers the agreed audit scope on time and on budget. If you fail to do this, you fail as a lead auditor. Optimizing the value to the auditee of the deliverables is also an important mark of an effective lead auditor.

An audit's purpose is to provide a level of *assurance* (i.e., either high confidence or low confidence) around the effectiveness of the BCF around the sampled significant risks, and to *alert* to areas of control weakness. In some audits, if agreed and necessary, an audit can also assist management to understand how to achieve control by providing *advice*. You may recall I called these audit objectives the *3As*. However, the responsibility for exactly how and when (or indeed whether) greater control should be achieved lies with the relevant audit committee. The audit team is *not* responsible for implementing its recommendations (unless this is agreed otherwise).

> When I report on my audits, I sometimes feel like the boy who cried wolf; but I always remind auditees that the wolf does show up at the end of the story . . .
>
> Dr Stephen Asbury

It is important that every audit team, as it prepares for an audit, thinks about what their journey may look like over the agreed time frame, in terms of the stages of preparing audit activities, then conducting, reporting upon, and concluding the audit. And close supervision of the audit team's work by the lead auditor is necessary throughout the process. Time lost is (virtually) impossible to recover.

To do all this effectively, every lead auditor must develop a high-level timing plan that allocates an appropriate amount of time to each stage of the audit. Figure 5.5 shows my suggested plan, which allocates 20 per cent of the total time available to preparing

FIGURE 5.5 *The Audit Adventure*™: Planning the division of time.

> **A-FACTOR 32**
>
> A lead auditor has the authority to decide how to use the audit time. For example, if they are leading a relatively inexperienced audit team or working in an area of the business they do not know well, they might increase the time for the preparation stage above the usual 20 per cent of overall time available. This extra time for preparation will reduce the time available for conducting the audit, as at least 20 per cent of the overall time available must be retained for the reporting stage.

audit activities (i.e., one day out of a five-day audit), 60 per cent to conducting the audit ('fieldwork': three days out of five), and the remaining 20 per cent to reporting on and concluding the audit (the final day out of five).

We will now look at each of the steps in *The Audit Adventure*™ in turn.

Prepare for the Audit

The lead auditor gets ready for a successful audit by carefully preparing both the auditee and their audit team. By 'auditee', I mean the senior member of the organization's management with responsibility for the scope subject, area, or theme covered by the audit.

The main objectives of the Prepare stage:

- Meet the auditee (possibly for the first time) at an opening meeting, to describe for them how the audit process will work (this includes 'selling' the benefits of audit and agreeing the terms of reference) and to listen to any concerns they may have.
- Ensure that the audit team is familiar with their roles, having read (and understood) the background documents selected by the lead auditor and knowing what needs to be done to deliver against the terms of reference.

In Chapter 6, you will find my tried-and-tested tips and techniques for conducting the opening meeting.

The main outcome of the Prepare stage is a detailed plan for the audit team's Conduct (or fieldwork) stage. This will include the specification for the fieldwork to be undertaken by each auditor in the team. The overall document produced is called the audit *work plan*, which will focus on a sample of inherent risks independently selected by the audit team as critical to the achievement of the organization's objectives.

I strongly recommend a pre-audit site visit, where time and budget allow. Ideally, the lead auditor—and possibly some members of the audit team—should visit the location where the main audit scope activities are performed for orientation, to see operations first-hand and to be walked through the processes. This is extremely helpful and will aid the preparation immensely.

As the planning progresses, a more detailed plan can be developed. Figure 5.6 shows a typical 20%, 60%, 20% division of the total audit time—two weeks in this example—between the three main on-site stages of a management system audit using the approaches in *The Audit Adventure*™ and ISO 19011:2018.

A-FACTOR 33

Monitoring by the lead auditor of progress against the audit timing and work plans ensures that the audit is completed on time, using those resources available to provide a level of assurance concerning the control framework within the auditee's area of responsibility.

Chapter 6 provides a detailed explanation of how to prepare audit activities effectively.

Conduct the Audit

The main objectives of the Conduct stage:

- Complete an independent review through (horizontally) the design of the management system being audited for each potentially significant risk selected for the audit team's work plan.
- Sample by verifying and testing the application and efficacy of the main controls within key elements of the management system for each selected potential risk in the work plan.

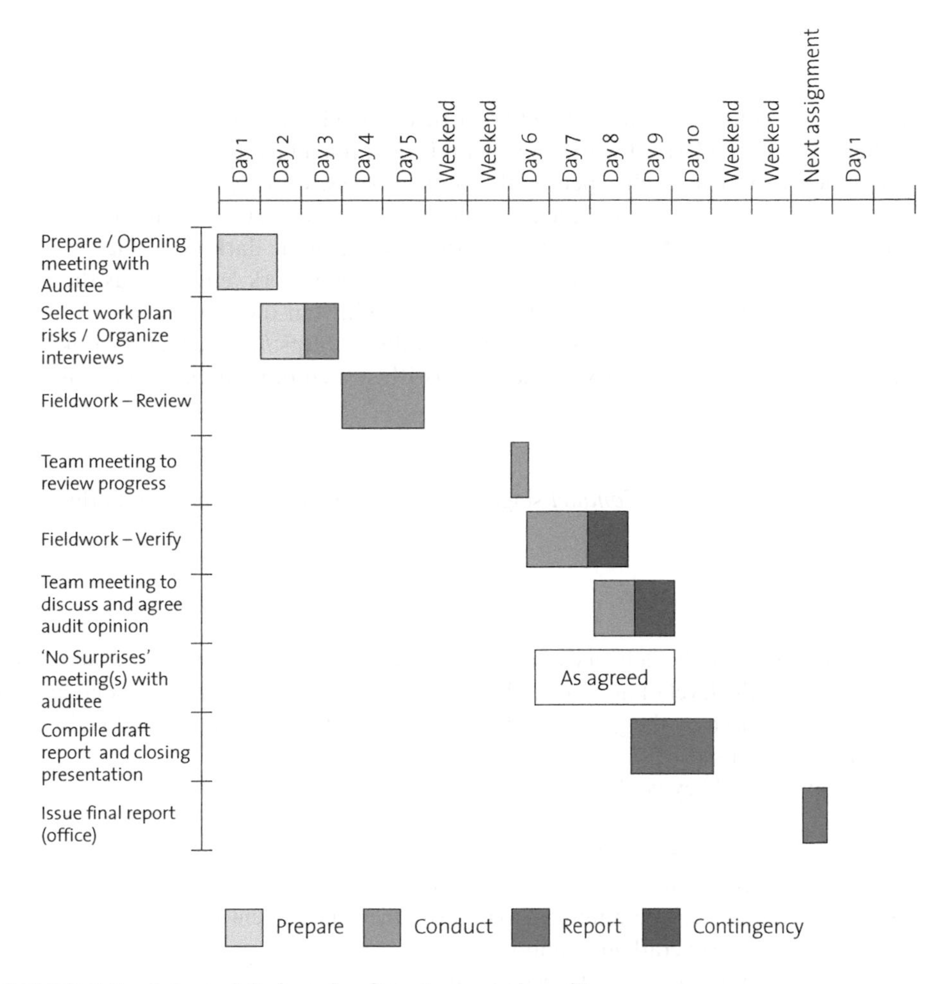

FIGURE 5.6 A typical timing plan for a two-week audit.

Effective audit interviewing will be the main technique used to gather information during the review stage, supported by examination of documentation and other evidence. Verifying (or 'testing') involves carrying out interviews to corroborate what has been said or learned elsewhere, but there are also five other techniques, including making observations. These techniques are described in Chapter 7. Some of these can be quite time-consuming, and thus require careful planning to be fully effective.

A-FACTOR 34

While the Conduct stage of an audit follows a logical sequence of Reviewing and Verifying, the methods used, and the substantive content, will involve significant repetition. This is especially true of interviewing, with similar enquiries being made in different interviews at successively finer-grained levels of detail, across various lines of enquiry, and possibly across different control frameworks.

TIP

When an audit team has more than two or three members, it becomes more difficult to share A4-sized working papers. I recommend using A1-sized flip chart sheets covering each aspect of the audit process, which can be wall-mounted. Everyone in the audit team can see them, and discussions about *the facts* are made easier. Metaphorically, as in a photographic darkroom, the image of the effectiveness of the business control framework 'emerges' as the audit progresses. A set of such wall charts for use in your own audits can be downloaded from the book's companion website (https://routledgetextbooks.com/textbooks/_author/asbury/). Their use is illustrated in Figures 9.7 and 9.8.

Chapter 7 describes the *Conduct* stage and shows you how to approach it effectively.

Conduct and Report the Audit

Reporting provides for the delivery of an audit opinion that compels the necessary actions, based on the level of assurance (high or low) the audit team has provided and the level of concern raised by the gravity and quantity of alerts. Reporting should be high-level and future-focused.

There are several key goals for the lead auditor to achieve during this stage:

- Complete and assess the audit team's findings (both positive and negative findings).
- Discuss and agree the main findings with the auditee (*nemawashi*—see page 282).
- Obtain the auditee's commitment to improve the management system (if possible).
- Determine the overall audit opinion.
- Present a draft report in advance of the closing meeting.
- Finalize and submit the final audit report within the agreed time frame.

Chapter 10 describes how to write great audit reports that compel improvement action.

A-FACTOR 35

Information about the future is much more valuable and interesting than information about the past. An audit team able to set their report in the context of future performance is more likely to be esteemed by management.

The lead auditor should ensure that there is sufficient time at the end of an audit to communicate a full understanding of the results and to explain the key findings to the auditee (and their line managers if they are present; perhaps at the closing meeting). It is generally best if the findings and recommendations can be agreed before the audit team

CASE STUDY

AUDIT REPORTS—SHARED OR SECRET?

As an HSE practitioner delivering internal audit services around the world, I hear disappointingly often—when discussing how successful audit reports can be instrumental in effecting change within an organization—that some line managers (even after they have contributed a lot of their own and their staff's time to assisting auditors) do not see the audit report or the recommendations.

LEARNING

Audit reports are not always shared with your interviewees. You cannot promise that they will be. This is not an issue for the audit team until the next audit, when the organization's communication can be questioned if necessary.

A-FACTOR 36

The main deliverable of *The Audit Adventure*™ is a well-presented audit report that triggers improvement in control of the most significant risks to the achievement of the organization's objectives (the *Big Rocks*).

A-FACTOR 37

People support what they have assisted to create.

leaves the site, as in my experience the advice and recommendations are much more likely to be implemented when there has been input and agreement from the auditee.

Preparation and submission of the report—draft, then final—will complete your audit assignment. Where necessary, your invoice may be submitted.

Chapter 9 describes in detail the steps needed for you to *Conclude* your audit effectively.

Test your knowledge with the next chapter assessment test (see Appendix 4).

Prepare for the Audit

6

Time spent on reconnaissance is seldom wasted.

—Dr Stephen Asbury

If you fail to prepare, you prepare to fail.

—Dr Stephen Asbury, adapted from Benjamin Franklin (1706–90, Founding Father of the USA)

INTRODUCTION

For a summary of this chapter, you can listen to the author's Microlearning™ presentation, which you'll find on the book's Companion Website. Then, delve deeper by reading on for further details on the Companion Website. (See Appendix 4.)

The remaining chapters in this book, Chapters 6, 7, 9, and 10 are set out as if you, the reader, have been appointed as lead auditor for an auditing assignment. These chapters follow *The Audit Adventure*™ and ISO 19011:2018, and take you step-by-step through your risk-based audit:

- Prepare for the audit—Chapter 6
- Conduct the audit—Chapter 7
- Conclude the audit—Chapter 9
- Write the audit report and follow-up—Chapter 10

In addition, set in the heart of these, is Chapter 8, which provides my tips and techniques for getting the very best from your audit team. My focus is on developing strong teamworking skills, developing questioning techniques, and the conscious use of language. You'll also learn about *The Sausage Machine*.

Through each of these chapters, I'll describe the most important matters to think through, and show you how to carry out each stage of the process to produce a high-level, future-focused, and well-received result. And as ever, they are loaded with A-Factors, tips, and case studies to bring your learning to life!

DOI: 10.1201/9781003364849-7

A Clear View of the Route Ahead

In the same way that a successful athlete will visualize each detail of their event and know where they must deliver their best to win the race, a lead auditor too must have a clear view of the route ahead and know how to act at each critical moment before they commence their *Audit Adventure*™. If the lead auditor is not confident and prepared to lead the audit process, then both they and their audit team may underperform.

> **A-FACTOR 38**
>
> Lead auditors must have a clear view of the risk-based auditing process and know how to prepare for, and perform, in each stage (Prepare, Conduct, Report).

HOW TO PREPARE AUDIT ACTIVITIES—THE 'TOP-DOWN' APPROACH

Unless you've read straight through to here from Chapter 5, I suggest you take a *schneller blick* (a quick look; German) at Figures 5.4 and 5.5 on pages 173 and 175 to remind yourself where we are on our route map.

Efficient and effective use of the first 20 per cent of the total audit time will enable the lead auditor to significantly increase the likely success and usefulness of the audit's outcome. This *Prepare* stage covers the preparatory activities and the project planning that is necessary to ensure that the next 60 per cent of the audit time, used to *Conduct* the audit, is spent as efficiently and effectively as possible.

You Must Understand Sampling

Sampling and the Expectation Gap

Auditing is a process of sampling. We auditors cannot and do not check 'everything' (management is on site 365 days per year and probably does not check 'everything'!). Auditors must understand this and then manage the *expectation gap*—the possible difference between the layman's perception of the type and extent of work that goes into an audit, and the actual work that is done.

In 1974, Carl Liggio was the first to define the expectation gap as being the difference between the actual and the expected performance (Higson, 2002: 164). This definition was extended by the Commission on Auditors' Responsibilities (AICPA, 1978), who represented the expectation gap as the gap between the public's expectations and needs and auditors' expectations of what is achievable. The AICPA report is available to download from this book's companion website at https://routledgetextbooks.com/textbooks/_author/asbury/.

Monroe and Woodliff (1993), followed by Hian Chye Koh and E-Sah Woo (1998), define the expectation gap as 'the difference between the beliefs of auditors and those of the public concerning the auditors' responsibilities and duties'. Jennings et al. (1993) argue that the expectation gap represents the difference between the public expectations about the responsibilities and duties of the auditing profession and what the auditing profession actually provides. Porter (1993), meanwhile, defined the 'expectation-performance gap' as the gap between the expectations of society about auditors and the performance of auditors.

Moiser (1997) reports the presumption of belief in human nature; that humans will seek the truth unless there is sufficient to be gained by being dishonest. There has been a debate for years concerning whether an auditor/audit team is a watchdog or a bloodhound (The Law Times, 1896); a 'partner' or a 'policeman' (Marson, 1993; Siegel, 2002; Fadzil et al., 2005), or, as Morrin (2016) puts it, a 'guardian' or a 'watchdog'. The role of the audit team—having been appointed to be independent—is indeed to act exercising independence, applying correct methodology and with the courage to report the truth. It is then for the auditee's management to act in a timely manner upon that truth when it is reported.

Quick (2020) brings up to date the literature on the *expectation gap* from 1974. He advises that the most frequently identified gaps refer to fraud detection, though other gaps also persist. Education and the expansion of the auditor's report to explain the gap are two response strategies proposed.

A-FACTOR 39

Be ready to explain the 'expectation gap' to the auditee, as well as clarifying it within the audit team. There is no expectation or possibility that you will audit 'everything'; the audit will be based on a structured sample.

The Audit Sample

In a risk-based audit, the sample we focus on is the audit team's independent selection from the most significant inherent risks identified in the audit scope. We select our sample from at-risk activities, products, and services—including *Black Swans*—which could have the greatest impact on the organization's objectives if they were not properly controlled.

Selection of a sample of potential risks for our review and verification during the Conduct stage of the audit is made by the lead auditor and their audit team members

TIP

It is unwise to allow the auditee or any of their team to overly influence the selection of risks for the work plan. An auditee seeking a positive audit opinion (from their perspective) may tend to steer auditors towards well-managed areas of the operation, while an auditee seeking further resources—such as additional recruitment or budget—may steer the auditors towards less-well-managed areas of their operation. The audit team can of course take account of site risk registers, minutes of meetings, and things they hear during early interactions with site staff, but the selection of risks for the work plan should be their own. This is the essence of acting independently.

together, and independent of the auditee's/auditee's organization's view, although, of course, some of the team's selection could concur and be like the site's own view.

So, you'll be focussing exclusively on the *Big Rocks* . . .

—Auditee to lead auditor at an opening meeting

A-FACTOR 40

The audit sample should be selected independently by the audit team (and no one else).

MAIN PREPARATION ACTIVITIES

The main activities of the *Prepare* stage are to

- obtain, read, and understand the audit's terms of reference (ToR);
- pursue background information from the auditee;
- develop your overview of the scope of the audit;
- make a pre-audit site visit; and
- meet the auditee at an opening meeting.

The main deliverables of the *Prepare* stage are

- an audit time and resource plan;
- a risk-based work plan identifying the risks to be sampled in the audit;
- an interview strategy;
- preparation of the *right* questions to ask the selected interviewees based on the expected business control framework (BCF); and
- an indexed file in which to retain documents relating to the audit.

Figure 6.1 illustrates the main activities of the Prepare stage of *The Audit Adventure*™.

Obtain the Audit's Terms of Reference (ToR)

At this stage, the ToR is a particularly important document. Both auditors and auditee must understand it, as it sets out the parameters, the boundaries, for the forthcoming audit. In essence, the ToR is the contract for the service to be provided to the auditee by the lead auditor and the audit team.

The draft ToR document will usually have been prepared at the time the audit was added to the organization's audit plan. As this may have been a while ago, it is important to check that the objectives (*3As*), reference framework, and scope remain correct, and that, for example, there have been no acquisitions or divestments of significant assets since it was prepared. If the audit ToR need to be revised because of such changes, the head of internal audit should be notified.

FIGURE 6.1 Activities of the Prepare stage.

A-FACTOR 41

If the audit reference framework or audit scope are significantly changed, the head of internal audit should be advised, so that they may amend the audit plan to reflect these changes.

TIP

The auditee is normally the most senior manager at the location being audited. However, in certain organization models, for example in a matrix organization (discussed in Chapter 1), functional or regional managers may be accountable for particular activities at the location, bypassing the senior manager on-site. In such situations, where an appropriate auditee cannot be clearly identified, the lead auditor should refer to the head of internal audit or the internal audit committee. An audit must not commence without absolute clarity concerning the auditee's accountability for the activities within the audit scope.

Note that ISO 19011:2018, 3.13, defines 'auditee' as 'organization as a whole or parts thereof being audited'. I say that the auditee is the most senior manager representing that part.

By obtaining the ToR as soon as possible after their appointment, the lead auditor will see the 'who', 'why', 'what', and 'where' of the forthcoming audit. Depending on the author, audit ToR come in many styles of content and presentation, but they should always include the following:

- Who—The name of the auditee and their manager, and the audit's sponsor if that is not one of these people
- Why—The objectives (*3As*) for the audit provide management with:
  1. An opinion on the effectiveness of the whole BCF (a level of assurance)
  2. Identification of any weak control(s) resulting in or causing exposure to unacceptable levels of risk, inefficient consumption of resources, or failure to benefit from business opportunities as they arise (alert)
  3. Development of appropriate actions for addressing the identified weaknesses and improving the overall strength of the BCF (advice)
- What—The reference framework should be specified (for example, the organization's own HSEQ-MS, ISO 9001:2015, ISO 14001:2015, and/or ISO 45001:2018)
- Where—The scope should clarify the processes and sub-processes at specific location(s), named facilities, etc. on which the audit should concentrate, including a description of the key interfaces with departments or sites either upstream (that is, suppliers to the auditee) or downstream (customers of the auditee).

Figure 6.2 shows the main features of a typical Terms of Reference. Completed ToR documents may also include the lead auditor and audit team members' names, logistical

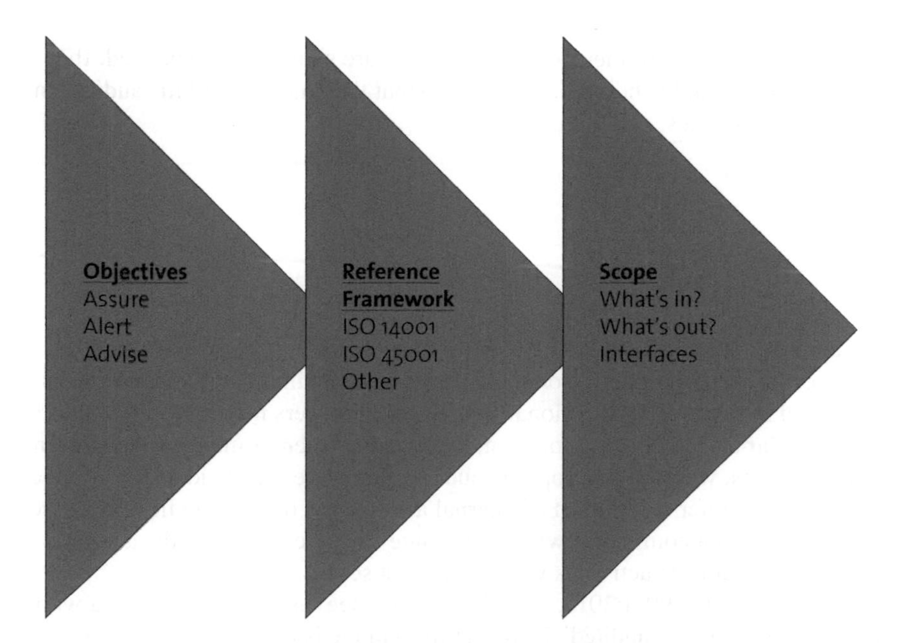

FIGURE 6.2 The main features of a typical terms of reference document.

information such as planned start and finish dates, and the dates for the presentation of the agreed deliverables (closing presentation and reporting) to management.

A-FACTOR 42

The audit terms of reference are generally not negotiable. They have been approved by the internal audit committee as one of their 'jigsaw pieces'. The scope as specified needs to be considered in full, without exclusions.

CASE STUDY

TO INCLUDE OR EXCLUDE?

The well-prepared distribution manager of a busy oil storage and distribution depot with a 400-metre frontage onto the Caribbean Sea suggested to the lead auditor, at the audit opening meeting, that there was little point in including the deep-water jetty operations within the forthcoming audit scope, since an audit team of marine specialists was scheduled to audit a nearby ship-bunkering facility in a few months and would cover this area.

The lead auditor declined this suggestion, saying that it was within the audit scope and that there would be sufficient time to include this potentially high-risk activity (i.e., from marine spillage) within his work plan.

The audit fieldwork included the jetty operations and identified that coastal tankers to be offloaded at the jetty often arrived before suitable spillage containment booms were available to be deployed. While sometimes offloading was delayed, and penalty demurrage charges paid until the equipment arrived, unloading of the most recent delivery had begun without this critical environmental protection control in place. It was discovered that this situation arose because the stock level of product in the depot storage tanks was below that needed to supply the next shift's deliveries, and undue pressure had been brought to bear on relatively junior operations personnel by the marketing department.

Fortunately, by retaining jetty operations within the scope of the audit, the audit team was focusing on this risk area during the period that the non-compliance occurred, and the lead auditor was able to alert senior management's attention to this unacceptable level of risk.

LEARNING

The scope is not a negotiation.

Obtain Background Information from the Auditee

As I explained in Chapter 4, the lead auditor should request to receive appropriate background information. As lead auditor, you are entitled to ask for 'anything' that is reasonable. This may be hard copy documents sent by post or email, files sent to a large file transfer service, or some may be available for download from the auditee's intranet.

In most organizations, there will be offices, drives, and archives filled with files, documents, and records. And don't forget that the whole of the World Wide Web is available for your review as required. Accordingly, the lead auditor must be highly selective in their requests for background information, or they risk being buried. I will advise upon this shortly.

TIP

Identify a file transfer service, so that large files (typically larger than 15MB which cannot be emailed) may be forwarded to you, and that you can use to forward large files to your audit team.

I like WeTransfer (www.wetransfer.com), which is free for up to 2GB single transfers.

The lead auditor should seek information that will enable analysis of the business environment (context) in which the organization is operating, and its business objectives. These are often found in the auditee's business plans. The lead auditor will extract the key information from the organization's plan(s) and from departmental objectives and confirm their alignment to each other.

TIP

Create an extract of the corporate and departmental business objectives, and give each one a unique reference number. For example, C1, C2, and C3 may be the first, second, and third corporate objectives, while O1, O2, and O3 are the first, second, and third operations objectives, and T1, T2, and T3 are the first, second, and third transport department objectives. Using this referencing system, create a list of all the objectives and give a copy to each member of the audit team. Get all your team members to use these references when you are developing the audit work plan, and in their working papers and team discussions. Then everyone will be clear about precisely which business objectives may be at risk; this really helps to focus everyone's minds on the things (the *Big Rocks* and *Black Swans*) which matter.

Before membership of the audit team is finalized, the lead auditor should be liaising with a nominated senior contact within the auditee's department to make the arrangements to carry out the audit at the work site(s). The lead auditor should check that the auditee has a copy of the draft ToR and request selected background information to be sent to them in advance, so that they are able to understand what must be audited and how to address the logistical issues. This also provides background and preparation materials for the audit team.

Some examples of background information to be sought are

- the organization's business plan and objectives;
- descriptions (and/or diagrams) of major processes;
- organization charts at each level of the organization;
- the relevant subject area manual (for example the health and safety manual, for a health and safety audit), or at least its table of contents;
- the location's risk registers;
- details of any external certifications or awards; and
- directions to the site(s), local area map, and site plans.

Other items a lead auditor may consider requesting, since they are very useful pre-audit reading, include

- recent management self-assessment (MSA) of business risks;
- levels of authority manual;
- a recent report of financial and operational performance; and
- recent surveys and reports by other interested parties (stakeholders), such as reports from regulators.

My suggested template letter for advising of the forthcoming audit and requesting background materials is provided in Appendix 3. This is also available to download, for use in your own audit preparations, from the companion website (https://routledgetextbooks.com/textbooks/_author/asbury/).

The lead auditor should review the organization's website, as well as arranging to review the auditee's intranet. Of course, there may be a range of online articles on the organization, and these are also useful pre-audit reading. News and other developments often feature there. Preparation by the lead auditor includes keeping an eye on relevant news and specialist sources. Information of this type can significantly affect the conduct and findings of an audit by informing the selection of risks for the work plan, as illustrated by the three case studies that follow:

CASE STUDY

READ ALL ABOUT IT!

A lead auditor, one day before the start of an audit, reviewed the auditee's website. A news release stated that the company was in the preliminary negotiations of a management buyout. This information was invaluable the next day at the opening meeting with the auditee—the audit team could present itself as very well informed.

LEARNING

It's called 'news' for a reason!

CASE STUDY

KEEP ABREAST OF LEGAL NEWS

While preparing for an audit, a lead auditor noted that there had been a recent change in national legislation concerning radiation safety. During the audit, the audit team observed that a batch of radioactive isotopes for use as X-ray sources for non-destructive testing of oilfield pipelines was at a level of radioactivity now above that allowed by the new laws. The preparation allowed the audit team to respond appropriately.

LEARNING

Use the news!

CASE STUDY

USE ALL THE SOURCES OF INFORMATION AVAILABLE TO YOU

A lead auditor read in an online trade journal that a major contractor used by the auditee had won a separate large contract that was likely to stretch its ability to deliver service in the short term. The audit team reprioritized their audit sample to take account of this new information.

LEARNING

New information can have immediate impacts on the organization!

A-FACTOR 43

As a lead auditor, it is important to encourage your team to be speculative. Think ahead about the business environment (*Context*) of the audit setting and how your auditee will be managing their organization considering its foreseeable changes and challenges.

TIP

Request the documents you require from the auditee's organization as early as you can. As mentioned previously, my advice is to commence three months in advance of the audit. Depending upon the priority the auditee gives to this audit, you may not get a response to your first request. You may need to send a second or even a third request. My 'three-two-one' rule works well: if there is no response to your first request three months out, a second request two months out is often fruitful; if not, a 'final' request one month out almost always does the trick. Whatever timescale you decide on, do give yourself time—remember you will need to read the documents obtained yourself, then create a pack of selected items to send to your audit team: 'Seek, Sort, Share' (see Figure 4.1).

TIP

While I have suggested here and in Appendix 3 the key information that you may wish to obtain about the auditee's operations, an excellent lead auditor will develop their own master list, updating and improving this over time.

Distribute Selected Background Information

When the pre-audit reading documentation has been received by the lead auditor (and note that sometimes it does not all arrive together), it should be read, sorted, and shared with the audit team. The lead auditor needs to extract relevant information from the total received. They will often sift through considerably more information than is eventually given to the team members for their familiarization. The aim is to identify the key information to circulate to the audit team. Packs of selected key documents, or extracts from them, should be prepared and sent to each member of the audit team, with instructions for reading and understanding the information provided. The intention is for each member of the audit team to gain a good overview of the auditee's organization, the site(s), and business processes to be audited in advance. You might be surprised to find out how often this does NOT happen!

TIP

In an ideal world, the best time to send pre-audit reading packs to your audit team members is about two weeks before the start of the audit. Earlier than this and the information may have been read and forgotten by day one. Later than this and team members may not receive the packs in time to read all the information, or even any of it (particularly if the packs must travel through international or company postal systems).

Be pre-warned that even this ideal approach does not always mean that all your team will have actually read the materials sent to them. I have come across several auditors who think they'll just turn up and catch up!

A-FACTOR 44

Key to successful audit is a thoroughly prepared audit team.

Develop an Overview of the Scope of the Audit

From the moment they are assigned to participate in an audit, all your auditors need to build their familiarity on how the auditee is managing the area in the audit scope.

Appointed auditors will receive information from their lead auditor, and it is important that it is read thoroughly. Over the years, much time has been wasted by audit teams when one or more of its members arrive poorly prepared. Auditors should therefore aim to assimilate the required information quickly, by concentrating on understanding the business challenges confronting the auditee, identifying key personnel, and generally getting a feel for the activity to be audited. This includes studying the company and department organizational structure—who does what job, who reports to whom, where do people work, how many positions are vacant, and so on—and confirming the physical locations where work takes place and what the work comprises.

The audit team will usually meet for the first time, in respect of a particular audit, on or just before the first day at the auditee's site. In Chapter 8, I describe and provide tips for how these initial meetings should be conducted for maximum effect. Briefly, the lead auditor should

- build rapport during discussions with each member of their team;
- finalize the team composition as regards subject matter expertise and knowledge;
- hand any last-minute background reading materials to their team members; and
- lay down the ground rules, such as 'We're working as a team'.

Make a Pre-audit Site Visit

Such a practical overview will help to build understanding and can result in a more realistic work plan. It will benefit the audit team if (at least) the lead auditor can visit the site(s) where the business activities take place as part of their familiarization. It allows auditors to meet and listen to personnel and ask questions about the technical aspects of the work being done, confirming the extent to which personnel are confident both in their knowledge of the business processes and how best to manage these processes. The auditor should try to get a feel for the maturity of the overall control framework and the extent to which it is respected and utilized by line management.

Your eyes, ears, and brain, however, should be switched to 'receive' mode when making a pre-audit site visit. Too many auditors want to make recommendations already, and the audit hasn't even started yet! Avoid too much detailed investigation by asking the right questions. I suggest that, unless life or limb is in immediate peril, you do not react to what you hear or see, and certainly do not jump to conclusions or suggest different ways of doing the work, at this very early stage of your audit. If you do, it's possible that operational staff may interpret this as arrogance or take your comments as criticism. That would be a poor way to start an audit!

Meet the Auditee at an Opening Meeting

It is essential to meet the auditee face-to-face at an early point, to get the audit moving on the right lines and to build the auditee's confidence that you and your audit team can really help to achieve their business objectives in the future. Meeting the auditee's line manager too would be a bonus, since you would then obtain another perspective on the quality of control in the area being audited and possibly some hints as to where there might be pressures in the system.

TIP

Gaining initial thoughts from the auditee and/or their line manager on their perceived level of control in the subject area will alert you to any likely difficulties in 'selling' your major findings and final audit opinion later.

A formal audit opening meeting is best. Give the auditee reasonable notice of the meeting and agree a firm date, time, and place to meet. Contact the auditee again shortly beforehand to discuss your preferred agenda and confirm the arrangements in advance. The main benefits of having a formal opening meeting with the auditee are

- ensuring that everybody who needs to know knows that the audit is 'on' dates x–y;
- finalizing the terms of reference (I usually ask the auditee to sign the document and to circulate it within their own team—it helps to underscore the importance of our forthcoming fieldwork);
- providing an explanation of the sampling, risk-based methodology you will be using for the audit and how this process will deliver the audit's objectives;
- confirming the auditee's knowledge and use of the reference framework;
- checking with the auditee the availability of their and their subordinates' time for interviews and *nemawashi* meetings (this latter is explained in Chapter 9);
- obtaining an explanation from the auditee on how risks should be controlled in the scope areas;
- explaining the purpose of the closing meeting to the auditee and agreeing a firm date, time, and place for it;
- starting your review of the auditee's understanding and use of the reference framework.

Establish Credibility

I've found that some auditees can be quite apprehensive, even worried, about audits—especially the audit opinion because of the message it may send to their boss. They may have a fear of finding out (FOFO) something about their process, or a fear of being found out (FOBFO). It is easy to understand why an auditee may be worried, especially if they believe it may threaten their career prospects, the viability of the operation they are managing, their performance scorecard, or even their bonus!

Your role at this time is to persuade them that you and your team have the necessary knowledge, skills, and experience to undertake an evidence-based analysis—using a recognized (this) methodology—of their operations. This will help them to identify with and address risks which may not be as well controlled as they thought. Your primary interest, you might tell them, is in ensuring their future success, not dwelling on the past.

TIP

Working to really understand the background information received, and how it relates to the business and the audit, before meeting the auditee enables the audit team to demonstrate considerable credibility.

Communicate the Audit Objectives and Approach

Describe how the audit will progress in practice, taking the auditee through the work you and the team have already done and then describing the next stages. Use the ToR document—which the auditee should already have seen—as an agenda, to confirm their use of the reference framework and to discuss the major challenges to their overall objectives and the possible significant risks in each of the main business processes.

TIP

Explaining the purpose of the opening meeting to the auditee beforehand may encourage them or their senior staff to provide a presentation describing some of the issues currently affecting the area being audited.

Confirm the Terms of Reference

Up to this point, the ToR document will usually have been a draft. Once the audit objectives and approach are clear, it is usual for the auditee and lead auditor to sign it to signify their respective agreement of the ToR.

Since the audit committee is responsible for the overall assurance plan, as they see the 'big picture'—the metaphorical jigsaw puzzle, in which each piece needs to be audited at the agreed frequency, with no overlaps and no gaps—you and the auditee cannot exclude (or include) scope areas willy-nilly. If the auditee has a real problem, you must refer the issue back up your line of authority (probably to the head of internal audit) before any changes can be contemplated. The ToR is usually formally agreed at the opening meeting. As I said, it is helpful if the auditee can be persuaded to circulate a copy of the final document to their line staff so that everyone is clear that this audit team's work is 'approved'.

Agree a Provisional Timetable

Depending on when the opening meeting is held, the audit team may be able to present the auditee with a detailed overview of the timing plan. Referring to the division of time within the plan (i.e., my suggested 20/60/20 division) helps to show the auditee that the audit team has already spent time studying the background materials supplied and how their fieldwork time will be spent, reassuring them that the result will be a well-researched audit report including, where possible, appropriate and agreed recommendations.

Agree Progress Meetings

Surprisingly, auditors do not know everything! It is wise for all auditors to remember this. Agreeing a programme of short meetings with the auditee—perhaps every day or two—to discuss early findings and observations is invaluable. It goes without saying that if these are not needed, they can be cancelled. Agreeing with the auditee—if you can—what the audit team has found at an early stage will also be helpful during the audit *Conclusion* stage.

Confirm Auditee Involvement

As mentioned, the auditee may be apprehensive. Make it clear that the audit team will value their input at every stage of the process and anticipate active interest and involvement throughout. I suggest you consider giving the auditee an open invitation to visit your audit team room at any time to see what is going on.

A-FACTOR 45

You only get one chance to make a positive first impression. Be sure to take it!

A sample agenda for an audit opening meeting is available for download from the companion website (https://routledgetextbooks.com/textbooks/_author/asbury/).

CASE STUDY

THEY THOUGHT THEY COULD BLAG IT!

A small audit team was commissioned by a first-tier supplier to audit one of its strategic second-tier suppliers in North London. A total audit duration of four days was planned, and this was communicated to the auditee in advance.

The opening meeting was attended from the auditee's side by the director of operations and the health and safety manager.

Immediately after the meeting, the audit team discussed management systems documentation for the selected risk areas in their work plan. The health and safety manager assured them that the systems were well understood but generally not documented: 'There is no need for all that paperwork in a family firm like this one. We've all worked together for years and know inside-out how to operate.'

The audit team decided to test some of the described informal systems and quickly confirmed that they did not work with any consistency.

At lunchtime on the first day, the audit team conducted a 'no surprises' meeting and explained their early opinion. The director admitted that he did not realize that the audit would be so thorough, and that he had decided in advance to 'blag it'. The audit was terminated, as there was no benefit to the first-tier supplier (the client for the audit) in gathering unstructured low-level evidence.

LEARNING

An audit reliably reveals the presence of structured means of control, which bring consistency to operations and, consequently, fewer losses. Likewise, it reveals the weakness or absence of frameworks very quickly.

PREPARE STAGE DELIVERABLES

Audit Logistics

The lead auditor should assume responsibility for communicating with and organizing the audit team as soon as they have been appointed. Simple things can be overlooked or not dealt with in time when trying to get everybody together to start an audit. Here are some examples of things that could go wrong:

- Team members don't know the travel dates or the start and finish dates, and/or they have conflicting assignments.
- Work or country entry visas have not been applied for in time or at all.
- Not all the audit team have had the mandatory vaccinations for travel or entry.
- Flights have been booked to Austria, not Australia; Budapest, not Bucharest; Charlotte, not Charleston.
- Local accommodation is not available because it is fully booked at the time of the audit (due to a large trade fair, for example).
- The flight is scheduled to arrive late at night, and local taxis will not accept GB pounds sterling, euros, or US dollars.

A detailed list of items for the lead auditor to consider as a part of the preparation and planning work is provided in Appendix 2. The list is also available on the book's companion website (https://routledgetextbooks.com/textbooks/_author/asbury/).

TIP

Consider applying for a second passport. In many territories, it is legal to have two passports, on the basis that you may need to send one to a foreign embassy for a visa while you are travelling. In the UK, your employer will need to endorse this request.

Details for obtaining a UK passport are provided in Appendix 2.

Time and Resource Planning

Detailed planning is critically important if an audit's objectives are to be fully met. Even in the shortest audits, a timetable provides the best estimate of time use, but it will need regular updating to make the most of the resources available and cover the work plan risks. It will also allow for planned contingency time to be allocated to best effect.

The timetable should include the activities for each member of the audit team. It should be presented to, and discussed with, the auditee at an early stage to secure their support for the logistics, interview schedules, and key meetings.

Figure 6.3 shows an example of an audit time plan at overall, location, activity, and resource levels.

Once you have confirmed the allocation of approximately 60 per cent of the time available to the fieldwork stage, as a broad guideline, reckon on spending 20–25 per cent of the total audit time reviewing the adequacy of the auditee's control framework, and 20–25 per cent on verifying and testing the application and effectiveness of the controls. This leaves a contingency of 10–20 per cent, which will be allocated by the lead auditor after they have finished their formal supervisory review to confirm the quality and sufficiency of the audit team's review and testing.

Figure 6.4 illustrates this suggested tie allocation, along with the lead auditor's review of each auditor's work. It is typically carried out around three quarters of the way through the fieldwork. Its purpose is to

- check and challenge work done by each audit team member;
- confirm the quality of the auditor's findings and how well they have been documented; and

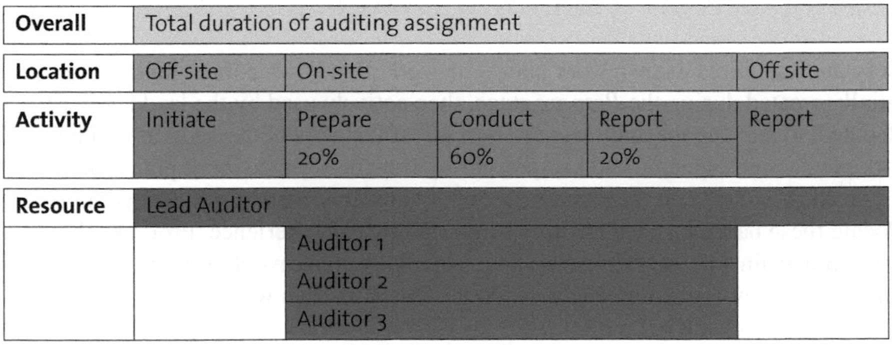

Overall	Total duration of auditing assignment				
Location	Off-site	On-site			Off site
Activity	Initiate	Prepare	Conduct	Report	Report
		20%	60%	20%	
Resource	Lead Auditor				
		Auditor 1			
		Auditor 2			
		Auditor 3			

FIGURE 6.3 Audit time plan showing the allocation of on-site and off-site time, and the division of time for the work at the auditee's site (Prepare, Conduct, Report).

Conduct the audit: 60% of total time

Review
20–25%
of total time

Verify
20–25%
of total time

Contingency
10–20%
of total time

Lead auditor's review

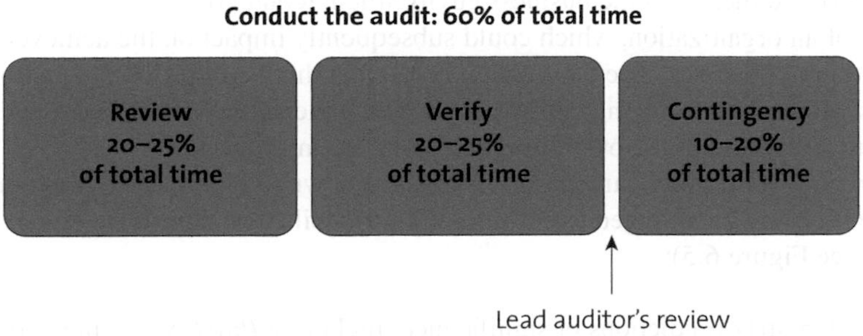

FIGURE 6.4 Scheduling the lead auditor's review and determining the use of planned contingency time.

- confirm how the individual auditors will spend the remaining time available in the fieldwork stage.

The planned contingency time is allocated by the lead auditor to such activities as additional sampling to bring greater clarity and strength of meaning to the results of samples already taken.

Scheduling Interviews

Identify the individuals you believe will be able to contribute to the audit, and obtain their contact details (telephone numbers and email addresses). As soon as you have agreed the dates for the audit fieldwork, contact the people on your list to introduce yourself and the audit, and suggest to them your preferred time for an interview, telling them how you believe they can assist with the audit. You should try to get as much of your schedule agreed as you can before you arrive on-site. Failing that, do it as soon as you arrive on-site.

Think about the logistical requirements and the travel time in getting from A to B, in terms of developing realistic interview schedules.

Develop a Work Plan for the Audit

Every audit requires its own work plan. The work plan is a separate, tangible deliverable initially created during the *Prepare* stage, then actively used by the lead auditor to allocate and manage the audit team to best effect throughout the audit assignment. The work plan assigns specific audit responsibility (for the selected work plan risks for review and verification) to each member of the audit team. It is normal for the lead auditor to allocate these based on each member's knowledge and experience. Progress should be monitored against the work plan by the lead auditor, in terms of both the quantity and quality of the review and testing work done. The work plan will keep the *Conduct* stage of the audit on schedule.

Creating a work plan that will be the focus of the audit team's efforts throughout the fieldwork stage is an analytical and speculative process in which the whole audit team should participate. The lead auditor will use their own and the team's understanding of the corporate, departmental, and operational risks and opportunities to identify situations that may arise, either from the business environment (context) or within the operational activities of an organization, which could subsequently impact on the achievement of the organization's objectives. The team must then select those sub-processes and business activities within the area being audited upon which the success of both the auditee's area of accountability and of the overall organization are most reliant.

There are well-known families of risk, well known to have caused previous losses, where auditors are encouraged to consider the probability of significant risk at this location (see Figure 6.5):

- Risk matrix predictions of significance, including *Black Swan*/high-impact, low-probability (HILO) risks (aka 'organization killers')

Risk matrix predictions

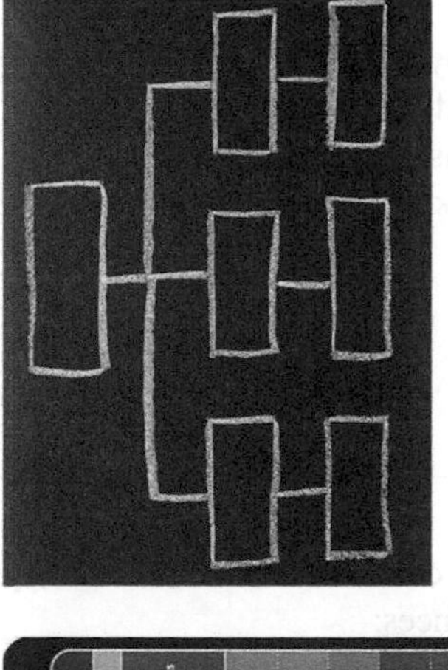

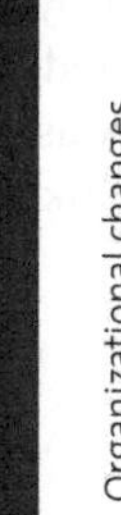

Organizational changes

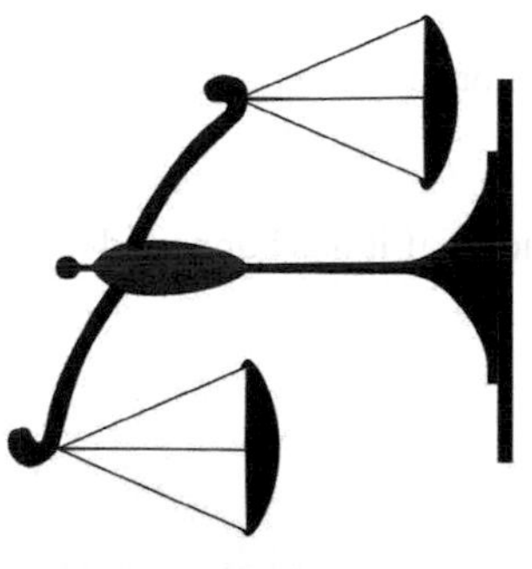

Imbalance in required controls

Priority areas of risk for consideration

Complex links and interfaces

Stressful, repetitive operations

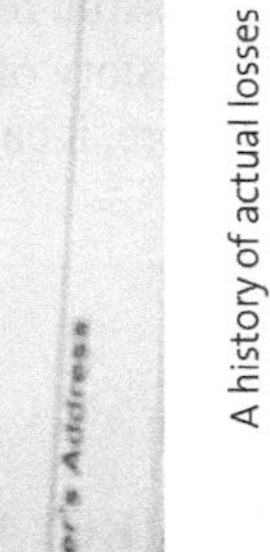

A history of actual losses

FIGURE 6.5 Six well-known risk families.

- Where there have been organizational or other significant changes (people, processes, or working methods)
- Imbalance in required controls identified (too many, or too few)
- Complex links and multiple interfaces
- Stressful, repetitive operations
- A history of actual losses (lightning *does* strike twice—if it has happened before, it can happen again!)

CASE STUDY

NAKED ROYALTY

In 2012, photographs of two members of the British Royal family, showing them without their clothes, were published in newspapers and on the internet. Kate, the Duchess of Cambridge (now the Princess of Wales), was shown sunbathing topless while on holiday in France; and Prince Harry (now Duke of Sussex) was shown unclothed while partying in Las Vegas.

LEARNING

Even the most unlikely lightning can strike twice.

In addition, experience has shown that circumstances can increase the probability of risks (or opportunities) arising, or magnify their consequences:

- Circumstances can arise where even a slight deviation from the control design would result in a disproportionately major loss (for example signals passed at danger on a railway network).
- Wherever a supply chain or business process has a link or interface, there is potential for failure due to omission (for example failure to take responsibility for a critical action) or due to duplication of effort (doing the same work twice).
- In parts of a business undergoing organizational or process change, it is more likely that appropriate maintenance or amendment to the control framework is not being undertaken.
- The sheer complexity and the speed of change in business today (aka VUCA—volatility, uncertainty, complexity, ambiguity; often led by external events and/or determined by senior management) creates stresses on people and on systems at all levels.
- Frustration can affect the attitudes and actions of personnel when ineffective or excessive controls are not identified and/or are not changed or eradicated in a timely manner.
- Management's knowledge of and commitment to correcting control failures is often well illustrated by how they have reacted previously to the discovery of poor performance or operational losses. Failure to assist in this identification, measurement and mitigation of such poor performance or loss would be a cause for concern.

The business activities selected by the audit team for sampling in the work plan must always be challenged by the lead auditor, for whom the question is 'Why will the results of reviewing these activities form a significant and relevant part of my audit team's assurance to management?' This type of challenge process will result in the identification of the most important business processes for review and testing in the *Conduct* (or fieldwork) stage.

The coverage in every audit is restricted, in terms of the number of discrete business activities that can be included in the work plan, by the time available (the *audit intensity*) for reviewing the reference framework and testing the underlying controls. To be clear: there will always be more potential risk areas than can be fully reviewed and verified in the time available (this is the *expectation gap*). Keeping this in mind, the content of the work plan must be finalized by the lead auditor before the in-depth analysis begins. This finalization includes a review of the sample selected to ensure that the scope specified in the ToR is adequately covered.

TIP

Assess each risk to be sampled by validating it in terms of how the main business objectives will be directly, or indirectly, impacted should the potential risk be realized. To do this, list the objectives affected using the reference codes for the corporate, operational, and transport objectives exampled earlier (i.e., 'C1', 'O1', or 'T1') against each work plan item. This will reveal where auditors' time should best be allocated.

A-FACTOR 46

The audit work plan is a living document, which is continuously referred to and updated by the lead auditor. It should be adapted as necessary to account for discoveries made by the audit team in the Review stage. This may occur if the team identified a new and significant area of risk not previously considered, which may then replace an item in the plan if the remaining time allows for this.

Figure 6.6 presents an example of a work plan, with seven risk areas selected for review and testing. The question marks ('?') show the selected areas where the presence (or absence) of expected controls would provide the audit team with a level of assurance that the sampled risk was adequately controlled (or not).

A-FACTOR 47

Time constraints and the need for efficiency mean that auditors should not set out to ask questions about *every* control element of the reference framework. They need to decide which of the control elements are critical as the basis for good risk management of the activity being audited. Their selections for review and testing will be shown on the work plan (as exampled in Figure 6.6).

	HSE-MS Sections → WORKPLAN	Leadership	Planning	Support & Operation	Performance evaluation	Improvement	
1	*Fire*	?	?	?	?	?	
2	*Security*		?	?	?	?	
3	*Spillage*		?	?	?		
4	*VOC exposure*		?	?	?		
5	*Work at height*		?	?	?		
6	*RTC / MVA*	?	?	?	?	?	
7	*Asbestos*		?	?	?		

Decide which are the critical HSE-MS elements to review and verify in each selected risk area

FIGURE 6.6 An example audit work plan showing seven selected risks (numbered 1–7).

CASE STUDY

PREVENTING COUNTERFEIT, FRAUDULENT, AND SUSPECT ITEMS FROM ENTERING A SUPPLY CHAIN

We've probably all seen *Adibas* sportswear, *Louis Vuton* handbags, and *Bolex* watches. Counterfeit, fraudulent, and suspect items (CFSI) are those that have misleading marking of the material, label, or packaging, or misleading documentation (e.g., age, material, composition, or testing/certification) but are represented as being original. They can create incidents/losses in an organization's supply chain.

A functioning quality management system (QMS) might identify CFSI as a specific significant inherent risk, with control measures typically to reduce the probability or likelihood of occurrence (once they are in a supply chain, it can be harder to reduce the consequences or impact of their presence).

A few years ago, I worked with an international organization determined to control CFSI risks. It briefed its managers as follows:

- CFSI items are being discovered all over the world in all industries.
- The use of substandard, nonconforming parts or equipment can affect safety, processes, and operations within our company and our clients.
- Mitigating the risks of CFSI entering processes and facilities has become a major focus of concern for our customers due to the possible consequences.

- Requirements for suppliers like us to build awareness through robust policies and procedures have never been greater.
- Customer tender documents now ask how we apply CFSI awareness. Everybody in the business should be vigilante for CFSI items.
- If CFSI parts are identified, they should be actioned in accordance with procedure QP1014 (Control of nonconforming product) and QP1015 (Quarantine).

Now we'll use the case study as an opportunity to understand an *expected* control framework. Let's imagine that your team has selected CFSI as one of several samples of inherent risks in its work plan for a QMS audit at one of this organization's sites. What might the appointed auditor expect from the QMS? If those expectations are met during her review and verification, will she be able to provide a high level of assurance of control? Here is an example of what she might expect:

- Leadership: Recognition of the CFSI issue by senior management, with clear policy to purchase only directly from manufacturers, or from recognized approved suppliers.
- Planning: Clear technical specifications. Procurement strategy of developing and maintaining a strong supply chain with strategic suppliers. Supplier credentials checked and verified at prescribed intervals in accordance with a selection, evaluation, and re-evaluation process.
- Support: Purchasing staff trained on CFSI. Aware of the ways CFSI may be identified—shorter lead times than other suppliers, sudden availability of obsolete or hard-to-source goods, lower than expected prices, slight differences in markings, packaging, labels, or data sheets.
- Operation: Incoming parts supplied with appropriate certification. Orders picked by workers highly aware of CFSI and empowered to declare them as nonconforming pending an investigation.
- Performance evaluation: Goods-in conformity inspections. Customer feedback sought/collected. Should a CFSI be discovered, it is identified as nonconforming and quarantined to prevent re-entry into the supply chain.
- Improvement: Evidence of learning from prior CFSI incidents (if they have occurred); for example, changed status of approved supplier list.

Once the auditor has created this *expectation* of control, she can progress to develop an interview plan—who will she need to speak with to determine the *actual* control framework? She can then think about the right questions to ask to those people.

Develop an Interview Plan

For each of the risk areas included as a work plan item, it is necessary to do the following:

- Identify the people who will likely involved with the relevant activities. It is usual to schedule interviews in a top-down fashion to understand how managers manage, how supervisors supervise, and how workers work.
- Decide which documents or records are needed to give reasonable assurance of the application of the ToR reference framework (ISO standards call documents and records together *documented information*).
- Be clear on how, when, and where critical activities are carried out.

A useful audit tool for lead auditors and their audit teams is to map each of the potential interviewees to a list of the risks to be sampled. Of course, we cannot be certain of their actual involvement until we meet them, but the likelihood of them being able to contribute to the audit team's review of each risk area will usually be identifiable from their position in the organization chart or their job title. This process not only helps the audit team to cover the most ground possible at each interview but helps to allocate responsibility for work plan items efficiently to individual auditors. An example of such mapping is presented in Figure 6.7.

The figure has the work plan items (R1 = Risk #1 from the work plan) along the top, and the desired interviewees by job title down the side. Each tick represents a probable requirement for an interview. After doing this mapping, the team can look horizontally across each line to identify the areas to be covered at each individual interview. This mapping creates an agenda for each selected individual in advance of your meeting. You

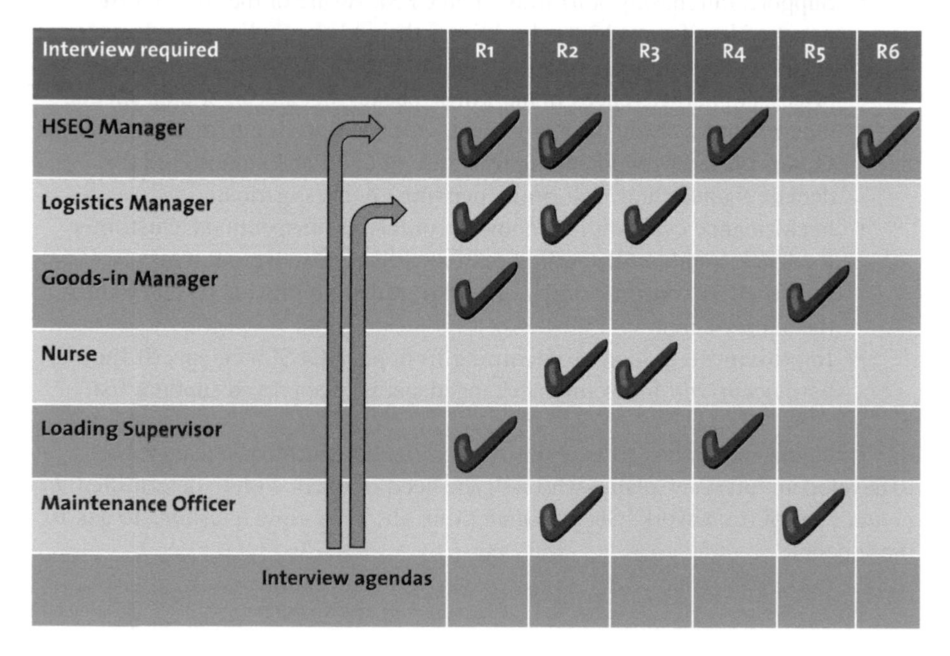

Interview required	R1	R2	R3	R4	R5	R6
HSEQ Manager	✓	✓		✓		✓
Logistics Manager	✓	✓	✓			
Goods-in Manager	✓				✓	
Nurse		✓	✓			
Loading Supervisor	✓			✓		
Maintenance Officer		✓			✓	
Interview agendas						

FIGURE 6.7 Mapping work plan items to interviewees creates agendas for each interview.

may send this in advance. This is a very powerful tool, meaning that a lot of ground can be covered in each interview.

Preparing the Right Questions to Ask

Once each auditor in the team knows which work plan item/s they have been allocated by the lead auditor, they can start to create effective interview agendas and questions. Since each work plan item must be reviewed and tested individually, the responsible auditors need to decide the best approach to take to obtain the most out of each interview.

For example, health and safety on a construction site using ISO 45001:2018 as the reference framework is reliant upon an intentionally selected continuity (horizontal) of control expectations including the following:

- Leadership:
  - Is there clear policy of H&S leadership, communicated to all workers and third parties?
  - Are site roles, responsibilities, and accountabilities clear?
  - Is there consultation with, and participation of, the site's workers?
- Planning:
  - Risk assessment—Have hazards been identified and assessed? Moving vehicles? Work at height? Dropped objects? Excavations? Trespassers, including children?
  - Review of legal requirements—Is the site aware of the required standards (legal and others) to which it subscribes?
  - H&S objectives and targets—Are objectives for significant risks set with action plans? Communicated to workers? Periodically reviewed?
- Support and Operation:
  - Training and competence—Have workers been trained and licensed to the required standards? Is a training needs analysis carried out at appropriate intervals? Are workers aware of site hazards and to control them?
  - Communication—Is there evidence of meaningful communication of H&S information with workers (employees and contractors) and site visitors?
  - Documented information—How are documents controlled? Site passes returned? How are data and records protected?
  - Operation—How are entrances and exits controlled? How is access to trenches, scaffolds, ladders, and other work equipment restricted? How are contractors managed?
  - Emergency preparedness response—Is the site ready for foreseeable emergencies?
- Performance evaluation:
  - Monitoring and measuring—Is there active monitoring of the specified control measures? Ladder and scaffold inspections? Vehicle checks? PPE checks?
  - Compliance—How does the site know if it is compliant with its legal and other requirements?
  - Audit—Is anyone reviewing the H&S-MS arrangements independently?
- Improvement:
  - Incidents—Are all incidents recorded and investigated? Including near misses? How are corrective actions progressed and implemented?

- Learning and continual improvement—How is management involved in developing H&S learning and improvement?

By approaching each work plan item in this manner, an auditor will be able to prepare themselves in an organized way to do the following:

- Create closed questions (that is, questions which can be answered only 'yes' or 'no') to establish which controls in the selected elements of the reference framework are expected to be applied to manage the risks arising. Answers will often be provided by corporate policies or procedures, legal requirements, or statements made at senior levels of management.
- Prepare a series of more probing follow-up questions (often starting with 'What/Why/When/How/Where/Who') for each of these expected controls to confirm the extent to which they are implemented and effective.
- Frame these questions in an appropriate way to individuals at various levels within the organization. The answers will give the auditor information about whether these key controls are known about and how well they are understood, implemented, checked, and reviewed within the organization.
- Confirm whether these controls are being applied as designed and to what extent their application is effective. You'll achieve this through further interviewing, examination of documents, observation of activities, and verification of assets.

There is specific help on questioning techniques in Chapter 8 and *Microlearning™ 8: Conscious Use of Language*, this latter you will find on the book's companion website at https://routledgetextbooks.com/textbooks/_author/asbury/.

By preparing a set of questions for each risk in the work plan, interviewing individuals at different levels of the organization, and at different stages of the work cycle, and then inspecting documents, the auditor will be able to establish the extent to which the expected reference framework has been adopted, and whether it is effectively supporting the achievement of the business objectives.

The ToR specifies the reference framework the audit is being carried out against; and the lead auditor and audit team should be fully familiar with it. For each selected work plan risk, the auditors need to ask themselves, 'How should this auditee be using the reference framework to manage these risks?' In answering this question, they will create a view of what I call the *expected control framework*, which you will use in the *Conduct* stage of the audit. Put simply, this means that each audit has an expectation for control of the risk they are sampling. It is far, far different to organizing interviews and wondering what might crop up.

TIP

Each auditor, and the audit team together, needs to understand what constitutes effective control in each element of the reference framework, and how the elements interact with each other in controlling risks within the auditee's business processes and activities.

As lead auditor, you should consider asking each member of your team to verbalize their expectation of control before they commence their interviews.

A-FACTOR 48

Before the time available for the *Prepare* stage expires, each auditor in the team should have prepared an agenda for each of their first interviews, together with lists of appropriate questions. This will enable them to commence the next stage of the audit—the *Conduct* stage.

Audit Finding Working Papers and the Audit File

The lead auditor and the members of this audit team need to develop an organized approach to recording and retaining the information obtained during fieldwork. It is normal that the rules for retention of audit working papers and final audit reports will be specified by the organization, by national legislation, or in audit codes adopted in the organization or country where the audit is being carried out. An example from the UK is provided in Table 6.1, which shows an example document retention schedule, which was developed by the UK National Archive. It sets out minimum periods of retention.

As soon as the work plan for the audit has been approved by the lead auditor, each auditor in the team should open what I call an 'audit finding working paper' (AFWP) for each risk in the work plan for which they have been assigned responsibility. AFWPs will be actively used throughout the audit by each auditor as the place where they record their collected facts and analyses (aka audit findings) gathered during their work at the client's premises.

A sample AFWP appears in Figure 6.8. This document is a well-structured form which has been used by me in over 1000 audits, as well as in auditor training classes. It is recommended for your own use, particularly if your own organization does not

TABLE 6.1 Audit Working Paper Retention Guidelines

DOCUMENT TYPE	*DISPOSAL AFTER*
Reports	
Audit reports with examination of long-term contracts	6 years
Fraud investigation papers	6 years after legal proceedings
Other audit reports	3 years
Undertakings	
Terms of reference	3 years
Programmes/plans/strategies	1 year after last date of plan
Correspondence	3 years
Minutes of meetings and related papers (including audit committee)	3 years
Working papers	3 years
Other records	
Internal audit guides	When superseded
Procedure manuals	When superseded

Source: UK National Archive

AUDIT FINDING WORKING PAPER (AFWP)

Box 1 – Work plan item *(Descriptive title for the area of risk)*

Box 2 – Risk(s) (Explain risk, referring to business objectives which may be affected)

Box 3 – HSE-MS framework expected *(Main expectations, according to reference framework)*

Box 4 – Identified and proven status of control (Both +ve and –ve examples from fieldwork *review and testing stages – refer to documents, interviews and examples)*

+ve

–ve

Box 5 – Specific Impact – Significance of any weaknesses (in terms of the effect on relevant *business objectives – may be N/A)*

Box 6 – Failing HSE-MS element *(may be N/A)*

Box 7 – Root cause *(may be N/A)*

Box 8 – Recommendation/corrective and/or preventive action *(SMART)*

Summary of auditee's response/'No surprises' *(This is where the auditee writes their response,so leave this blank)*

Prepared by (Auditor):	Date:	Reviewed by (Lead Auditor):	Date:

FIGURE 6.8 An example of an audit finding working paper (AFWP).

have its own audit recording forms. This AFWP is also available for download from the companion website (https://routledgetextbooks.com/textbooks/_author/asbury/).

The AFWP in Figure 6.8 presents the following fields (boxes) for completion:

- Box 1: Work plan item—a descriptive title for the work plan risk being audited. Usually numbered 1–10 (or whatever) to form a series of AFWPs
- Box 2: Risk(s)—your justification for including this risk/business activity in the work plan
- Box 3: Control framework expected—which BCF (or MSS) controls are expected in each part of the reference framework (Leadership, Planning, Support and Operation, Performance evaluation, Improvement)
- Box 4: Identified and proven status of control—this is where auditors record the details and evidence of strengths and weaknesses obtained from their review and verifications
- Box 5: Specific impact—this is where the auditor spells out the specific impact of any identified control weakness(es) upon the organization's objectives
- Box 6: Failing BCF element—if possible, specify here the basic management system cause of any weakness(es) identified
- Box 7: Root cause—if identified, specify the root cause of the basic control weakness(es)
- Box 8: Recommendation/corrective and/or preventive action—the audit team's SMART recommendation for action by the auditee
- A space for response or comment from the auditee
- Sign-off by the individual auditor, countersigned by the lead auditor, both with dates

Each audit will have its own master file containing all the records relating to that audit. This should be retained securely for an agreed period after the audit. The file should be structured in such a way that auditors (and anyone following you) can find documentation easily from every stage of the audit. Individual auditors will have preferences for naming the various sections of the file, but a simple document referencing system, with an index at the front of the file, should be sufficient.

A-FACTOR 49

Each audit has its own master file containing all the audit records, including interview notes, critical documents, and other audit evidence. This will be retained securely for an agreed period after the audit.

Test your knowledge with the next chapter assessment test (see Appendix 4).

Conduct the Audit

7

A *gemba* attitude means going to the source to check the facts to arrive at well-informed decisions.

—Dr Stephen Asbury

Successful *nemawashi* enables change to be the result of consensus among the parties.

—Dr Stephen Asbury

INTRODUCTION

For a summary of this chapter, you can listen to the author's Microlearning™ presentation, which you'll find on the book's Companion Website. Then, delve deeper by reading on for further details on the Companion Website. (See Appendix 4.)

As I explained in Chapter 1, *risk* can be defined as anything—an event, occurrence, interruption, threat, activity, or opportunity—with potential to impact on the achievement of an organization's objectives. Risk concerns the effect of uncertainty on objectives (ISO, 2018c). Successfully addressing an uncertainty is positive and helpful to the organization—grasping an opportunity represents the 'upside' of risk. Alternatively, uncertainty is a negative exposure to be avoided—representing the 'downside' of risk. I refer to these opposing faces of risk as reflecting 'value creation' and 'value protection'.

Audits must consider the significance of risks arising from the context of the host organization's business environment (*Context*) and from needs and expectations of interested parties. As we have seen, the context is turbulent—'more things will change . . . in the next ten years than in the previous 100' (Zakaria, 2006). This likely permanent state of *white water* is the dynamic of mostly uncontrollable changes in the political, legal, economic, social, demographic, technical (PEST), organizational (SWOT), and other environments.

Audit teams must identify a representative sample of potentially significant inherent risks (*Big Rocks* and *Black Swans*) for their audit work plan. This is done by estimating the significance of identified risks from their relative probabilities and consequences. A risk matrix is a common way of presenting the *relative* significance of each risk exposure.

DOI: 10.1201/9781003364849-8

I say (Asbury, 2005) that three questions can assist auditors (as no doubt they may assist management) to decide the significance of identified risks:

- How often will this happen (the probability, likelihood)?
- How big could the impact be (the consequence, severity)?
- Who might be affected by an occurrence (which interested parties, stakeholders)?

My recommendation is that HSEQ/OI auditors focus on the *relative* position of risks on the risk matrix—a useful approach is to focus on, say, the top six to twelve risks. The number of risks sampled in any particular audit will depend ultimately on the audit *intensity* (auditors x audit days) available to the lead auditor and their team.

While there are many quantitative risk assessment, measurement, evaluation, and estimation methodologies, along with software packages and other toolkits galore to assist with this, my experience from over 1000 completed audits suggests that, unless it is necessary, it is wise to avoid the 'numbers game'. Quantitative methods are usually better suited to use by HSE and risk managers within organizations charged with recording the significant risks and impacts and prioritizing these for subsequent improvement. They have more time to research, use, and (try to) understand them.

CASE STUDY

RISK RATING OF 103.1

I remember some years ago being told by a health and safety manager in an engineering composites company that his risk assessment had revealed a risk rating of 103.1. I asked him what this rating meant, how it compared with other health and safety risks in the organization, and of any resulting priority for additional control.

He told me that he had multiplied some factors together in accordance with a procedure and that this (103.1) was the result. He believed that he had complied with the legal requirement to 'assess risks'.

LEARNING

There is more to conducting a risk assessment than arriving at 'some number' and filing it.

TIP

Remember that 62.5 per cent of all statistics are made up (including this one!).

FIGURE 7.1 The work plan keeps the audit team on track.

However the sample of inherent risks for the audit work plan are selected, the principle is that the greater the risk (that is, the greater the value of the opportunity or the greater the value to be protected), the more resilient and reliable—and more emphatically set at the heart of the organization—any framework for risk control should be expected to be.

We learned (Chapter 3) how to initiate an organization-wide audit plan to provide levels of assurance (high or low) to senior management, via the head of internal audit/ internal audit committee. Chapter 5 provided an overview of *The Audit Adventure*™—the vision an effective lead auditor should have in their mind throughout an audit; the satnav for their audit assignment. And Chapter 6 explained how to *Prepare* for a single audit, including creating a risk-based work plan. This audit work plan guides the auditors throughout their fieldwork and keeps them on track and avoid being side-tracked (see Figure 7.1).

We can now move on to the *Conduct* stage of the audit, comprising the *Review* and *Verify* sub-stages. I'll show you how to reliably and effectively evaluate (*Review*) the actual system for business control in the auditee's operations, and then to test (*Verify*) how well it works.

Figure 7.2 provides a representation of your thought processes in this stage of *The Audit Adventure*™. Training course participants like this model, as it shows the whole of the process in just one illustration, highlighting the relationship of the *Review* sub-stage (shaded darker blue) to the *Context* (the whole in which the model is set), the objectives, and risks.

In this chapter, we enter the steeper part of the planned descent of *The Audit Adventure*™. This is the 'Top-down' approach to auditing.

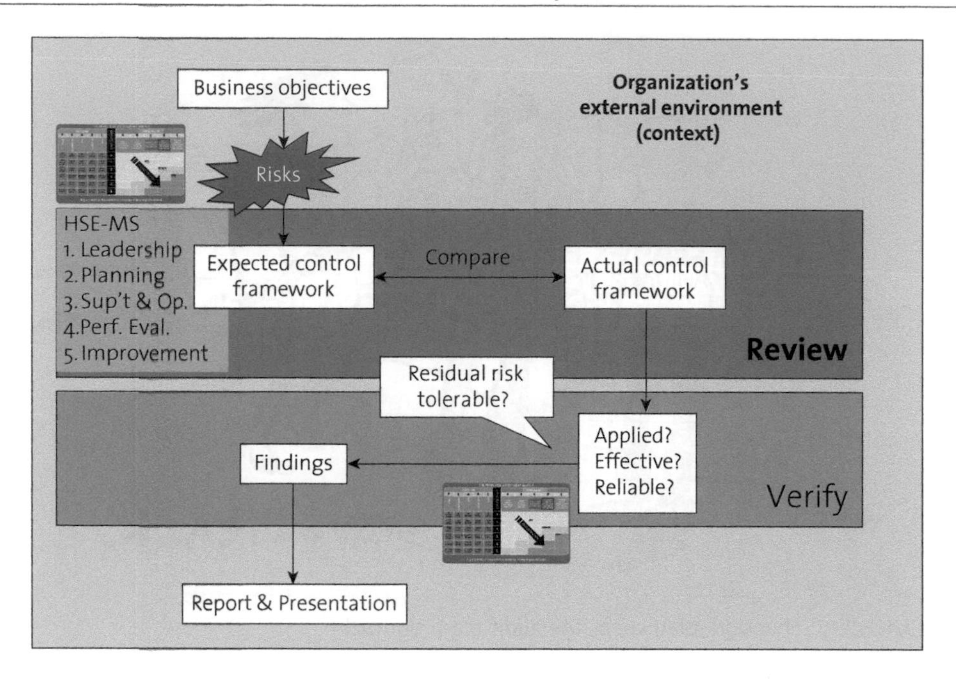

FIGURE 7.2 Audit thought process, with the Review sub-stage highlighted.

TOP-DOWN—CONTROL THE DESCENT

The model for risk-based audit used throughout this book is *The Audit Adventure*™, which is based on the metaphor of the topography of Poldhu beach (see Figure 5.1 on page 170). Auditors need to try to 'stay at the top' for as long as has been planned (*c*.20 per cent of total audit time) in the *Prepare* stage, even though their enthusiasm 'to get started' tends to draw them downwards. It is like an invisible but powerful force, akin to gravity, pulling them inexorably down from their high-level view of the organization into ever finer levels of detail.

Lead auditors should tell their audit team that they will have plenty of time (*c*.60 per cent of the total audit time) to look at the details later—presuming they continue to plan for and use their time well. The controlled descent is very focused on the risks selected for your work plan—it is as though we are guided by the well-trodden pathways of Poldhu to keep us on track.

A risk-based audit is not a low-level compliance check—there is no checklist, and no box-ticking. For each work plan selection (i.e., for each selected significant inherent risk), the audit seeks evidence of how leadership, planning, support, operation, performance evaluation, and improvement (the high-level structure from ISO Annex SL) is used as a structured means of control THROUGH the Plan-Do-Check-Act or PDCA cycle. This is shown in Figure 7.3 and contrasts significantly with the siloed approach described in the Preface and shown in Figure P.5 on page xlii.

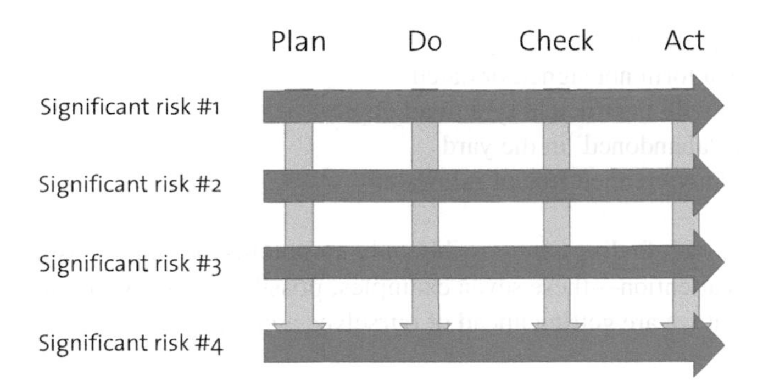

FIGURE 7.3 PDCA: How management systems should be implemented and audited.

Applying My BCF (business control framework; see Figure 2.12 and Table 2.1) means seeking evidence of demonstrable

1 leadership and worker engagement related to the sampled work plan risk;
2 planning for the systematic control of the specific significant risk;
3 support including resources, competence, awareness, and communication;
4 operation of the selected control framework, including preparedness for emergencies;
5 performance evaluation how these controls work in practice (for example, through effective supervision, evaluation of compliance, active and reactive monitoring); and
6 improvement(s) based on identification of nonconformity and taking corrective action.

This linkage THROUGH the PDCA continuity within the HSEQ management system is exactly as expressed by the primary heads of My BCF:

- Leadership
- Planning
- Support and Operation
- Performance evaluation
- Improvement

A controlled descent, progressing into the management system facts 'top-down' maintains the auditor's focus on the work plan, rather than on trivial risks which can present themselves as symptoms of basic loss of control (aka basic causes) during the conduct of the audit, often spotted during a plant tour or walkabout.

It is all too easy for the inexperienced auditor to report the facts found as a long list of low-level findings, amounting to hazard spotting. In my judgement, findings such as these are not audit findings. They are indicators of the past performance rather than the future. They could and should have been detected and corrected by the site's own performance evaluations:

- Worker(s) not wearing eye protection, hard hat, or other PPE
- Trash in a recycling container, or stuck by wind to the boundary fence

- Unacceptable housekeeping
- Box on a form not signed or dated
- Vehicle with insufficient tyre tread depth
- Ladder 'abandoned' in the yard
- A bund not pumped free of rainwater

In my opinion, such findings are possibly only symptoms of a larger issue for management's attention—these seven examples, possibly being symptoms of ineffective supervision. But we are getting ahead of ourselves again . . .

CASE STUDY

SOMETIMES IT'S THE SMALL STUFF THAT MATTERS . . .

An audit at a Middle Eastern oil and gas company identified a mutual aid arrangement agreed among regional public services as well as with its competitors for deployment in the event of a tier-3 emergency, for example a major fire or spillage.

The arrangements advised the auditors that firefighters would attend the affected site by helicopter, with all their equipment ready to respond.

In *reviewing* and *verifying* this emergency mutual aid system, it was discovered that some connectors for refilling self-contained breathing apparatus (SCBA) tanks were incompatible. After a 30–45-minute helicopter transfer, the only air available for the firefighters was that in the tanks brought to site—possibly 15–20 minutes' worth—as refilling by host organizations was not possible.

LEARNING

Sometimes small details really matter.

CONDUCT STAGE TIME PLANNING

Review, Verify, and Contingency

An audit is like any other project, and it should be managed as such, with a carefully thought-through time plan. As I have said, lead auditors should plan for the *Conduct*—the *Review* and *Verify* sub-stages—of an audit to take about 60 per cent of the total audit time. So, for a two-week audit (ten working days), the *Conduct* stage will take about six working days. It can be further broken down into the following elements, with approximate time allocations as shown (as Figure 6.4 on page 197):

- Review: 20–25 per cent of total audit time (2–2.5 days)
- Verify: 20–25 per cent (2–2.5 days)
- Contingency: 10–20 percent (1–2 days)

The planned allocation of contingency time is important. It gives the lead auditor time to direct their team to go back and check details of the management system to confirm matters about which an auditor is not sure, or to take a larger sample—for example, to speak to more people or review further records—to provide a more reliable assurance (or a louder alert!) to management.

A-FACTOR 50

The lead auditor should ensure that an audit is managed as any other project, with careful time and resource planning, including a provision for contingency.

The case study which follows highlights the link between a management system and its audit—the mirror and its reflection (a reminder of Figure 3.1 on page 102). In the example, a one-plan OH&S-MS implementation, and a one-line audit work plan. Success in both might provide a 67 per cent improvement in overall safety performance . . .

CASE STUDY

WHEN TWO-THIRDS OF ALL INJURIES HAVE A COMMON CAUSE . . .

In 1991, I joined GKN as its safety manager, a role I held for five years and one day. GKN plc was a multinational automotive and aerospace business with headquarters in the West Midlands of the UK. GKN could trace its origins back to 1759 and the birth of the Industrial Revolution. It was acquired by Melrose Industries in April 2018 for £8.1bn; this case study and my associated recollections and experiences of the company predate this.

In the months leading up to my appointment, my mentor had suggested to me that I needed a *proper challenge* to test and validate my emerging thoughts on the operation of PDCA-aligned OH&S management systems for controlling hazards. Through my contacts as chair of a ROSPA-affiliated health and safety group, I spoke with the Midlands head of UK regulator Health and Safety Executive. I asked 'who' had a poor H&S record. 'Who must have statutory notices served, and who do you prosecute?' It was that enquiry that led me to GKN.

On my first day, I met my new boss who introduced me to my (small) team. One of my team was 'buried' in paperwork, mostly related to civil legal cases brought against GKN by our 6000 or so workers in the UK. He would take them from the solicitor's envelope, dutifully log them onto a spreadsheet, and post them onward to our employer's liability insurance company. There were hundreds of such claims. And it seemed to me that we settled them all. My first job was to review the causes of these injuries, and subsequent claims, to create a profile of these.

In a nutshell, 67% of all recorded injuries in the prior three years arose from cuts and burns to the fingers, hands, and wrists. The purchasing manager had recently been mentioned in despatches for negotiating a reduction in our purchase price for protective gloves by 5p (to 22p per pair). You might only imagine the build quality and the (low) level of protection that such gloves might provide!

A view I had then, and which I continue to believe in today as a safety management system implementer, is to identify a smaller number of actions and then to implement them superbly, rather than taking too wide a focus and failing broadly. In my first three months at GKN, we had just one health and safety action plan—to reduce this large quantity of cuts and burns to hands and wrists. I remember feeling that my job (and my views on how to apply structured control) depended upon showing early success.

We reviewed case laws on what constituted *so far as is reasonably practical* (Health and Safety at Work etc. Act, 1974) insofar as protecting hands was concerned. Over three months, I organised series after series of glove trials in all our departments. Gloves for welding, gloves for painting, gloves for handling sheet metal, gloves for injecting wax into fabricated chassis, and so on. Different manufacturers, different materials, different sizes, different fits, different cuffs, different colours. The objective of our trials was to identify and specify the *world's best* glove for every relevant manual activity. We engaged with as many workers as we could. We let them contribute feedback on what fitted, what was comfortable, and what protected them best. When we were done, we launched our *Glove Guide*. It was an immediate success. Let me explain what I mean by 'an immediate success'.

The main GKN production site in the UK was in Telford. Amongst the site buildings was a fully staffed medical centre, led by an occupational health Sister with a team of nurses and visiting specialists.

Two months in, Sister telephoned me to enquire 'whether I was trying to put her team out of their jobs. We spend much of our time stitching workers hands and treating their burns before escorting them to the Princess Royal Hospital', she continued. 'Suddenly, we have almost nothing to do'.

I explained to her that there would be lots of new, more interesting occupational health programs in the future, such as well-man and well-woman clinics, occupational monitoring, and other ill-health prevention programs. And so our OH&S-MS implementation progressed.

In the months that followed, we went on to select progressively other significant risks to focus upon—fire safety, chemical safety, electrical safety, noise, and vibration exposure, and so on. I'm certain that the approach we took of focusing on the most significant risk and attending to it well was the correct approach.

Just before the fifth anniversary of my joining GKN, the company was awarded a RoSPA Gold Award for health and safety. When I'm asked how long it takes to implement an OH&S-MS, I usually say 'with concerted effort, five years'.

Again, we'll use a case study as an opportunity to understand an *expected* control framework. Let's imagine once again that your team has select cuts and burns to hands and wrists as a sample of inherent risks in its workplan for a OH&S-MS audit at the Telford site. What might the appointed auditor expect from the OH&S-MS? If those expectations are met during her review and verification, will she be able to provide a high level of assurance of control? Here is an example of what she might expect:

- Leadership and worker participation: Clear leadership of the drive to reduce hand and wrist injuries. Increased budget for protective equipment (gloves). Senior engagement with workers and their representatives on the plans with opportunities to participate in the decision-making. Sharing of information as the implementation progresses.
- Planning: Identification of hand and wrist injuries by type, task, location. Identification of protection suppliers and product ranges. Plans to make initial selections of gloves for trial, and deployment of controlled trials.
- Support: Workers briefed on the plans to improve protection of hands and wrists. Questions encouraged, received, and responded to.
- Operation: New gloves deployed according to plan.
- Performance evaluation: Monitoring of injuries reported. Incidents investigated.
- Improvement: Feedback on gloves received and responded to. Alternative protection sourced for further trials. Continual improvement of provision as the *Glove Guide* developed.

I know (because I was the Safety Manager) that the organization will highly value the findings of your audit team. Your independent assurances and alerts will enable GKN to focus even more closely on the details of its hand safety implementation.

I'll now explain the two parts of the *Conduct* stage—*Review* and *Verify*.

REVIEW

The *Review* sub-stage is based on the audit work plan, the sample selected by the audit team of potentially significant (inherent) risks. The work plan subsequently guides all auditing activities: it is this sample of risks which are reviewed and verified against the ToR reference framework, and which then provides the audit team with the basis for their overall audit opinion.

A-FACTOR 51

The audit team will base its overall audit opinion on the effectiveness of the structured means of control for the sample of risks in its work plan.

CASE STUDY

LIFEBOATS IN THE DESERT

An HSE audit team was flown 250 miles by helicopter into the Omani desert to conduct an audit of contractors' health and safety performance.

In selecting their work plan and conducting the audit fieldwork (*Review* and *Verify*), one of the auditors was shown a very detailed HSE case for a land rig in the oilfield. (An HSE case is a structured argument, supported by evidence, that a given system is healthy, safe, and environmentally acceptable).

Reviewing the HSE case, the auditor was initially impressed by the overall quality and detail included. The auditee believed that it was one of the better-developed HSE cases in the company. However, on a middle page, a description of a series of emergency controls included this line: 'lifeboats must be provided inside accommodation blocks'.

It was immediately clear that the case had not been developed for the land rig in question. On further review, the auditor felt that it had probably come from the North Sea and had merely been given little more than a new front cover prior to approval by local management.

LEARNING

Don't be too impressed too soon. Things are not always as they seem.

Audit Finding Working Papers

The *Review* commences with developing an audit finding working paper (AFWP) for each risk selected for the work plan. The document is prepared by the auditor allocated to the risk.

A blank example AFWP appears in Figure 6.8 at the end of Chapter 6. This can be freely copied by you for use in your own audits. Its structure—the number of boxes, the section titles used, and so on—is not prescriptive, but I have found in my own work as an auditor that this format works extremely well, as it provides for recording a logical (i.e., structured and repeatable) auditing process and gives a complete and detailed record of the auditing work conducted on site. A template for this AFWP is also downloadable from this companion website (https://routledgetextbooks.com/textbooks/_author/asbury/). This download version enables you to make amendments or adjustments to suit your own preferences.

The information for Boxes 1, 2, and 3 of the example AFWP—the name of the work plan item, the risks identified, and the auditor's *expectation* for control—will already be to hand on your audit wall charts if you have decided to use them (see Chapter 5 for more details). Box 3 is where the expected management system controls will be set out, and these will frame your questions in the *Review*, as described later.

AFWP—Box 1

Give the AFWP a 'working title', as per your audit work plan. You may decide to revise this at a later point.

AFWP—Box 2

The second box on the AFWP is the place to confirm to management *why* you have selected this risk as significant and worthy of a portion of your (and the auditee's time) for review and verification. This second box is important. For those readers with an insurance background, this is akin to a possible maximum loss (PML) scenario, where the 'worst case' is considered. Other readers may have experience of emergency response exercises. This is also like the disaster scenario selected as the trigger for response and recovery.

As an example, I might write the following in Box 2:

> A small fire in the tank farm will disrupt operations and may lead to adverse media reaction. A large fire may trigger explosions which would terminate operations on this site—perhaps permanently—and will impact significantly on neighbouring industrial and residential occupants and the company's own deliveries to its customers. Significant, possibly international media reporting is foreseeable, and this may impact on other group companies and their operations.

Keep it credible, and explain the scenario clearly. It is unlikely that any management team (or internal audit committee) will believe 'Thousands may die if machine guard fails'.

You could refer to the impact on business objectives, for example, 'C3—to avoid adverse media coverage'. (In this regard, to save time, consider using your own set of reference codes for business objectives, as outlined in the *Tip* on page 188. In addition, consider revisiting relevant parts of Chapter 6 to review the need for familiarizing yourself with the auditee's business objectives and their significance in determining risk.)

AFWP—Box 3

The next step, in Box 3, is to set out clearly the considered expectation for the structured means of control of the work plan risk. Doing this will create 'the questions' for your audit team to ask. It explains to the auditee how you came to your opinion about the state of control. Start by selecting and describing a key set of controls, without which, in your opinion, there would be inadequate control. How this is expressed will depend on the framework being used in the audit, but the general framework of PDCA will guide us.

I have given numerous examples of this expectation, including the case studies on pages 202–3 (Quality/CFSI), 217–9 (OH&S/Hand safety), and 268–9 (Environment/pollution of controlled waters).

I will provide two further examples here. For the first, if *Fire and explosions in the tank farm* was selected as a significant risk, the expectation, based on a PDCA BCF, might be as follows:

- Plan
  - A specification for the tank farm based on national/international design codes
  - Safe systems of work based on detailed risk assessments

    - Trained, competent staff and supervision
    - Fire alarms and fixed fire suppression systems
- Do
    - Safe systems of work implemented
    - Operational controls and management of change
    - Emergency preparedness and response
- Check
    - Active supervision of workers on all shifts
    - Incidents, accidents, near misses and losses of containment reported
    - Incidents and non-compliance identified, investigated, and rectified
- Act
    - Regular management review of performance
    - Learning incorporated into future plans and targets

This is a basic example, in which I do not claim to have identified every expectation, but the reader can extrapolate this approach for interconnected series of expected controls for other risk areas, either against specific management system frameworks or against PDCA.

The audit terms of reference (ToR) will identify the framework you should use to create this expectation of control. Subsequently, I have presented a further example, this time using ISO 45001:2018 for the work plan item *Maintenance team exposure to asbestos*. Again, I do not claim to have identified every expectation. Note that the formal headings from the standard can be aligned (in an auditor's mind) to Plan-Do-Check-Act.

5. Leadership and worker participation
    - Leadership commitment to preventing asbestos exposure
    - A written statement of health and safety policy, signed by a member of top management within the last two years, which has been communicated to all potentially affected workers and contractors, expressing awareness of asbestos as a possible health risk to workers, and committing to eliminating or minimizing this risk.
6. Planning
    - An asbestos management plan which is up to date and identifies all possible asbestos exposures. This may include laboratory sampling to identify crocidolite, amosite, chrysotile, and other asbestos-containing materials (ACMs). In the absence of sampling, the plan should identify *possible* ACMs, and these should be treated as though they *are* asbestos until the contrary is proven.
    - ACMs scheduled for inspection to identify any deterioration in their condition.
    - Selected ACMs which could deteriorate or be damaged (for example those in locations near vehicle entrances) and those which have already deteriorated scheduled for safe removal by approved contractors.

7/8. Support and Operation
    - Asbestos awareness training/re-training completed for relevant workers—records retained.
    - Contractors briefed on the location of known ACMs—records retained.
    - OH&S policy clearly communicated to workers—records retained.

- Asbestos management plan is up to date.
- ACM waste taken from site by authorized contractors to approved disposal sites—records retained.
- Waste disposal records retained for required period.

9. Performance evaluation
   - Regular ACM inspections—records available.
   - Periodic review of records for all potentially affected staff and contractors to ensure training or briefings have been given.
10. Improvement
    - Planned review at regular intervals by top management on how the organization is delivering on its commitment to asbestos safety, as set out in its OH&S policy.
    - Corrective actions addressed as necessary—records available.

A-FACTOR 52

The *expected* control framework should be identified before fieldwork commences. It is essential for focusing the auditor's questions about the structured means of control during the *Review* sub-stage.

It can be seen from all these examples that each *expectation* is validated by the MSS/BCF identified in the terms of reference.

Figure 7.2 shows the audit thought process. It provides a useful pictorial representation of the comparison made during the Review stage between auditors' expectation—based on the applicable BCF or reference framework—and the actual control framework that site management has established for its operations.

Differences of Approach

At this point, I must highlight the possible differences in approach between

- ISO 19011:2018 type 3 certification audits, where the auditor's *Review* stage compares the expectation set out in the applicable clause of the standard, and the organization's arrangements; and
- This risk-based approach, where the focus is on structured control of selected significant risks using a reference framework.

ISO-Type Audits

This type of audit can become a tick-box, yes-or-no process, typically with a lone auditor working through the standard, clause by clause, within a day or two, asking, 'Do you have a policy?', 'Do you have a register of legal requirements?', 'Do you record and investigate incidents?' and so on to determine suitability for (re)certification.

I suggest that ISO-type standards can provide so much more to an organization when they are used intelligently by auditors with big brains and adequate time to use them.

Risk-Based Audits

In these audits, the *Review* process can become more intellectual (and, for an auditor, much more personally rewarding). It requires the skilled use of judgement on the part of the audit team to decide on the right balance in the selection of elements of control and then to consult with the auditee to agree on this balance.

I think that to audit an intentionally selected risk THROUGH the structured means of control—an ISO standard or any other BCF—provides a much more powerful and repeatable audit opinion than simply checking for the presence or otherwise of stand-alone controls specified by an ISO standard or any BCF. Risk controls should be based on the organization's own significance tests and are thus themselves risk-based operational controls. This illustrates that the real purpose of controls—and of auditing—is not regimented compliance or standardization but the effective management of risk.

Sources of Information

As we have seen, the well-prepared auditor will already have much relevant information gathered, read, and organized during the *Prepare* stage. The importance of developing and then using an organized filing system for audit paperwork cannot be overstated. Being able to find particular documents when they are needed is vitally important. Few things look worse than an auditor who cannot trace the source of facts being presented.

The auditors may have seen, for example,

- business plan and objectives,
- minutes of meetings,
- subject-specific manuals and procedures,
- training records,
- job descriptions, and
- work methods and risk assessments.

Collectively, these documents may begin to show how the site has thought about its risks and addressed these with a structured means of control. Of course, much more information will usually be available to the audit team once it is actually on-site. It will probably have access to

- the organization's records of performance evaluations for, say, two to three years;
- people performing the work activity to observe; and
- workers from all levels of the organization to interview.

FIGURE 7.4 Get to the level of detail you need.

These three resources reflect three approaches to gaining information—reading, observing, and asking. Each has pros and cons and provides different information on management's desired approach to control.

Figure 7.4 highlights the importance of knowing what you are looking for and adopting a *gemba* attitude.

> **A-FACTOR 53**
>
> A *gemba* attitude means going to the source to check the facts for yourself, to arrive at well-informed decisions (*gemba* means 'place' in Japanese).

Reading

Reading documents—if they are up to date, and not too long (!)—can be a very illuminating source of information at the *Review* sub-stage.

A key drawback is that in some organizations documentation lags current practice. I do not mean non-compliance per se, which is discussed later. Rather, the documentation may not reflect the newer practices within the organization. This may be the case because the 'continual improvement' so prevalent within those organizations ('change is the only constant') means the written system has not kept pace with the work in the field. Another common drawback is that in some organizations the drafting of the system has followed a 'why use one page when 105 pages will look better' approach. If this is the case, even at this early stage, we may have found a possible area for useful comment to management!

> **TIP**
>
> If you are given a 105-page document to read, look first through the table of contents and the section headings. You'll get a feel for the content and probably be able to spot the sections relevant to your work plan more quickly.

Observing

Observing the owner of a process can be a very powerful method of learning how work *should* be done. A useful question for an auditor to ask is 'Can you show me how you should do X?' A trusted member of the organization's staff will then usually demonstrate the designed system in its application.

A key pitfall here is the potential for slipping into a verification and deciding too early that someone is doing something correctly or incorrectly. At this stage, we are simply trying to compare (and write down for later verification) management's chosen system against the expectation we had developed in Box 3 of our AFWP.

> **TIP**
>
> Don't be frightened to ask someone to 'do it again' when learning by observing. During the *Review* sub-stage, it is important that auditors gain an understanding of how control of the work has been designed.

Asking

Asking can also be a very powerful method of learning how a task *should* be completed. 'Do you have a procedure for X?' or 'What method have you been trained to use for Y?' are good questions at this stage—again, we'll be verifying and testing this in application later.

If one person from the management team gives us a 'wrong' answer—that is, an answer different from our expectation—this does not in itself give rise to an audit finding. When we later test the system in application, we can decide whether this 'wrong' or different system still provides reasonable control.

A-FACTOR 54

Remember the 'ten-foot rule'. People within 10 feet (3 metres) of the operation will likely have very good knowledge of the work processes there.

A-FACTOR 55

The more you know about a subject, the more you can see both sides.

TIP

And this is one of my favourite tips . . . Remember that auditors are 'from Missouri'.

Let me explain. In the US, each of the states has a nickname—Florida is the *Sunshine State*, Texas is the *Lone Star State*, New York is the *Empire State*, and so on.

Missouri is known as the *Show-Me State* (see Figure 7.5). There are several different explanations for this. Two of these are summarized later.

- Version 1:

Missouri became known as the *Show-Me State* in 1899, when Congressman Willard D. Vandiver said: 'I come from a country that raises corn and cotton and cockleburs and Democrats, and frothy eloquence neither convinces nor satisfies me. I'm from Missouri. You've got to show me.'

FIGURE 7.5 Missouri 'Show-Me State' licence plate.

- Version 2:

'Show me' is a term of ridicule and reproach, coined in the mining town of Leadville, Colorado, in the mid-1890s. A miner's strike had been in progress there for some time, and miners from the lead districts of southwest Missouri were brought in to take the place of strikers. They were unfamiliar with Colorado mining methods and required frequent instruction. Pit bosses began saying, 'That man is from Missouri. You'll have to show him.'

Whichever version you prefer, remember to ask—at the Review *and* at the Verify sub-stages—to see ('Show Me') the relevant practices, procedures, methods, documents, and records you are told are important to the effective control of a particular risk. This can sometimes feel repetitive and has even been known, at times, to make auditees annoyed, as they feel they are not being believed.

TIP

A great way to put your interviewee's mind at rest—and a conversation starter—is to briefly explain the Show-me/Missouri story at the start of an audit interview. This will help interviewees to understand why, as an auditor, you ask them to 'show me'—to see evidence of their control framework and related documentation.

It may sound like motherhood, but unless you ask the right questions, you won't always get the most useful or actionable information. For example, too many companies use satisfaction surveys that are constructed to answer questions based on their own hierarchy of needs, i.e., 'What do we want to know?' rather than 'What does the client see as important?'

Carey Evans (2006), Relationship Audits & Management, London

TIP

When preparing questions for audit interviews, know what you want out of the interview beforehand. You and your interviewee are both busy, and keeping to the point is essential. A well-structured agenda that shows which risks are to be discussed can really help, particularly when meeting with senior and middle managers. If this is provided in advance, the interviewee may often be better prepared for you and your enquiries.

CASE STUDY

FIRST-CLASS HOME

In 1998, I led a team conducting an environmental audit at a Malaysian airline seeking external certification to ISO 14001. We spent two weeks in-country looking at airport, airside, and aircraft maintenance operations.

In our work plan, we had decided to conduct a 'water balance', because water (even in 1998) was viewed as a scarce resource in the country. It was selected as an example of a sustainability opportunity.

In our review, we hypothesized that the water purchased as a raw material by the airline would approximately balance the water ultimately leaving the organization as waste. Of course, aircraft cooling systems and toilets are filled with water and then fly away, but a corresponding number of aircraft are filled with water elsewhere prior to flying back. A water balance is a recognized tool, and it can be applied in most audit settings.

By looking at paid water supply invoices, we were quickly able to calculate the volume of water purchased. We set about tracing its path through the organization to its disposal points. We looked at aircraft washing, aircraft systems, catering, toilets, and cleaning. Within a day, we knew that a lot of the purchased water—more than 90 per cent—was 'missing'.

We considered various options:

1. Were there leakages in underground pipes?
2. Had there been a site subdivision, or was someone consuming water without paying for it?
3. Were the water supply bills incorrectly totalled?
4. Were airline employees stealing it and taking it home?
5. Were our calculations incorrect?

We asked a lot of questions to resolve this; I will merely summarize the outcome of this part of the project. The answers to our five questions:

1—Probably, but this would not account for losses on the scale detected.
2, 3, 4 and 5—No.

What we discovered in our *Review* and *Verification* sub-stages of the water management system was that the incoming water was measured by a linear flow metre, approximately twenty years old. There was no evidence that it had ever been calibrated, and now when one cubic metre of water passed through it, it spun like the office fan! Approximately 500 per cent over-readings were the result, equivalent to an annual saving of approximately USD 20,000 for our client. We were also able to provide the underlying information that resulted in a negotiated settlement for a one-off refund of USD 20,000 to cover all past costs errors.

When we checked in at the airport to come home, we had been upgraded to the airline's first-class cabin. Clearly, on this occasion our client had appreciated the audit!

Three months later, we were also pleased to hear that the client achieved external certification to ISO 14001.

I wish you similar 'luck' in the field . . .

LEARNING

A *gemba* attitude—persistence in getting to the facts—pays off!

Possible Review Outcomes

There are probably two main outcomes of the *Review* process for each risk selected and included in the audit work plan:

- The framework's design appears to provide reasonable assurance that objectives will be achieved.
- The framework's design appears not to work as intended—there may be a gap between the actual framework in use and a reasonable expectation for control. Objectives may not be achieved.

Either of these outcomes (our audit opinion) is good information. Remember that audit is *not* about finding things wrong—remember 'I Will Audit' (IWA). Audits provide:

- I: *Independent* assessment of the control frameworks deployed to control significant risks;
- W: W*ell-balanced* control framework between the levels of risk and the cost of further control; and
- A: A*ppropriate* approach for this organization.

Returning to the AFWP, recall that, at this point in the audit process, the first three boxes should have been filled in, at least in draft, to show

- a title for the work plan item describing the area of risk to be sampled,
- an explanation of the risk and the reason for its selection for this audit, and
- the auditor's expected framework of controls based on the TOR BCF/MSS.

As a reminder, an example AFWP appears in Figure 6.8, and this is also available as a downloadable template from the companion website at https://routledgetextbooks.com/textbooks/_author/asbury/.

As you progress through the *Review* sub-stage, making your comparison between the expectation for control created by the reference framework and the organization's actual controls, you can start to populate Box 4 on the AFWP. As indicated in the example AFWP in Figure 6.8, auditors should report both positive (+ve) and negative (–ve) facts about evidence of control, based on the information gathered and making cross-references to other working papers as necessary. Box 4 is discussed in further detail later.

TIP

Do not refer to individual interviewees' names in audit working papers or reports. Firstly, you must maintain the confidentiality you may have promised respondents in line with the principles of auditing set out in ISO 19011:2018. Secondly, there is no need to divulge 'who told you' that something works or does not work. Using the same methodology as you have, 'anyone' would be able to follow in your footsteps and come to a similar conclusion. Refer to documents (by title, reference number, etc.) and, where people are concerned, stick to 'The audit team was told that . . .' If 'everyone' told you something—good or bad—a useful phrase is, 'Everyone we spoke to told us that . . .'. I have also numbered my interviewees and noted 'Interviewee 6 advised that . . .'.

A-FACTOR 56

Remember IWA—I Will Audit (Independent, Well-balanced, Appropriate) in a manner which meets the needs of the auditee's organization.

VERIFY

Once management's framework for control has been established by the *Review*, the next sub-stage for the auditor is to *Verify* the control provided—to test it—in operation.

Figure 7.6 shows how there should be a very clear linkage between the results of *Review* and those controls or BCF elements which go on to be *Verified*. As auditors, having established management's preferred (or chosen) control framework, it is very important that you can *Verify* that it actually works as intended and that it is consistently applied, effective when needed, and reliable in mitigating impacts.

If your tests reveal that the controls are not applied properly, or are ineffective or unreliable, then you should try to assess the significance of the residual risk to the achievement of the business objectives. It is upon this verification that the audit team will base their assurance or alerts to the auditee in their *nemawashi* ('no surprises'), at the audit closing meeting, and to the head of internal audit and the internal audit committee in their final report.

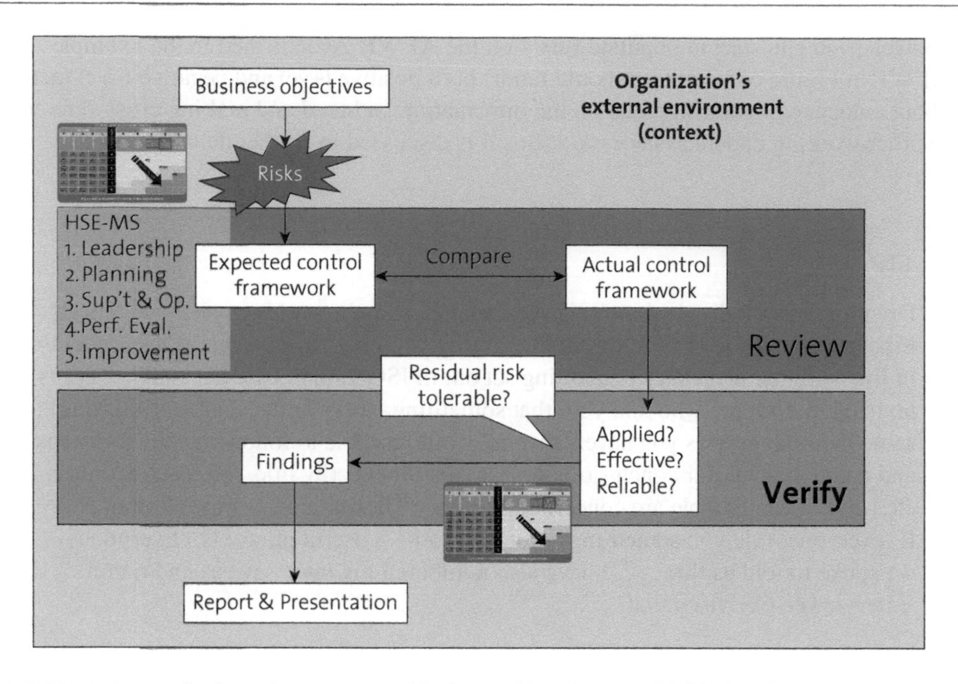

FIGURE 7.6 Audit thought process, with the Verify sub-stage highlighted.

Possible Verification Outcomes

There are probably three main outcomes of the verification process for each risk area:

- The activity is in a controlled status as designed.
- Control is not implemented, leading to unauthorized exposures and/or inadequately controlled risks.
- Control is implemented as designed, but it is not sufficiently effective in controlling the specific risk adequately.

Any of these three outcomes is 'good information' to progress the audit. Remember, auditing is not aimed just at finding things that are 'wrong'.

Lead auditors and their teams need to consider the depth of the detail they need to go to, at the bottom of *The Audit Adventure*™, as this will differ depending on the results of their *Review* of the expected BCF. In areas where the expected controls are present, the journey will continue downwards towards the beach, where, metaphorically, the auditors may even dig holes in the sand to get to ever-deeper levels of detail; while in areas where the auditee either confirms or reluctantly accepts that that the necessary framework for control is not designed well, or is completely absent, then *The Audit Adventure*™ curve may be rather shallower, as shown in Figure 7.7.

FIGURE 7.7 Decide the level of detail necessary to Verify each risk.

Outcome—In Control as Designed

Audit verification may show, based on the samples the audit team has taken, that the intended system has been communicated to operators and works as intended.

> **TIP**
>
> As an auditor, remember to explain to your auditee that you are not giving a 100 per cent guarantee that this system works and will always work with absolute and perfect reliability –you are not! But you can advise that it works as tested. We call this *reasonable assurance.*
>
> The more time we give to testing, the more confidence we can have in the assurance provided.

> **TIP**
>
> We can leverage our testing sample by reference to, and reliance on, others' audit work, in which case we need to verify their processes.

Outcome—Control Is Not Implemented So Does Not, or May Not, Work As Intended

Alternatively, your verification may show that while authorized procedures are in place, they are not applied, or are not applied as designed. This may or may not give cause for concern, since the actual control may be more cost-effective, or the structured means of control may provide a compensating control which makes up for the apparent control failure.

TIP

When such compensating controls are identified, lead auditors should satisfy themselves that the auditee both knows about the situation and has an action plan to either reinstate or revise the authorized control framework. Often compensating controls are found in the supervisory control element because 'loyal' employees recognize the shortcomings in implementing the approved BCF, and, out of professionalism or the 'goodness of their hearts', they do things to ensure a successful outcome—for example, that employees are not injured. Clearly, in these circumstances the audit finding must be reported for urgent management attention.

TIP

In my experience, telling an auditee that their staff are working well and doing a good job (provided this really is true!) is always well received—sometimes they glow with pride! If they have thought about their work and have improved it, all that may be necessary is to bring the documented system into line with what is actually happening—at low or nil cost in most circumstances.

CASE STUDY

THE RUSSIAN CONNECTION

During a property protection audit in St Petersburg, Russia, an audit team was advised by technical staff in the maintenance department of an automatic wet-pipe sprinkler system which was said to protect all areas in the facility. Automatic sprinkler systems provide fast response in the event of fire. Insurers' data shows that 68 per cent of fires in protected buildings are controlled by the activation of just one sprinkler head, and 95 per cent by the activation of six heads or less.

In every part of the facility, the audit team saw sprinkler heads, including an in-rack sprinkler protection system in the warehouse areas. This was an important layer of protection, as the site held very large quantities of flammable materials, including PET preforms (used to make PET bottles), pallets, shrink-wrap, and juice concentrates.

The auditors were advised that fire risks had been assessed as very low, and the automatic sprinkler assurance statistics were quoted by the auditee. Close to the end of the audit, the lead auditor—noticing the freezing weather conditions outside—wondered how the system was protected from frost, and asked to see the sprinkler pumps so that he could perform a flow test.

After a little resistance from the auditee, the auditor was eventually shown to the sprinkler pump room. Coming down the wall was the main sprinkler feed pipe, which terminated in fresh air. There was no water tank, no connection to the city mains, and no reservoir provided.

LEARNING

Follow all systems (both physical ones and management ones) to their origin to make sure they are 'connected'. This is a good example of auditors exhibiting a *gemba* attitude to get to the source of the facts.

Outcome—Control Is Implemented as Designed, But Not Effective

The audit verification may show that management's controls are implemented as intended. However, to give reasonable assurance, an audit should be able to report whether this expected implementation, done 100 per cent correctly, can provide adequate control of the identified risk. For example, an annual check that hearing protection is worn is probably too infrequent to enable a meaningful assurance, whereas an annual check of the lifting cables and mechanisms on a goods lift may give adequate assurance.

TIP

Tell it as you see it. Sparing someone's feelings does not enhance your credibility; it can only undermine it. But of course, there are techniques for delivering bad news without shutting down the audit (or being escorted off-site by security).

A-FACTOR 57

The *Verify* sub-stage involves checking that expected controls are correctly implemented, and that those controls are effective at controlling the sampled risk. It can also involve verifying, in the absence of expected controls which are not considered appropriate or necessary by management, the acceptability of the residual risk exposure.

CASE STUDY

IT'S ALL DOCUMENTED, BUT WOULD IT REALLY WORK?

The workers at an offshore oil terminal in the Arabian Gulf were male Sikhs. The terminal served a 'sour' field, where a known hazard was hydrogen sulphide (aka H_2S, sewer gas, swamp gas, stink damp, sour damp), an extremely flammable and highly toxic gas.

Emergency alarms and procedures were established, and all the employees had been provided with, and trained to use, H2S respirators.

Male Sikhs generally do not shave. The respirators of the type provided would almost certainly not seal properly against a bearded face. One hundred per cent of employees remained at risk of exposure in the event of a leak. A potential risk of multiple fatalities.

LEARNING

Verify that the control framework will REALLY work as intended, otherwise it's just 'paper safety'.

AFWP—Box 4

Whatever we find—positive, negative, or both—during *Review* and *Verify* sub-stages of the *Conduct* stage of the audit, we should record our findings in Box 4 of the AFWP as evidence of control.

It is very important that auditors should give a full and factual account of what they have found, as this will establish the facts that provide an audit trail leading to their opinion.

TIP

Tell the truth. Don't be afraid to describe precisely what was found, but don't embellish a minor shortcoming. Honesty is always the best policy.

CASE STUDY

BENEFITS OF HEALTH AND SAFETY AUDITS IN A MEDIUM-SIZED PUBLIC SECTOR ORGANIZATION

As a part of a new initiative to support young people, a UK government-funded agency was established to support teenagers through a transitional stage in their lives, working across four local authority areas, with 160 employees and an annual grant of £11 million GBP. With funding accompanying a statutory requirement to deliver services, this new organization had to recruit staff from a range of youth work, voluntary, and private sector backgrounds to meet its targets.

As a part of its business plan, a health and safety procedures manual, a training program, and a health and safety audit plan were introduced. Many of the staff had never experienced audit interviews before, and the range of backgrounds and lack of organic growth meant that the concept of auditing was met with scepticism and concern.

The health and safety audit plans focused on an inspection of the site to ensure office hazards were being correctly managed, and a structured selection of staff was interviewed, including frontline staff, administrators, team leaders, and managers. Managers were given a copy of the audit report and were re-audited after three months to ensure that the recommendations had been implemented. The number of non-compliances was used as a key performance indicator of effectiveness of the safety management system.

Once the first round of audits had been completed, it was possible to identify a range of benefits for the organization, as follows:

- Non-compliances were actioned, where previous initiatives had failed.
- The audit encouraged compliance because staff knew they would be checked.
- The audit reinforced agency policies and procedures.
- Compliance with statutory requirements could be demonstrated to regulators.
- Local and organization-wide problems could be identified for action.
- Resources and training could be targeted to where they were most required.
- By signing the report, managers understood health and safety was their responsibility.
- The auditing encouraged all staff to consider their own responsibilities for safety.
- The audit interviews provided an opportunity to refresh training provision.
- The audit process improved the safety culture.
- Staff reported that the process demonstrated the organization was really concerned about their and their clients' health and safety, which improved morale.
- Key performance indicators provided a proactive, quantitative measure of the effectiveness of the safety system.

LEARNING

Audits can deliver continual improvements even in relatively low-risk organizations such as this one.

CONFIRMING FINDINGS

Confirming Findings with the Lead Auditor

It is important that each auditor discusses and confirms their preliminary and developed findings with their lead auditor and then shares these with the rest of the audit team. I'll discuss some techniques for doing this in Chapter 8, Teamwork and the Conscious Use of Language.

Confirming Findings with the Auditee

As areas of strength and weakness emerge, the lead auditor will wish to hold periodic *nemawashi* or 'no surprises' meetings with the auditee, to confirm issues that have arisen, and/or to seek guidance.

Nemawashi is a Japanese word which translates literally as 'going around the roots'. It indicates an informal approach to preparation for change or new projects within an organization.

This is a powerful technique for auditing, which I commend to you for the following reasons:

- Auditor errors and misunderstandings are identified at an early stage, when they can be corrected—that is, facts are confirmed.
- Missing documents, awkward interviewees, and other day-to-day issues can be resolved while the auditors are on-site.
- The audit team can seek early buy-in from the auditee to findings as they arise; this delivers on the pre-audit 'sale' of the process and demonstrates powerfully to the auditee that audit is not a 'secret' process.
- Auditees like it when they can tell their own boss the facts and the progress and share the good news or start to develop corrective actions at an early point, as appropriate to what has been discussed.

TIP

Use the powerful *nemawashi* approach, even on a short audit. If the audit lasts even just one day, have lunch or coffee with the auditee, and talk with them about what has been found and how the work is progressing—you'll be glad you did!

A-FACTOR 58

The best recommendations an audit team can ever make are those agreed in advance with the auditee. The best chance of gaining agreement comes from bringing the auditee onside at the earliest possible opportunity.

In Chapter 9, I'll show you how to gather and group together your detailed findings to present to the auditee to *Conclude* your audit. This is the 'Bottom-up' stage of *The Audit Adventure*™.

SAMPLING

If we wished to know how many residents of the state of Kansas watched the last Football *Super Bowl* on television, would we ask all its near three million inhabitants? The answer is assuredly 'no'. To establish how many are likely to have watched the game, we would ask a sample of inhabitants whether they watched it, and extrapolate the results to the wider population. The same approach is used for opinion polls about politics or likely Oscar winners.

Similarly, if auditors wish to know whether all effluent discharged is within the limits set by a permit, we do not have to spend a month at the discharge point or down the drains! We can organize a series of independent effluent samples to be taken and analyzed. Or we could also look at the organization's test results, or an external regulatory agency's results, and base our opinion upon this information.

As auditors, we have a variety of sampling tools and techniques at our disposal, and in the following sections, I will comment on the merits of each. In reality, you will likely use a mix of these techniques in each audit. In addition to the techniques described here, there is a broad range of mathematical, statistical, and analytical tools, which tend to go beyond what is required in a risk-based HSEQ audit. You will also find suggested reading on statistical techniques in the *Further Reading* section including Witte and Witte (2017). A few years ago, I participated in Six Sigma training, and I found the approaches to measurement interesting. I commend it to you.

Sampling Techniques

In this section, I will explain six sampling techniques available to auditors to assist them in considering whether the management system as prescribed is effective and how to intrusively (but selectively) *Verify* it in operation.

I use all six of these techniques personally and can recommend them to you (see Figure 7.8). The mnemonic C-COVER may also help you to remember them:

- **C**orroborate.
- **C**onfirm independently.
- **O**bserve for yourself.
- **V**erify physical evidence.
- **E**xamine records.
- **R**ework the system's results.

FIGURE 7.8 Six sampling techniques for auditors: C-COVER.

Corroboration

The first of my six testing techniques is to ask several people, perhaps at different levels in the organization hierarchy, the same question and compare their responses. It is a powerful testing technique and provides (or not) reliable verification of effective implementation.

For example, a corroboration question asked might be, 'How well was the recent fire evacuation practice conducted?' Consistent opinions from several people that the practice was conducted well verifies that this was the case. If nine out of ten responses are positive, this may also represent verification (depending on what number ten says!) However, if most people describe chaos, it probably was!

> **TIP**
>
> Make notes during all interviews. You will surely not remember all the details of what was said afterwards. For note-taking, work as a team if you can, with one person asking questions and another writing down the responses—it is difficult to ask, listen, and write at the same time. Decide in advance who is doing what: it is probably best if the 'techie' asks the 'techie' questions, for example. If the reply is 'techie' or important, make sure the note taker has an accurate record of it—ask them to read it back for the agreement of both the interviewer and interviewee.

TIP

When you are auditing an activity which is carried out by a close-knit group of colleagues, do not be surprised if you get the same response to the same question from each person in the work group. The closer the group, the more there will be a feeling of 'us' and 'them'. In these circumstances, you will need to check what they say against substantive evidence, such as transactional documents, procedures, or reports. If you need to rely on another person's word, then that person needs to be independent of the influence of the group upon which you are relying.

Confirm Independently

Sometimes you may need independent confirmation of something's status. For example, that

- a composite wall panel is non-combustible
- a passenger lift winding cable is in a safe condition,
- an electrical transformer does not contain PCBs (polychlorinated biphenyls), and
- a water sample contains suspended solids below a specified level.

A broad range of laboratories and specialists offers analytical services that will provide independent confirmations of this type. Similarly, independent confirmation of legal points may be needed, in which case the lead auditor should arrange access to an appropriately qualified solicitor or another lawyer.

TIP

If independent expertise can only be provided from outside the auditee's organization, it will probably have to be paid for. Make sure you have the necessary budget provision before committing the audit department to such expenditure.

Observe for Yourself

This is straightforward verification—watch the activity taking place, and compare it with the standard as reviewed. Does it match?

TIP

Auditing is not a covert exercise. Auditors should tell the auditee which activities they would like to observe and at which stages of the process. You should ask the auditee to advise the appropriate process supervisor that you will be attending at a particular time on a particular day. As soon as possible after you arrive at the work site, appropriately clothed and with the requisite PPE, you should introduce yourself to the foreman or supervisor and confirm that they will be carrying out the activities which you would like to observe. There should be no reason why you cannot describe your test plan to the foreman and, if there is an appropriate opportunity, to the operators. Once this is done and you have checked the necessary paperwork (such as permits to work or operators' licences), try to become as inconspicuous as possible so that the supervisor and operators forget you are there observing them.

Verify Physical Evidence

The fourth testing technique is to verify physical evidence, for example, checking that the bund wall is completed, or that the spill kits are fully stocked. Much can be learned by an auditor by actually 'seeing' the risk in control.

TIP

As an auditor, always carry a camera. Remember that 'a photograph is worth a thousand words'. Naturally, you should check the site rules in advance of taking photographs (are there issues of security, or of flammability and ignition?), and I suggest asking people for their permission before you include them in your shot. If this permission is not (verbally) forthcoming, ask them to step out of your shot.

Examine Records

Organizations have many records (ISO MSS calls records *documented information*). All these are available to you for sampling if required by your work plan, including

- purchase orders;
- invoices;
- inspection reports;
- incident and investigation records;
- minutes of meetings;
- contractor records;
- performance appraisals;
- training plans, training attendance, and training feedback records;
- interview and recruitment notes; and
- competency evaluations.

Examining a sample of applicable documented information for evidence in support of your work plan is (virtually) always a good idea.

Rework the System's Results—A 'Brown Paper Exercise'

Reworking a complex management system as a 'brown paper' (or desktop) exercise commences, if taken literally, by covering a large wall or tabletop with brown paper. Of course, you can use whatever colour you like, but many stationers sell wide rolls of brown paper. Onto this, map out a work system—a document flow, for example—to show the origination and the route along and through which each document passes and where it ends up.

Reworking a system in this way shows an auditor what to expect in the records of a department. When you have completed the brown paper exercise, follow the system for the sample set of transactions in practice to see if the outcomes are the same.

CASE STUDY

LISTEN FOR THE WHISTLE!

During an audit in Moscow, an auditee told an auditor (at least) three times that the site did not have all the necessary building and planning consents for its buildings.

On reflection, the auditor was apparently too focused on his questions concerning construction materials and fire separations and, as a result, missed the critical point that this 'whistle-blower' was trying to make.

When the message was finally heard, the result was the involvement of senior management from headquarters and local site management, which led to an action plan to acquire the necessary consents before the local authorities stopped work on the site or began prosecution proceedings.

LEARNING

LISTEN! (see page 244)

TIP

Before you start a Brown Paper Exercise, ask the auditee whether a flow chart of the relevant business processor activity already exists. If it does, save yourself time and use it. If it is only available in digital form, to be viewed on a computer screen, there may still be value in carrying out the exercise described, to increase the transparency (to you) of the controls.

A-FACTOR 59

Learn about the management system by listening closely. There is more to hear in any answer than just 'yes' or 'no'!

TIP

LISTEN

L: Look interested—it may rub off on your interviewee. Enthusiasm can be highly infectious!
I: Inquire with questions—relevant to your interviewee's area of responsibility and/or competence.
S: Stay focused—on your interview agenda so that you can cover the ground required in the time available.
T: Test your understanding—of the facts by asking supplementary questions and summarizing.
E: Evaluate—whether you need to continue verifying to a greater depth of detail an agenda item and, if so, where to probe.
N: Neutralize your emotions—do not be distracted by apparently good or bad news, the interviewee's irritation, or their inability to find the record you seek.

Sample Sizing

In any sampling technique, the auditor should select the sample size before their sampling starts. The sample size might be measured as, for example,

- the number of interviews to be held,
- the number of samples to be taken and verified,
- total observation time (and at what time those observations should be made), and
- the number of records to be viewed.

Table 7.1 suggests sample sizes for use where a specific population is being sampled. Where the population is small, it is obvious that you can take all of them. As the population size increases, it is also obvious to reduce the number of samples.

TABLE 7.1 Suggested Sample Sizes for Any Size of Population

SIZE OF POPULATION	*SUGGESTED SAMPLE SIZE*
2–10	100%
11–25	50%
26–100	25%
101–500	10%
501–1000	5%
More than 1000	1–2%

An alternative statistical approach, easy to remember and calculate, is a *square root sample*. If there are (say) 144 records (or workers), a reasonable sample size is the square root of 144, which is 12 records or workers.

Auditors should remember to take the individual records within the sample from different places in the sequence (that is, to avoid the most recent twelve or the earliest twelve, and so on). Random number generators (provided on some mobile telephones and calculators) or random number tables can be used to identify a random sample, but in most cases this is unnecessary. As an example, if examining training records, the sample might be structured in the following way:

- The newest starter
- The longest server
- The first alphabetically
- The last alphabetically
- The most senior employee
- The most junior employee

Then use numbers 1, 4, 9, 16, 25, 36, 49, 64, 81, 100, and so on (square numbers) to make up your planned sample construction.

It is a good idea to focus (or skew) your sample towards current or recent items (1, 4, 9) since it generally provides a more reliable picture of the future than older or historic sampling (64, 81, 100). As you can see, using square numbers provides such a focus on recent records or events.

As you become more used to determining appropriate samples, I recommend that you 'correct' sample size by increasing or reducing it; for example, if you are sampling the output of a process, you might consider adjusting the sample size as follows:

- If the job is done by lots of different people, you may choose to take a larger sample to ensure broad coverage.
- If there is only one operator for the process, and you have met them and been impressed by their knowledge and efficiency, you may decide to take a smaller sample.

DETERMINING AND REPORTING FINDINGS

Whichever blend of sampling approaches is selected by the auditor, there will tend to be a variety of outcomes to our verifications. These can be quantitative, statistical, or qualitative.

Quantitative Findings

An audit finding can be given in quantitative terms—for example, 'We checked 20 inspection records, and 19 were okay' (or whatever).

TIP

Sample sizes will often be determined by the time available for sampling and the cost of carrying out the tests given the audit scope and objectives. A key to controlling the audit effort at this stage is to know how significant the results of the work already done in a particular area are and how much more information is necessary to gather through detailed testing.

Using Statistics

Where a sample meets the criteria for use of statistical methods, the data collected can be tested, and the results evaluated in terms of statistical significance—that is, in terms of how likely the finding is to be a 'real' finding, rather than merely due to chance. If you have the right resources—including appropriate statistical software—you can decide on an appropriate statistical test, collect suitable data, and calculate the test statistic to find out, as a matter of mathematical probability, how confident you can be in your findings.

You will find a guide to a user-friendly statistical sampling software package, *TurboStats*, in the Further Reading section (Hart, 1993). While *TurboStats* is a relatively low-cost product, I have found it easy to use and highly effective, particularly its reporting functionality. If it is out of print, you may find it on eBay or similar sites. Other products are available.

Qualitative Findings

An audit finding can be given in qualitative terms—for example, 'Everyone we spoke to told us that . . .'

TIP

Whether the testing approach is quantitative or qualitative, the actual selection of items within the sample can be skewed towards factors which the auditor judges to be important—for example, transactions which are a representative cross-section, items of higher value and high-risk exposure, or activities carried out at times of high pressure (such as shift handover or emergency response)—and which reflect current or recent time periods (unless the intention is to test the management system over a longer period of time, for example the effectiveness of a steering committee for a major project).

A-FACTOR 60

Whatever you decide to use as the sampling strategy, record the sample size and how the sample was derived, and then report the results—what the sample told you—in Box 4 of the AFWP.

TIP

Use the planned 15 per cent contingency time for additional testing as required.

CASE STUDY

A DAY AROUND THE POOL—ALTERNATIVE USE OF OUR CONTINGENCY TIME . . .

In 1998, I was a member of an audit team at one of the largest and most complex audits I have ever participated in. Our audit scope was the 'Canal Zone'. This was the prior name for the secured operating location of the Panama Canal, which runs approximately north to south for fifty-two miles through the isthmus of Central America. Until 31 December 1999, it was a US federal agency; ownership was transferred to the Panamanian government on 1 January 2000.

The Panama Canal Commission was a huge organization, with around 70,000 employees involved in operations, administration, security, maintenance, and other work. In fact, the Canal Zone quite literally did everything for itself. If it wanted a nut and bolt, it manufactured them. And it managed its own waste, including the operation of two of its own landfill sites.

This was a major audit using the then newly published ISO 14001:1996 as its reference framework. A large audit team worked over several weeks. Nothing of special note happened during our preparation or fieldwork, or at the start of preparation for our audit conclusion. But with three days to go, our lead auditor was asked if it were possible to bring the closing presentation to senior management forward by one day.

Throughout our audit, the lead auditor had kept the team working closely to time in accordance with the audit plan. With this careful planning, we felt as a team that we could accede to the request without impacting the overall quality. We did just that, and our report was received very well.

Given the title of this case study, you can probably guess what we did with the day we inherited . . .

LEARNING

Careful planning and timely execution of all stages of an audit give auditee and auditor assurance that the audit can be delivered on time or, exceptionally, ahead of time.

Test your knowledge with the next chapter assessment test (see Appendix 4).

Teamwork and Conscious Use of Language

8

I keep six honest serving-men; They taught me all I knew.
Their names are What and Why and When, And How and Where and Who.

—Rudyard Kipling; journalist, poet, novelist (1865–1936)

I know you are wondering . . . and it's good to wonder . . . because . . . that means you are learning things . . . and all the things . . . that you can learn provide you with new insights and new understandings. And you can, can you not? One can, you know. And it's more or less the right thing. You are sitting here, listening to me [reading my book] and that means your unconscious mind is also here, and can hear [read] what I say. And since that's the case, you are probably learning about this and already know more at an unconscious level than you think you do . . . Do you feel this is something you feel you understand?

—Dr Tad James and Dr Adrianna James (2009)

INTRODUCTION

For a summary of this chapter, you can listen to the author's Microlearning™ presentation, which you'll find on the book's Companion Website. Then, delve deeper by reading on for further details on the Companion Website. (See Appendix 4.)

In Chapter 4, I discussed the importance of creating and maintaining good working relationships with individuals *outside* the audit team and gave tips and examples for doing this. However, the lead auditor also bears the principal responsibility of ensuring that the individuals in their audit team work well together.

Over the years, I have learned some powerful techniques that have helped me as an auditor and as a lead auditor to fulfil this responsibility. I'd like to share some of these with you. In Chapter 8, I'll share these helpful approaches, before concluding with my thoughts on *cabinet rules*, and playing as a team—in, and out, of school.

These seven approaches have proved themselves invaluable in countless applications:

1. The sixty-second rule
2. First meeting with the whole team

DOI: 10.1201/9781003364849-9

3. The sausage machine
4. Conscious use of language
5. Questioning techniques
6. Framing
7. Metaphors

However, before we review these, we must recognize that quite a lot of audits are undertaken by lone auditors. In that circumstance, the auditor has only themselves to manage. This can have its own challenges and difficulties.

THE CHALLENGES OF AUDITING ALONE

Many audits are carried out by a lone auditor, even in quite large organizations. This is not an approach that I support or recommend. Audits require analysis and synthesis of many and various strands of information from a range of sources to arrive at an opinion about the effectiveness of one or more BCFs or MSSs. Doing this on your own can be tough. My advice is that there should be at least two auditors on-site, with access to support for peer review. I raised this advice in Chapter 3, Initiating Audit Culture, so that those with responsibility for the audit plan can take account of this resourcing requirement at the earliest stage (and see A-Factor 22).

By the time an auditor has completed half to three-quarters of the *Conduct* stage (i.e., *Review* and *Verify*) of *The Audit Adventure*™, they will likely have generated a large amount of detailed information about the way in which the auditee and their managers have established and are using the elements of the BCF. The auditors may have decided which controls they wish to *Verify* (test) for application and effectiveness and may already have the results of some verifications. After doing all this work there is a need to stand back and reflect on what it all means, especially in preparing for their *nemawashi* ('no surprises') meetings with the auditee.

In this situation, it is extremely difficult for a lone auditor to be able to arrive at a balanced assessment that considers both positive and negative indicators. Having at least two auditors is thus preferable. This promotes reasoned discussion and (professional) argument which is more likely to deliver the balance required to provide auditee management with relevant assurance and improvement messages.

POWERFUL TECHNIQUES THAT CAN HELP AUDITORS

1. The Sixty-Second Rule

Back in the day, when I used to walk back into my home after work in the evenings, I would mumble 'Hello' to my wife, who would shout something back at me as I headed for my study to mess around with my email or LinkedIn or something else. I wondered why we sometimes lacked rapport. And then I learned the sixty-second rule—yes, a simple tool that takes precisely one minute to deploy successfully.

Now when I arrive home, I find my wife straight away, kiss her, and ask about her day. I ask about what she has been doing; I ask if her parents, sister, or friends have called. I look her in the eye and listen intently as though she were the most important person in the world to me (actually, she is). I do this for sixty seconds, before we break and do our separate life stuff. And our rapport is so much stronger. A similar approach (probably without the kiss) can assist with building rapport in an audit team.

In larger organizations, it is unlikely that the lead auditor will know all the individuals assigned to their audit team personally. Sometimes they may not know any of them. The size and composition of the audit team will vary depending on how the audit plan was initiated, on the complexity of the audit to be performed, and the audit intensity required/level of assurance required by top management and other interested parties.

The lead auditor should use the sixty-second rule when meeting each of their team for the first time for any audit, whether they know them already or not. It seems auditors notice this, and it gives rapport building a flying start.

As well as building rapport, the lead auditor must confirm for themselves that they have been given enough people with the right competencies to carry out the assigned audit. Where necessary, additional resources may need to be sought or arranged, usually by contacting the head of internal audit.

Once the team composition has been finalized, the lead auditor should arrange to have an early one-to-one meeting or call (of perhaps a quarter to half an hour) with each member of the audit team. The aim is for the lead auditor and the team member to get to know a little more about each other in a non-pressurized situation. The lead auditor can identify the level of personal and professional commitment the team member has to their assignment, their operational readiness, and whether they are likely to be distracted during the period of the audit by either work or personal issues. They can also check the level of auditing experience and whether they have any subject matter expertise which will be useful when allocating specific tasks.

Each team member should be encouraged to ask questions at this informal meeting, since they should understand that the next meeting will be with the whole team on-site where the emphasis will be on getting on with the job, with less time for personal and social interaction.

A-FACTOR 61

The lead auditor's first action to promote effective teamworking is to have an early discussion with each team member—and remember the sixty-second rule!

2. First Meeting with the Whole Team

This will be a session at which the lead auditor will want to show that they *mean business*. Lead auditors know that they *only get once chance to make a good first impression*. Having recently met with each member of their team, they will be able to make this a very business-focused interaction.

Two weeks ago, the lead auditor will have sought, sorted, and shared (see Figure 4.1 on page 154) their selection of audit-relevant background reading with each auditor and a copy of the draft terms of reference (ToR). They will have asked that the team acquaint themselves with this information, and accordingly at this initial team meeting, they should be well prepared to discuss the auditee's *Context* and objectives, and some of the potential significant inherent risks.

Figures 8.1 and 8.2 show two useful forms that I have used to record the products of this first meeting with my team. I use the first one to record

- the auditee's Context, Objectives, and possible high-level risks and opportunities; and
- the team's analysis of possible significant risks in each Scope area (there are six areas in this example).

When I am the lead auditor, I use the first meeting with my team to set out our ground rules.

Ground Rules for the First Meeting

The lead auditor needs to make it clear to their team at the first meeting that the best results will only be produced by an audit team where each auditor

- is clear about their personal responsibility for auditing correctly risks allocated to them from the work plan;
- executes their fieldwork (reviews and verifications) to a high standard, recording the information obtained on their working papers and sharing this information (analyzed across the elements of the BCF/MSS) with the other team members;
- is committed to giving and receiving feedback on the team's, and their own, work and behaviour;
- listens to their fellow team members as they explain the results of their fieldwork and the conclusions they have drawn from those results, before they challenge what their colleague has done;
- welcomes that the product of their fieldwork and the logic of their thinking will be (constructively) challenged by the other audit team members, as well as by you (the lead auditor); and
- is willing to seek assistance from other team members and the lead auditor to optimize the quality of their own work.

A-FACTOR 62

There is no space in an effective audit team for auditors to hold back from commenting on something they are uncomfortable with. Similarly, there is no space for petulance from a team member whose work is being (constructively) challenged by another member of the team.

Business Environment (political; economic; social and technological):

Corporate and key departmental objectives, strategies and plans:

Main company risks and opportunities:

FIGURE 8.1 Useful form (1): Initial review of the context, objectives, and risks.

INITIAL MAIN OPERATIONAL RISK ASSESSMENT:

Scope Area 1:
Scope Area 2:
Scope Area 3:
Scope Area 4:
Scope Area 5:
Scope Area 6:

FIGURE 8.2 Useful form (2): Initial operational risk identification.

3. The Sausage Machine

The floggings will continue until morale on the ship improves.

—Attributed to a Japanese naval commander in World War II

Imagine a machine that makes sausages. It has a hopper at one end of the device where the butcher places the sausage meat, a sausage-filling device in the middle, and a place where the linked and packed sausages emerge at the other end. If I were to ask you how the very best sausages could be made, you might well suggest that the quality of sausage meat introduced to the hopper would determine the quality of the resulting output.

So imagine now a lead auditor who pays little attention to their team, humiliates them and ridicules their 'stupid questions' at the first meeting, and is badly prepared and allows meetings to drag on and on. Should the same lead auditor be surprised or annoyed by the poor output of that meeting, or the poor work ethic or attitude of their team?

If a lead auditor desires the best outcomes, they must introduce the best inputs. And that's all I have to say on that.

4. Conscious Use of Language

My desired outcome from this section is to help you to make better choices in your use of 'language' to gain the best results. This works in your interactions within the audit team, as well as with the auditee, others in the organization, with third parties, and in your life generally.

Chunking Up and Down

Information is input and output by humans in all levels of detail. It can be distorted (intentionally or accidentally), it can be generalized, or there can be deletions from its content.

Information occupies a continuum from ambiguity to specificity, as illustrated in these two simple examples:

- Example 1
  Sport (ambiguous) → Football (soccer) → English Premiership → Manchester United → The game last Saturday → Outstanding ball control leading to a breathtaking goal by Marcus Rashford in the twelfth minute (specific)
- Example 2
  Training (ambiguous) → Health and safety training → HSE-MS training → Auditor training → Dr Stephen Asbury is teaching my class today → Session 1 is about risk management (specific)

Knowing where you are in terms of the level of detail being spoken or sought is important to auditors (and humans), irrespective of who you are interacting with. You may have noticed the similarity here (Top-down, Bottom-up) to the dynamics of *The Audit Adventure*™.

Language Patterns

James and James (2009) describe a *Meta Model* comprising five distortions, three generalizations, and three deletions in use in common language today. By noticing these language patterns and responding to them, an auditor can ask better follow-up questions. By understanding these language patterns, an auditor can be as precise or imprecise as they wish to be in their use of language in team and auditee interactions.

Distortions

1. Mind-reading—Claiming to know someone's internal state
   Example: An auditor says, 'I know you think I'm doing a bad job'
   Possible response from the lead auditor: 'How do you know I think you're doing a bad job?'
2. Lost performative—Value judgements where the person doing the judging is left out
   Example: Auditee says, 'It's bad that we haven't trained our staff this year'
   Possible response from an auditor: 'Who says it's bad?' or, 'Bad according to whom?'
3. Cause-effect—Where cause is wrongly located outside the self
   Example: 'You make me angry'
   Possible response: 'How does what I am doing cause you to feel angry?'
4. Complex equivalence—Where two experiences are interpreted as being synonymous; one thing causes another
   Example: 'She's always yelling at me, because she doesn't like me'
   Possible response: 'Have you ever yelled at someone you liked?'
5. Presuppositions
   Example: 'If my boss knew how much I suffered, he wouldn't do that'
   There are three presuppositions here: (i) I suffer, (ii) my boss acts in some way, and (iii) my boss doesn't know I suffer.
   Possible responses: 'How do you choose to suffer?', or 'How do you know he doesn't know?'

Generalizations

1. Universal quantifiers—Universal generalizations such as 'everyone', 'never', or 'no'
   Example: 'We never do any training at this site'
   Possible response: 'When was the last training session organized?'
2. Modal operators of necessity—Words which imply necessity, which often form rules of life; use of words and phrases such as 'should', must', or 'need to'
   Example: 'I must fill these forms in'
   Possible response: 'What would happen if you didn't?'
3. Modal operators of possibility (or impossibility)—Words which imply possibility, which may also form our rules of life; use of words and phrases such as 'can/cannot', 'will/will not', or 'may/may not'
   Example: 'I cannot tell my boss the truth'
   Possible response: 'What would happen if you did tell her?'

Deletions

1 Nominalizations—Process words which have been frozen in time, making them nouns
Example: 'There is no communication here'
Possible responses: 'Who's not communicating what to whom?', or 'How would you prefer to communicate with your colleagues?'
2 Unspecified verbs—Unspecified use of a verb
Example: 'He rejected me'
Possible response: 'How, specifically, were you rejected?'
3 Simple deletions—Failure to specify a person, thing, or comparator
(a) Lack of referential index
Example: 'They don't listen to me'
Possible response: 'Who is not listening to you?'
(b) Comparator deletion
Example: 'Company A is the best training company'
Possible responses: 'In what way is Company A the best training company?', or 'The best compared to whom?'

A-FACTOR 63

Avoid using absolute terms (no, never, or always) in meetings or reports, as they are seldom true.

5. Questioning Techniques

Now that you are more knowledgeable about language patterns (and you'll note I have just used a presupposition!), we can review the basic questioning techniques. Of course, these can be used when talking with your colleagues in the audit team and when interviewing. And if both you and your colleagues understand this conscious use of language in questioning, you will be better informed when (for example) taking notes during an interview.

There are (at least) ten basic questioning styles:

- Closed
- Direct
- Follow-up
- Probing
- Open
- Encouraging
- Linking
- Multiple
- Leading
- Overcomplicated

Mastering the use of the first seven of these, and avoiding (wherever possible) the final three, will help an auditor to be successful in talking with their team, in their audit interviews, and in their life generally. Let's briefly review each in turn.

- Closed
  - Example: 'Does your department hold health and safety committee meetings?'
  - Pro: Provides a short, concise answer, often 'yes' or 'no'
  - Con: Allows the interviewee to provide very little information or to be evasive
- Direct
  - Example: 'How often do you hold health and safety committee meetings?'
  - Pro: Obtains detailed information on the area questioned about
  - Con: None, but you do need to know why you are asking this question
- Follow-up
  - Example: 'You say you hold monthly health and safety committee meetings. Who are the usual attendees?'
  - Pro: Shows you are listening, shows you are interested, and enables a greater understanding of the area about which you are enquiring
  - Con: None, but you do need to know why you are asking this question
- Probing
  - Example: 'Can you show me the records of who attended the health and safety committee meetings in the last twelve months?'
  - Pro: Shows you are listening, shows you are interested, and enables a greater understanding of the area about which you are enquiring
  - Con: None, but you do need to know why you are asking this question
- Open
  - Example: 'Can you tell me more about the work of the health and safety committee?'
  - Pro: Encourages the interviewee to talk more freely and to express views and opinions
  - Con: Need to control long or rambling answers; need to make sure the interviewee stays 'on topic'
- Encouraging
  - Example: 'The excellent structure you have described must mean that management is strongly supportive of the work of the health and safety committee. Does this increase the support for safety initiatives?'
  - Pro: Shows your interest in the subject and encourages the interviewee to see the audit as a positive experience
  - Con: You may give a false impression of what the audit can deliver
- Linking
  - Example: 'How has the extra funding you have received for health and safety projects changed the culture for health and safety?'
  - Pro: Shows your interest in the subject and encourages the interviewee to see the audit as a positive experience
  - Con: None, but you need to know why you are asking the question

- Multiple (I recommend avoiding this approach where possible)
  - Example: 'How many times has the health and safety committee met in the last six months, who attended each meeting, who kept the minutes, and are all of the action items closed out?'
  - Pro: Theoretically possible to obtain lots of answers in a short time
  - Con: Easily confuses the interviewee, and often parts of the question go unanswered
- Leading (I recommend avoiding this approach where possible)
  - Example: 'So everything is going really well in your health and safety committee?'
  - Pro: None
  - Con: You can easily leave with 'yes' answers which do not reflect the reality of the situation
- Overcomplicated (I recommend avoiding this approach)
  - Example: 'Are you using the approach set out in ISO 9999 of 2019 and UN Standard 1234, section 2.22 for developing agendas for your health and safety committee, and Group Guidance Document 555, section 66 for OSHA, COSHH, DSEAR, blah, blah . . . ?'
  - Pro: None
  - Con: The interviewee becomes completely lost and probably thinks you are a pompous idiot

You must also remember Rudyard Kipling's 'Six honest serving-men', and use these consciously when you form your questions: What, Why, When, How, Where, and Who.

A downloadable copy of my interview skills checklist, as used on our auditor training courses and includes analysis of questioning techniques, is provided on the companion website (https://routledgetextbooks.com/textbooks/_author/asbury/).

A-FACTOR 64

In every questioning interaction, know specifically what you want from your question before you ask it.

6. Framing

Being able to 'frame' statements is another important aspect of your conscious use of language, and seven frames are useful for auditors:

- The agreement frame
- The purpose frame
- The 'what if' frame
- Using words that create positive representation
- Conditional close
- Tag questions
- The double bind

Again, we'll review each in turn:

- The agreement frame
  - Example: 'I agree with you that health and safety committees are a valuable addition to cross-departmental communication'
  - How it is useful: Facilitates agreement that health and safety committees are valuable for communication in the organization
- The purpose frame
  - Example: 'For what purpose do you hold health and safety committees on a weekly basis?'
  - How it is useful: Encourages the interviewee to see the purposefulness of any actions they are taking. Identifies gaps in their 'purpose'
- The 'what if' frame
  - Example: 'What would happen if you reduced the frequency of health and safety committees to monthly?'
  - How it is useful: Compels the interviewee to think about specific consequences and to verbalize these
- Using words that create positive representation (say how you want things to be; a group of five positives seems to work well, in my experience)
  - Example: 'People who attend health and safety committees often feel better informed, are more involved, are more aware of changes taking place, have better cross-departmental contacts, and are viewed more positively by top management'
  - How it is useful: Creates a positive internal representation for the interviewee of the health and safety committee and makes them more likely to identify with it
- Conditional close
  - Example: 'If I agreed to work with Rick to finish the first work plan item, would you work with Michonne to finish the second?'
  - How it is useful: Encourages the other person to commit to do what's been asked of them
- Tag questions
  - Example: 'This is something you're interested in, isn't it?'
  - How it is useful: Designed to displace resistance; it reveals interest on the part of the other person
- The double bind (gives the illusion of choice, when really none is available)
  - Example: 'Would you like to go to bed at 8 pm or 9 pm?'
  - How it is useful: The kids will be in bed by 9 pm

7. Metaphors

The purpose of a metaphor is to pace and influence another's behaviour through a story. I will describe briefly the four major points to be considered when constructing a metaphor, prior to describing two examples I have used in the last few years (with a summary of their respective contexts).

The four points to be addressed in constructing a metaphor are as follows:

- Displace the referential index from the other to a character in a story.
- Gain, hold, and pace the interest of the other by establishing behaviours and events between the characters in the story that are like those in the other's situation.
- Provide resources and ideas for the other within the context of the story.
- Finish the story with a sequence of events in which the characters resolve their conflicts and achieve the desired outcome.

Example 1

- Metaphor: I feel happy and secure working in the company I own. So much more so than when I worked for an insurance company where I could have been relocated or made redundant anytime. I think leaving and setting up my own company eighteen years ago was the best decision I ever made, and a great career move for me.
- Context: I had a good friend who wanted to start his own company but was nervous and unsure about whether he should resign from his employment. He has gone on to run a successful business for almost twenty years now.

Example 2

- Metaphor: Mrs Chipperfield was stuck in a dead-end job, regularly left at home while her husband travelled with a touring circus. The circus was a pretty good show, but often the big top was not full—people just don't go to circuses anymore. Mrs Chipperfield was full of ideas about how to encourage local businesses to offer tickets to their employees, so she left her job and joined the circus. The big top is now full almost every evening.
- Context: A friend of mine disliked her job and thought that her husband's printing business could grow to support the family better if only he could get better administrative support.

CONFIRMING PROVISIONAL AUDIT FINDINGS WITH THE LEAD AUDITOR

When auditors master these techniques—including the conscious use of language—they obtain better information with which to develop better audit findings. As each member of the audit team progresses their own part of the audit, it is a good idea to arrange regular meetings with you (the lead auditor) to report back on what has been found so far in the fieldwork. Lead auditors have discretion to do this one-on-one or involve other team members.

Periodic confirmation by the lead auditor with each audit team member that they are on track (or not!) and covering the work plan item/s allocated to them is a critical supervisory aspect of the lead auditor's role. It is both beneficial and essential, as,

- if a team member cannot convince the lead auditor that they have a valuable finding, they'll be unlikely to convince the auditee;
- the lead auditor needs to know at an early stage about the quality of risk management in the auditee's *Scope* and the strengths and weaknesses in the management system;
- the lead auditor will want to know whether any major or serious control weaknesses have been identified so that he can advise the auditee, if necessary, during their *nemawashi* ('no surprises') meetings.

PLAYING AS A TEAM

If the lead auditor is successful in engendering the styles of teamwork, use of language, and the questioning techniques described here, the result will be a high-quality and individually committed performance. I strongly encourage open and vigorous discussion among team members, including reasoned challenge and debate without irritation or rancour, so that in principle, every team member feels personally committed to every finding.

This team-playing approach is especially important during the *Conclude* stage of the audit (see Chapter 9), when the lead auditor will sign off or approve each auditor's audit finding working papers (AFWPs; see Chapters 6 and 7). They will initiate and progress sharing of information in the team through the BCF illustrated on the audit working papers and/or wall charts (see Chapter 5 for use of audit wall charts), so that the team together can see 'the big picture' and start to evaluate, group, and summarize their findings. This can only be done from a base of common understanding of the facts gathered on the BCF.

Each auditor needs to be prepared and ready to explain and support the facts that they have contributed to the wall charts. This process can often generate some *heat*, which the lead auditor will need to focus as *light* to ensure this time is productive.

At the end of this process, the facts upon which the audit opinion will be based will (ideally) have been agreed by the whole team. Similarly, each team member assigned responsibility for preparing a balanced assessment of an individual element of the BCF for inclusion in the management summary of the audit report will work more effectively, since they will have been involved in discussions which picked out examples of the most positive and negative indicators, which they will use in preparing their summary.

CABINET RULES—'OUR OPINION'

Members of any audit team must understand that the lead auditor is ultimately accountable for presenting the collective and balanced result of the audit team's work. Sometimes they may therefore have to make judgement calls based on their assessment of all the facts collected.

If the audit team has operated in the effective way I have described, it is considerably less likely that there will be any serious disagreement between its members, or between the team and its lead auditor. However, should there be such a disagreement, it is incumbent upon the lead auditor to remind the team members that audit reports do not include 'minority opinions'. Divided audit teams are generally *un*successful.

A-FACTOR 65

The only place for the audit team to disagree is in the privacy of their team room. Outside the team room, they must present a united face. There are no minority opinions in audit reports.

CONSULTING WITH EXTERNAL SOURCES

Lead auditors should remember that there will be experts outside their team to whom they can turn for advice or guidance. This is particularly relevant when the team is assessing highly technical areas of risk, or the possible impacts and consequences of the auditee's actions (or inactions) uncovered during the audit. The reality these days is that the audit report is likely to be seen by third parties including banks, shareholders, regulatory authorities, or sometimes the media. If there are errors, of either commission or omission that impacts them, this gives licence to challenge the audit results.

In sensitive situations, such as those affecting the occupational health and safety of employees, customers, or the local community (and especially if there have been recent incidents), the lead auditor must be ready to speak to specialist, technical, or legal experts. In the context of advice provided, it should be treated as having been given by a member of the audit team, and the source named within the report. For example, a lawyer might assist the audit team by providing appropriate language to describe a control weakness that needs to be strengthened to prevent the recurrence of incidents.

This should not be confused with external sources seeking to dilute, soften, or remove audit findings. This is always, of course, unacceptable.

Auditors can be held (and have been held) liable for defamation if an audit report is inaccurate and subsequently harms the subject of the report. Clearly, truth is the best defence against legal action(s). Where necessary, work with experts early on to clarify the facts found and the resulting opinions based on them. Particularly in sensitive areas, contact with expert sources should not wait until the audit report is being drafted.

Note that in some jurisdictions, expressions in an audit report referring to poor management practices and unacceptable levels of exposure to certain risks could lead to civil and/or criminal liabilities, which could result in substantial punitive damages and/or imprisonment for the managers involved.

PEER REVIEW

Towards the end of an audit, the lead auditor should seek (or be offered) access to another senior auditor who will either have experience of auditing a similar *Scope*, or will have been sufficiently briefed to enable a review the audit findings and draft audit report, to calibrate agreement with the draft or final audit opinion.

Such peer review will also help to iron out matters of presentation, which *fresh eyes* often see more clearly. A peer review process is always likely to make a significant contribution to the quality assurance surrounding an audit, but it is an essential part of any audit carried out by a single auditor.

A-FACTOR 66

It is good practice to provide the lead auditor with access to a competent colleague for peer review and to calibrate the draft audit opinion.

PLAYING OUT OF SCHOOL

No chapter on teamworking would be complete without reminding lead auditors to *let their hair down* occasionally and to take their auditors for dinner during the audit. Nothing expensive or flashy is needed—just a non-verbal reminder that you are a human being who is interested in other human beings.

Playing just a little 'out of school' makes for wonderful team working, rapport building, and lifelong-friendship-making. It should not be overlooked.

A-FACTOR 67

Rapport = building and maintaining positive engagement with other human beings. It should not be overlooked.

Test your knowledge with the next chapter assessment test (see Appendix 4).

Conclude the Audit 9

Think in advance about afterwards—whatever you decide, think ahead about the consequences.

—Dr Stephen Asbury

The more you know about a subject, the more you can see both sides.

—Dr Stephen Asbury

INTRODUCTION

For a summary of this chapter, you can listen to the author's Microlearning™ presentation, which you'll find on the book's Companion Website. Then, delve deeper by reading on for further details on the Companion Website. (See Appendix 4.)

Chapter 5 provided the 'big picture' of *The Audit Adventure*™, which provides a powerful metaphor for any management systems audit aligned to the key steps in ISO 19011:2018. Having journeyed through the *Initiate* stage in Chapter 3, the *Prepare* stage in Chapter 6, and the *Conduct* stage in Chapter 7, it is now time to *Conclude* the audit (this chapter) and *Report*; to write the audit report and understand its follow-up (Chapter 10). Your *Audit Adventure*™ now enters the final 20 per cent of its timed work plan.

The purpose of this final stage of the audit process is to *Conclude* the audit 'Bottom-up' (see Figure 5.4) in such a way that the audit team helps auditee management to understand the status of control of a sample of significant risks and, when necessary, how control may need to be improved.

Staying with the metaphor described in Chapter 5 of an audit being like a peak-to-peak walk at Poldhu, the concluding stage of the adventure commences just after your allocated time on the beach expires. The descent 'Top-down' was breathtaking, and the focus narrowed to the beach and its flora and fauna. Gravity has done its work, and after the pleasure of the sand and sea, and the resident beach life, tension has eased—until you see the steepness of the climb directly in front of you. But the speed of descent down to the beach, with the resultant flood of information into the audit team's *group brain*, normally provides the momentum to enable you to reach the top on the other side, at which point the audit team should be ready to deliver their audit report.

DOI: 10.1201/9781003364849-10

Figure 9.1 illustrates the concluding stage of *The Audit Adventure*™. This chapter and the next describe how to deliver this.

> **A-FACTOR 68**
>
> Momentum accumulated 'Top-down' provides auditors with a burst of energy for the journey 'Bottom-up'.

CONCLUDING THE AUDIT—'BOTTOM-UP'

However competently the lead auditor and their audit team have performed in the previous stages of the audit, it will all count for little or nothing if the audit work to be performed in this stage of the audit process is not fully understood and carried out with precision, imagination, and creativity. The challenge for the audit team throughout their work is to create an appetite for their audit findings and any consequent opportunities for improvement. The final proof of the audit pudding will be determined by how senior management relishes its eating!

The main activities involved in concluding an audit:

- Finish your fieldwork (*Review* and *Verify*) to finalize your facts.
- Agree areas of strength and any control weakness.

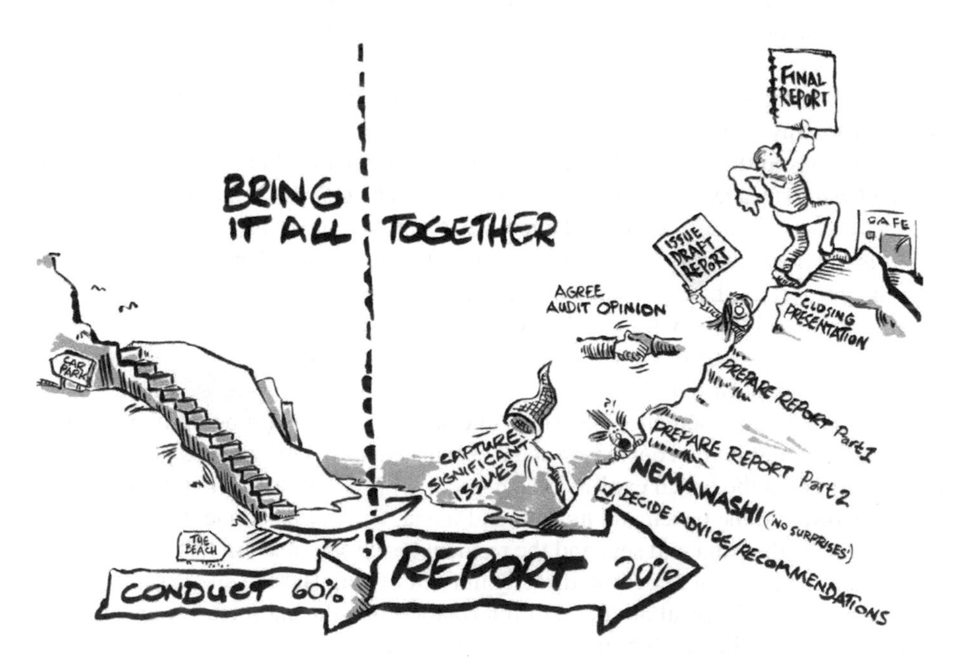

FIGURE 9.1 From detail to a high-level audit opinion; bringing it all together.

- Map results onto the BCF.
- Identify and focus upon the main findings/areas of significant *residual* risk.
- Evaluate the BCF.
- Discuss the findings with the auditee—*nemawashi.*
- Determine the overall audit opinion.
- Prepare and present your findings and written audit report.

These activities are not necessarily carried out in the exact sequence listed, since the best results generally come from using an iterative approach. That is to say, at the start of this final stage, auditors will have available to them a lot of detail regarding the individual controls and control framework relevant to each of the work plan items they have had responsibility for auditing. Auditors not only need to assess this information in terms of its impact on the risk exposures associated with each work plan item, they also need to review the results as a whole. It will then be possible to identify whether there are any common traits or patterns which can be reported as a summary finding. Each auditor must share the results of their audit work in a structured way, so that every member of the audit team can compare the detail of their own audit findings and determine on any such summary findings.

Getting a Clear Picture from the Mass of Information

The starting point for the *Conclude* stage of the audit:

- A completed set of factually-accurate audit finding working papers (AFWPs). There should be one AFWP for each risk selected and sampled in the work plan.
- An accurate understanding of the auditee's actual business control framework (BCF).

By this point, lead auditors should have ensured that they have challenged robustly the content, details, logical extension, and conclusion of every AFWP prepared by each member of their audit team. The underlying fieldwork (*Reviews* and *Verifications*) should demonstrate sufficient, relevant, and reliable evidence which is comprehensively cross-referenced to its source and to transactional documentation which is in turn linked to the organization's BCF. The audit's working papers should also be properly filed, as described in Chapter 6.

Lead auditors should also have ensured that the work carried out in preparing the AFWPs is clearly and directly traceable to individual or groups of work plan items. Only then can the lead auditor demonstrate that their team has fully audited the effectiveness of the BCF, as applied to the sampled significant inherent risk activities (including *Black Swans*) selected for the work plan.

Figure 9.2 shows the AFWP section headings (see Figure 6.8, Boxes 5–8 and confirmation signatures) associated with the results of the audit of a particular work plan risk—in this example, work plan item 2.2—and then the lead auditor checking this off as 'done' in the work plan.

Sometimes the audit team will be able to provide a high level of assurance as a result of their *Reviews* and *Verifications*, and an example of such an outcome follows.

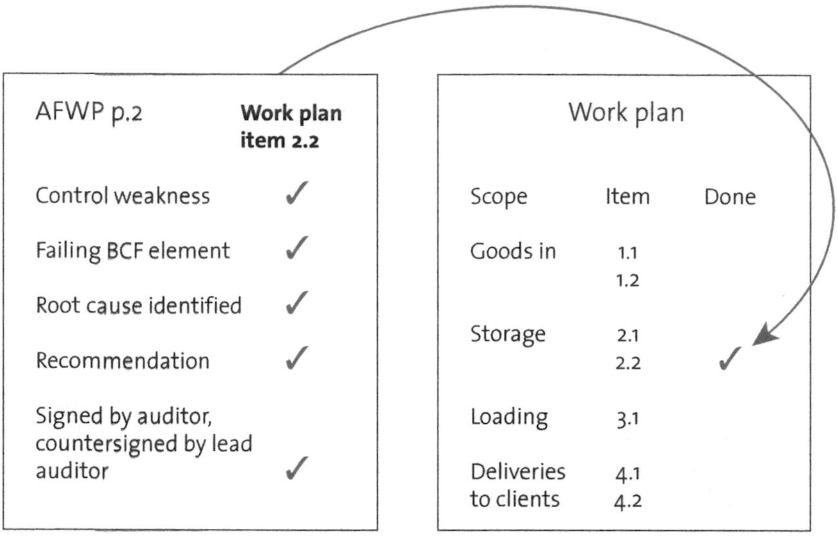

AFWP p.2	Work plan item 2.2
Control weakness	✓
Failing BCF element	✓
Root cause identified	✓
Recommendation	✓
Signed by auditor, countersigned by lead auditor	✓

Work plan

Scope	Item	Done
Goods in	1.1	
	1.2	
Storage	2.1	
	2.2	✓
Loading	3.1	
Deliveries to clients	4.1	
	4.2	

FIGURE 9.2 The lead auditor updates the work plan to confirm that 'Risk #2.2' has been audited (*Reviewed* and *Verified*), and the facts agreed.

CASE STUDY

EFFLUENT TREATMENT AND DISCHARGE INTO CONTROLLED WATERS NEAR AT A SITE OF SPECIAL SCIENTIFIC INTEREST

An UK industrial effluent treatment plant operated under the conditions of an environmental permit, as required by the *Water Industry Act 1991.* It treats wastewater from an industrial paint plant, and often has a foamy head until treated, settled, and pH-corrected. After treatment, it is discharged into a section of river, which feeds into an estuary four miles downstream. The estuary is a designated a Site of Special Scientific Interest (SSSI) due to the wading birds that inhabit the area.

There had been several previous breaches of permit conditions detected, one of which led to a prosecution in the Magistrate's Court. The organization pled guilty and was fined £20,000, plus costs.

Since then, new permit conditions require additional pollution prevention measures, including provision of pollution control booms to secure the estuary mouth in the event of a pollution event. Pollution response operations are notably difficult in rivers with strong current or river traffic.

The organization operating the treatment plant purchased a high-buoyancy inflatable boom system of size and performance recommended by its manufacturer. Its design makes it suitable for use in a wide variety of conditions from sheltered water to open sea waves. Each inflatable boom section length is segmented into individual inflatable chambers by way of internal bulkheads. That ensures that the boom will maintain positive buoyancy and continue to float in the event of any damage. Booms feature rapid inflation, and high buoyancy to weight ratios for use in waves and demanding conditions. The system also has substantial foam floatation sections along the top of the boom to keep it afloat should one of the chambers become damaged and deflate.

The system is compact for storage on boom reels. The manufacturer says that this makes it easy and fast to deploy and retrieve with a minimum amount of workforce. It is inflated using handheld or backpack inflation blowers. These integrated hydraulic blowers can be mounted on boom storage and deployment reels. The system sits unused in a transport container acquired for this purpose.

For the third time, I'll use a case study as an opportunity to understand an *expected* control framework, then review and verify its application. Let's imagine that my audit team has select pollution of controlled waters as one of several samples of inherent risks in its work plan for a five-day EMS audit at this organization's site. What might I expect from the EMS? If those expectations are met during my team's review and verification, will I be able to provide a high level of assurance of control? Here is an example of what I might expect:

- Leadership: Environmental policy signed/endorsed by top management includes commitment to avoid pollution.
- Planning: Plans to meet permit conditions, including ownership of suitable pollution control booms.
- Support: Training for the team appointed to inflate, deploy, and recover the pollution control booms in the event of an incident. Secure storage for the boom system.
- Operation: Able to correctly deploy the booms within the required time.
- Performance evaluation: Records of exercises, with time taken to deploy. Periodic inspection of the booms for damage/deterioration.
- Improvement: Learning from deployment exercises incorporated into future plans.

As it happened, while my audit team was on site, management agreed with me that a planned exercise to inflate, deploy, and recover the pollution control booms was desirable. The local authority gave notice to mariners of the closure of the river access for 36 hours. We were able to observe the inflation and effective deployment on the second day of our audit, as well as observing the recovery, deflation, and return to storage of the system on the fourth day. One yacht did cause an obstruction, but this was satisfactorily overcome. We took some photographs, a few of which a presented herein. Of course, these were also included as evidence in our audit report, which provide a high level of assurance of the effectiveness of this element of the environmental management system.

Identify Control Weaknesses

In her recipe for jugged hare, the celebrated British writer on domestic science, Mrs Beeton (1836–65), apocryphally said, 'First, catch your hare.' This is likewise important for auditors: first, they must have clearly identified a control weakness. Each step of the Conduct stage of the audit process enables you to do this:

- Your *Review* will identify whether there are any gaps in the design of the business control framework.
- Your *Verification* (or testing) will identify whether there is non-compliance with the expected controls, while your substantive testing will identify those controls that produce ineffective results.

TIP

The identification of control weakness needs to be followed by a process of analysis and synthesis as soon as possible if the audit team is to be successful in meeting the objective of minimizing the number (whilst increasing the quality) of audit findings.

At this interim stage, the audit findings will fall into three categories:

- OK: the audit fieldwork has provided the audit team with clear evidence of adequacy of control design, appropriate levels of compliance, and effective outcomes
- Not OK (i.e., control weakness): the fieldwork has identified evidence of inadequacy of design, varying degrees of non-compliance, and/or ineffective outcomes
- Doubt or uncertainty: final testing using the planned contingency time to confirm strength or weakness has not yet been completed (or an auditor is yet to make up their mind!).

As soon as possible in the *Conclude* stage, the lead auditor should have classified the level of control related to all the audit findings on the fieldwork which has now been completed as either OK (= assurance) or not OK (= alert, plus possible advice).

TIP

Exceptionally, if the lead auditor is in any doubt as to the sufficiency or reliability of the evidence obtained by the auditor leading on a work plan item, then that part of the fieldwork must remain inconclusive. If appropriate, the work in progress can be passed to a local internal auditor for further review.

As a general observation, I have found in most major organizations that there are significantly more risk controls classified as OK than as failing (not OK). Sometimes this irritates auditors, as some still think their job is only to find failing controls. Some even forget to record evidence of control strength on their AFWP as a result!

TIP

A competent auditor is one who makes a judgement, based on the evidence of strengths and/or weaknesses identified, in the significant risk area/s they have sampled. That final judgement is built up from incremental judgements based on the results of their audit fieldwork for each work plan item.

CASE STUDY

AUDITING THE AUDIT PROCESS

Throughout an HSEQ audit at a bottler of carbonated, flavoured soft drinks in Zagreb, Croatia, an audit team was accompanied as agreed by a representative from the auditee organization's London market insurer. The insurer wished to independently verify that the audit process was sufficiently thorough, and that findings and recommendations were commensurate with identified levels of residual (insured) risk.

The intention at the outset was that the insurer would send a representative to attend a series of such audits at locations across Eastern Europe prior to drawing a conclusion from all these.

The audit approach followed in Zagreb was as described in this book. Intelligent use of the *nemawashi* ('no surprises') approach resulted in plant management accepting all the audit findings and the improvement recommendations as proposed.

In the week following this first audit attended by the insurer, a telephone call and follow-up letter expressed satisfaction with the independence of the auditing process, the suitability of the opportunities for improvement identified, and the way these had been embraced by the site's management team. The letter advised that 'no further verification of the audit process was necessary'.

LEARNING

Done properly, auditing really does provide assurance to others.

Categorizing Control Weaknesses

The risk-based audit process described in this book progressively increases any audit team's knowledge and understanding of how well discrete, then progressively larger, parts of the auditee's business control framework function.

The lead auditor must optimize the *group brain* of the audit team through a structured process of analysis and synthesis, assimilating underlying facts and extracting similarities between common denominators within individual control weaknesses. This information can then be used as the basis for grouping or clustering detailed audit findings into higher-level audit conclusions.

For example, an auditor may note the following weaknesses:

- Employees and contractors not wearing the required PPE at all times
- Housekeeping falls short of expectations
- Site speed limit frequently exceeded
- Signatures and dates missing from sampled documents
- Planned inspection program behind schedule
- Routine supervisor absence at planned HSE committee meetings

In this example, I believe that these are not six individual control weaknesses but probably just one—you're just seeing the symptoms. The control weakness here seems to be related to the efficiency and effectiveness of the first-line supervisor.

There are no absolutely definitive categories an auditor should use for grouping or clustering their findings. However, possible categories which I have come across during my audit assignments are suggested in Figure 9.3. In practice, it is often best to discuss the preferred grouping (and subsequent presentation) of the findings with the auditee during *nemawashi* meetings.

By priority of importance

- Highest (Red)
- Intermediate (Amber)
- Lower (Green)

By cost

- Nil
- Low (Rev-Ex)
- High (Cap-Ex)

By organization

- Process 1
- Process 2
- Site A
- Department B

By reference framework element

- Leadership
- Planning
- Support and operation
- Performance evaluation
- Improvement

FIGURE 9.3 Ideas for grouping your audit findings.

I have provided four example groupings of audit findings, though, of course, there are many other possibilities.

Grouping by Priority

I have found that many auditees like to know which weaknesses need to be addressed first, and which may wait a while. A typical approach to clustering weaknesses is making priority groups, such as

- immediate attention required,
- up to three months,
- three to six months, and
- one year.

Another way to do this is simple colour-coding, with a traffic light approach—red (highest priority), amber (intermediate), and green (lower priority)—to show which matters require the promptest attention.

Grouping by Cost

Ultimately, remedying some weaknesses will require expenditure. Some may be funded from revenue expenditure (RevEx), and some from capital expenditure (CapEx).

A useful alternative to clustering weaknesses by priority is to cluster by cost, with nil- and low-cost items which can be funded from current operating budgets separated from matters requiring application to investors for authorization of capital investment.

Grouping by Organization

A third option is to group weaknesses under organizational or departmental headings—for example, by process, section, building, or site. This approach can also be used in conjunction with one of the others, for example clustering control weaknesses within each department by priority.

Grouping by Reference Framework Element

Throughout your audit, starting with the terms of reference, the audit will be approached with reference to one or more reference frameworks (BCFs and MSSs; see Chapter 2). A useful clustering could be under the elements of the relevant framework—PDCA, or whatever.

This is a useful approach, as it tends to further cluster the weaknesses according to their root causes. That is, a significant number of failings or weaknesses identified within any of the elements may become a signpost leading the audit team towards a higher-level root cause.

Any relevant category can, however, be used to group detailed control weaknesses into a few main audit findings. Suitable categories will become obvious to the audit team as they begin to organize the collected audit evidence.

TIP

Discuss with the auditee in advance how they may wish to see any control weaknesses grouped. I have observed several auditees dealing with audit findings by (literally) ripping pages from the report and handing them to the appropriate managers to attend to. Should a preference be expressed, and you have flexibility to prepare in this way, it would be a good idea to match those preferences.

Identifying the Root Causes of Control Weaknesses

A huge potential benefit of carrying out internal audits would be lost if each auditor, the audit team, and the lead auditor failed to ask the simple question: *Why or how was it possible for this aspect of the control framework to fail?*

Finding a significant control weakness and securing the auditee's agreement to fix the problem is certainly a good result for any audit. However, unless the auditors and/or the auditee unearth the *root cause* of the control failure, it is quite possible that the benefit of the fix will be short-lived; there may be a similar control, not included in the audit work plan, which is already failing or will be allowed to fail soon.

By identifying the root cause of a control weakness, an auditor may expose similar controls that have not been audited and which, if they were to fail, would have a similarly negative impact on the auditee's organization.

There are various established and well-documented techniques, proprietary software tools, and approaches available for assisting with root-cause analysis—I have listed several relevant titles in the Further Reading section (including Ammerman, 1998; Johnston, 2019; Roderich, 2021; and System Improvements, 2023). However, I believe that an effective way of finding the root cause of a control failure is to ask questions about how well other key controls in the same BCF are working.

For example, if the identified control failing is a procedure, then a competent auditor will ask, 'Was this failing identified by a supervisory control? If so, what was done about it?' If the supervisory control failed, then one needs to ask the question, 'Does an adequately designed supervisory control exist and was it operating properly?' If there was no suitable supervisory control, then one needs to ask the question, 'Are there sufficient competent personnel or automated resources available to carry out the necessary supervision?' Then one could ask the question, 'Were the risks associated with the control failure properly assessed? If so, what was the agreed risk response? If not, is there a fit-for-purpose risk assessment and risk control process in place?'

For as long as an auditor keeps asking questions about the existence and effectiveness of controls in the BCF, this approach will lead them towards the control failure at the highest level at which it occurs in the framework—for example, unearthing a failure to have clear objectives or direction. This approach is shown in Figure 9.4.

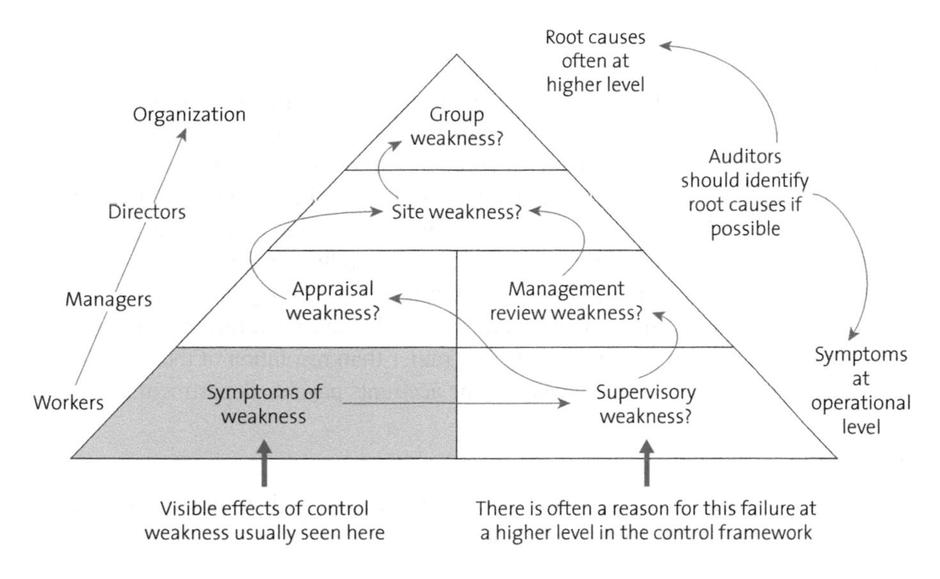

FIGURE 9.4 Tracking down the root cause of basic control weaknesses.

> **TIP**
>
> In addition to fixing the failing control and addressing the issues surrounding the root cause, another valid audit recommendation would be for the auditee to review the effectiveness of each of the other controls potentially affected by the same root cause.

CASE STUDY

KEEP ASKING QUESTIONS

An employee of a large organization cut the palm of his hand very deeply on the ragged edge of a large sheet of steel being handled in a power press shop. The investigating health and safety officer applied a technique of asking questions about how well other controls within the same BCF for hand protection were working, applying the style of '5-Whys' and 'Domino theory' (Boyle, 2002; Fuller & Vassie, 2004).

Instead of blaming the worker, the investigation chose to consider the possible reasons why the worker was not wearing the strong Kevlar-lined gloves that had been approved for use on that type of work. It was discovered that the gloves had not been available that day in the stores, and that the worker had 'done his best' and progressed the work wearing suboptimal gloves. Instead of blaming the storekeeper for this apparent stock-out, the investigation chose to consider the reasons the stores did not have the gloves. It was discovered that the glove supplier, having been chased five times for a delivery, had made five delivery promises to the storekeeper. Instead of immediately blaming the supplier, the investigation chose to consider why the delivery promises had not been kept. It was discovered that the supplier had not been paid for three months; rather, it had been promised five times that payment would be made that day. Each time, the payment had not been received, and accordingly, the account was not settled and normalized.

If the organization had 'blamed the worker' (as sadly remains so often the case), told him 'to be more careful next time', or sent him for 'training' (as in my experience organizations often do), the probable outcome would have been further cut hands and other injuries related to the absence of PPE, rather than resolution of the root-cause weakness, in this case in the diligence of the accounts payable department.

LEARNING

Each time the next proximate reason for control weakness is found, a good auditor looks for another 'Why' question to ask.

Prioritize Control Weaknesses

Before any member of an audit team tries to persuade the lead auditor that they have identified a control weakness, they should ask themselves, 'So what?', or 'Is the residual risk tolerable?' In other words, at an early stage, they should focus on control issues that may have significance of impact upon the objectives—not long lists of symptoms and hazard spots which just add to a mass of low-level findings in which senior management typically has little or no interest.

A-FACTOR 69

A significant audit finding is one for which the answer to the 'So what?' question represents a significant impact on the auditee's objectives. If the auditors cannot attach this significance, then it is unlikely that senior management will either.

TIP

The objective most likely to be impacted is the same one identified by the audit team when the work plan was selected and the AFWP created.

Along with the 'So what?' question, an auditor can seek answers to two other questions to help prioritize the control weakness.

- 'How easy is it going to be to fix the whole problem?' Their assessment of this would consider access to the required competence and sufficient financial resources.
- 'Why has it been left to an internal audit to discover this control weakness?' Here the auditor needs to realize that many audit findings are not necessarily 'news' to auditees. However, what may be 'new' is a perspective that provides a clear understanding of exactly how exposed either the auditee or the business will be if they continue to condone the status quo.

Control weaknesses can be prioritized using a variety of terminology to signify their relative importance. Generally, I go no further than splitting the control weaknesses into Serious, High, Medium, and Low categories. I have also seen auditors—particularly from standards certification bodies—use three categories:

- Major non-conformity
- Minor non-conformity
- Observation/opportunity for improvement (OFI)

Whichever method you select, the top-level categories indicate control weaknesses that require timely action from senior management to reduce or eliminate the risk of substantial and/or imminent negative effects on the achievement of objectives. In a nutshell, they indicate matters of urgency.

The lower categories are used for control weaknesses that may impact the achievement of the auditee's process or departmental objectives, without significant impact on the corporate objectives.

Very low-level weaknesses are generally those that affect the efficiency of the auditee's outputs. I often take these outside the context of the audit findings and report them in an 'audit memorandum'—a list of low-level findings sent separately to management in advance of the formal conclusion of the audit.

There needs to be some means of 'flagging' to management those situations where control weaknesses that have been identified and reported previously (whether through a previous independent audit or departmental monitoring activity) have not been properly and effectively reacted to by management and formally closed out. One way of handling such a situation is to prioritize the underlying audit finding at one level higher than the original rating. However, the delay may have resulted in the situation worsening further, in which case the control weakness may be prioritized at an even higher level.

Reduce the Number of Discrete Findings

One of the reasons many internal audit reports don't have a 'wow factor' is because often they include too many low-level findings. A mass of audit findings, while quite correctly identifying things that have happened which should not have happened, will probably not be attractive to most senior managers.

TIP

Senior managers do not expect to look at a plethora of findings and detail. They wish to be told how they can help the business. They expect lower-level managers to take care of all that detail! Remember who you are writing for!

The challenge for the audit team during the concluding stage is to interpret what the mass of information acquired during fieldwork says about the state of risk management throughout the auditee's area of responsibility. Their interpretation then needs to be expressed in high-level terms.

TIP

Whatever the number of discrete audit findings (symptoms) the audit team has produced as a result of the detailed audit fieldwork, the lead auditor needs to set a stretch target for the team to develop a maximum of, say, twenty main themes and a maximum of, say, five main messages for management. Figure 9.5 shows this process of clustering findings.

To achieve this step change in squeezing value from any audit—to increase the chances that senior management will get involved in putting right what needs to be put right—the audit team must use a structured approach towards grouping and clustering

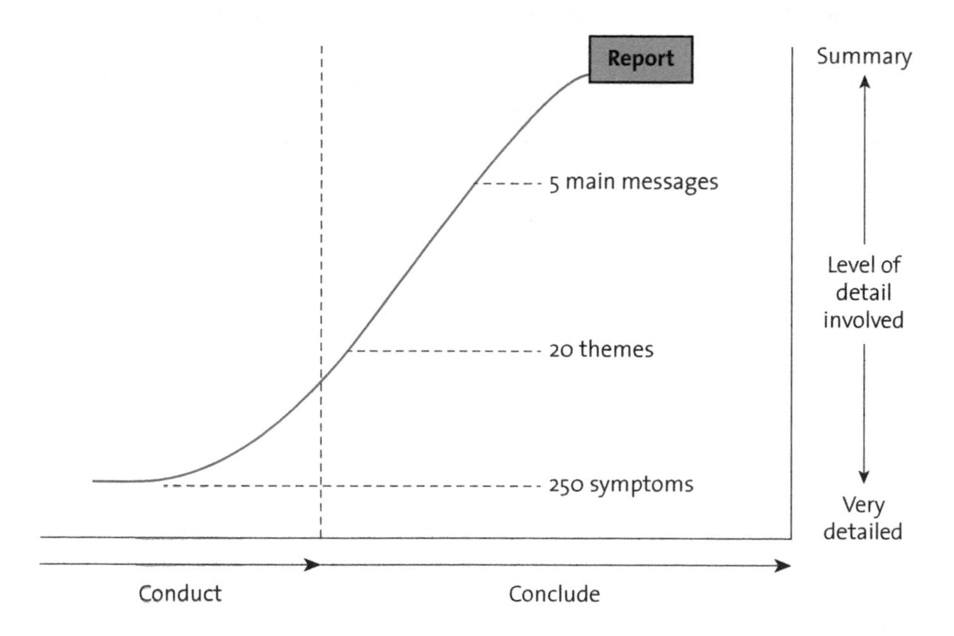

FIGURE 9.5 Consolidate the number of findings for senior management.

the lower-level audit findings. This will demonstrate the audit team's understanding of how the basic audit observations and findings can be built up into meaningful, substantial findings which are of interest to management because they relate to their efforts to achieve or even exceed their business objectives.

Sharing the Audit Findings

The success of this audit methodology depends on the ability of each auditor to share relevant information they have obtained during their part of the fieldwork. To do this, they need a means of recording key statements of fact onto a simple simulation of the auditee's BCF. To make the information coherent, they need to allocate the facts and audit evidence to each element of the BCF and to say whether the fact indicates strength or weakness.

In this process, the BCF in physical terms can be either on wall charts or a database in a computer. However, what is important is that the whole audit team can easily see at a glance every fact allocated to each BCF element and how these facts populate the overall BCF. Figure 9.6 shows how facts learned can be transferred from interview notes into the elements of the BCF.

For larger audit teams of more than, say, two or three members, it is more difficult to share A4-sized working papers. In Chapter 5, I recommended using A1-sized flip chart sheets which can be wall-mounted, enabling everyone in the audit team to see them at once and making discussions concerning 'the facts' easier. Figures 9.7 and 9.8 show the use of this approach during an audit in Angola, powerfully illustrating the process for every member of the audit team to see and refer to. A set of such wall charts for use on your own

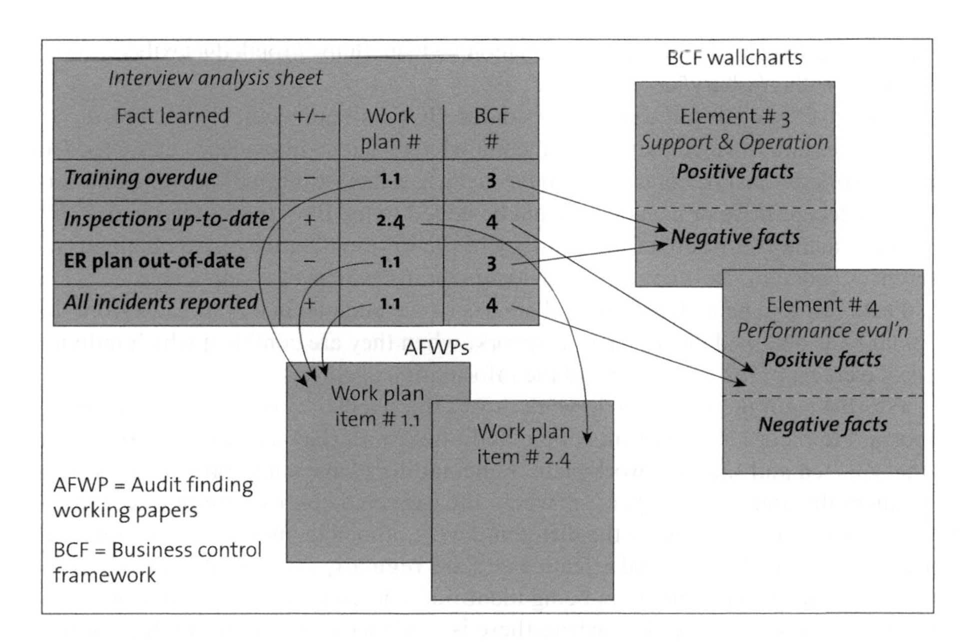

FIGURE 9.6 Allocating facts from each interview to BCF wall charts.

FIGURE 9.7 Wall charts help an audit team to share their information.

audits can be downloaded from the companion website (https://routledgetextbooks.com/textbooks/_author/asbury/).

These fact-loaded wall chart images of the BCF will quite quickly become a veritable treasure trove containing all the knowledge and audit evidence extracted from the fieldwork. It must be contributed to by every member of the audit team. New facts and evidence, positive or negative, should be added to the BCF from individual AWFPs as soon as auditors are sure of the accuracy of the information they have obtained, with references added to interview and hard documentary evidence, as applicable.

Facts can also be added to the wall charts as the audit team extrapolates from the original facts obtained during their fieldwork, when they are confident which individual control element of the BCF to record the information against.

For this audit methodology to work most effectively, it is of course critically important to record all the results of the audit fieldwork (facts) based on full coverage of the selected and sampled work plan. Some auditors have a natural reaction to skate over, mentally and practically, areas where the expected controls are in place; their purpose is well understood by the different levels of management, they are applied correctly, their application and effectiveness are regularly confirmed, and there is evidence of incorrect application being identified and corrected. In other words, auditors tend not to notice risks where there is evidence of strength. But they must learn to do this—their audit approach should be balanced and aimed at providing *both* alerts to weakness *and* assurance of strength!

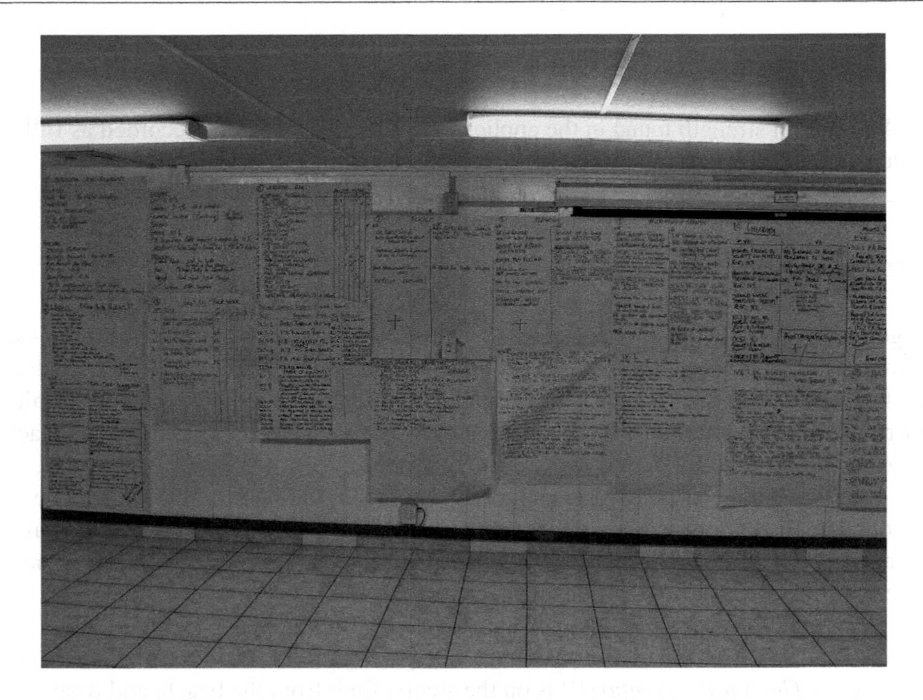

FIGURE 9.8 Records of the work done on the client's premises can be presented highly visually on audit wall charts. This photograph shows the walls of an audit team room for an HSE audit conducted in Angola in December 2012.

My thanks to our client and to my audit team for their permission to use these images.

> **TIP**
>
> Unless all the areas of strength found in the BCF are recorded as fully as any areas of weakness, there will not be an accurate weighting of strong controls to influence the audit team's deliberations when they come to determine the overall audit opinion.

> **TIP**
>
> It is important to be able to trace every fact (positive and negative) recorded on the AFWP and wall charts back to its original source, to the auditor who carried out the fieldwork, and, if relevant, to the work plan item to which the fact refers.

> **TIP**
>
> The aggregated information obtained during fieldwork, recorded on AFWPs, and allocated to the BCF elements on wall charts will be the factual and logical basis upon which the audit opinion is founded.

A-FACTOR 70

All areas of strength found in the application of the BCF must be recorded as fully as areas of weakness, so that there is accurate weighting of each.

Keep in Contact with the Auditee

Regular contact with your auditee at this stage of the audit can have many benefits—not least discussing in a low-key manner any apparent control weaknesses discovered during the fieldwork, and either having your misunderstandings clarified with the opportunity for re-auditing specific controls, or reaching agreement with the auditee about the facts, the extent of the weakness, and perhaps the root cause.

If auditees understand why an auditor is concerned, they will generally provide the opportunity and means for further testing, checking, and quantifying of exposures to sampled risk scenarios—particularly if the control weakness is likely to affect the successful performance of their department.

Such a consultative approach, particularly if practised by the lead auditor from the start of the audit, will often result in the auditee giving assistance to the audit team at a stage when *The Audit Adventure*™ is on the steep climb from the beach, and time is (as ever) critical.

Formulating effective remedial action is another task the auditee will be more eager to assist with once they have understood and participated in a joint analysis of a control weakness. Although you will have ideas as to what needs to be done to rectify the situation, it is better to encourage the auditee to suggest what can be done and what improvement they expect to achieve (see A-Factor 37).

TIP

The degree of interest in and commitment to making changes to the way the control framework operates, and the amount of cooperation in ensuring that the changes ultimately have the required effect, will normally be in direct proportion to the openness, constructiveness, and professionalism with which the lead auditor handles the auditee at this time in the audit process.

Nemawashi

The constructive and open approach described in the previous *Tip* is *nemawashi*. *Nemawashi* is a Japanese word which translates literally as 'going around the roots'. It indicates an informal approach to preparation for change or new projects within an organization.

Its original meaning referred to making a tree ready for transplantation; in its business meaning, it refers to close working and discussion with those people involved in and affected by change.

Organizations that employ the *nemawashi* approach see it as an important aspect of instituting change, aside from the formal arrangements to be made, and a way of achieving change through consensus. The practical application of this type of approach to audit is discussed later in this chapter.

Step Back and Analyze the Information Obtained

Every fact gathered—positive or negative—requires careful organization and recording. Every fact obtained should contribute to the moment when the audit team must make their final evaluation of each control weakness, each element of the auditee's BCF, and their overall opinion on the BCF.

TIP

The challenge for the audit team during the latter part of the fieldwork stage and the initial part of the concluding stage is to sort and use effectively the mass of discrete facts obtained during the fieldwork.

As AFWPs are completed, any underlying weakness ratings (major, minor, observation) should be recorded both on the AFWPs and against the appropriate business control element on the BCF wall charts.

Even though the analysis of the information gathered during the fieldwork may not have been fully completed, and the conclusions drawn about the controls may not yet have been entered on the BCF wall charts, the audit team should be encouraged to start paying more attention to the information as a whole. This is the point at which a methodology that uses wall charts, rather than a database hidden in a computer, comes into its own.

A-FACTOR 71

The audit team must be able to *see* the balance of the emerging facts if they are to apply their 'group brain' to what those facts mean for the control of significant risks. Large wall charts are a fantastic idea because they lend great visibility.

From a relatively cursory analysis of all the information (sorted into positive and negative impacts on each control element), each member of the team should be able to see the extent to which there is correlation between their audit findings (which have come from their focus on individual work plan items) and the aggregate information emerging on the wall charts. Looking for this correlation early in the *Conclude* stage will encourage the audit team that the dual approach to evaluating the overall BCF is working.

> **TIP**
>
> This holistic view may also help individual auditors to understand better the dynamics of the control framework within the areas they have been looking at, especially when working on the analysis of root causes.

Control Assessment Matrix

At this point in their process, the lead auditor can begin using a summary evaluation tool called a control assessment matrix (CAM). The CAM, rather than having a lot of detailed information as recorded on the wall charts, records only the resultant failings or weaknesses with their weakness levels and the root cause for each audit finding. Table 9.1 shows an example of a CAM.

As each AFWP is completed by a member of the audit team, and control weaknesses are identified and grouped into generic findings of control weakness, the information can be recorded on the CAM as levels of strengths and weaknesses, identifying the failed element and its root cause. I'll show you an example of a completed CAM in Figure 9.12.

> **TIP**
>
> The lead auditor needs to be aware of the source of the information to avoid recording both the results of discrete control weaknesses (or strengths) and the aggregate results of a group of the same control weaknesses—no 'double-dinging'.

Develop the Main Issues

The audit team, and especially the lead auditor, must now use the momentum created by *The Audit Adventure*™ to confidently develop the main issues their audit has revealed.

The most significant criterion for these main issues will be that they directly and significantly affect the achievement of the organization's objectives. The most important issues are those which relate to activities within the auditee's organization that are already having or may be predicted to have, with some degree of certainty, a significant negative impact on the achievement of those objectives, in both quantitative ('what') and qualitative ('how') terms.

This stage of the audit process is a great opportunity for the lead auditor to demonstrate their sound judgement, as well as their technical ability.

FINDINGS AND RECOMMENDATIONS

Prepare Summary Audit Findings

Continuing Mrs Beeton's recipe for jugged hare, now that you have your main ingredients, you can start to prepare them for cooking and serving. In this instance, this involves writing them up in a style which will lead to their eager consumption by senior managers.

TABLE 9.1 Control Assessment Matrix (CAM)

	OVERALL MANAGEMENT SYSTEM RATINGS				
REF.#WORK PLAN ITEM	*LEADERSHIP*	*PLANNING*	*SUPPORT AND OPERATION*	*PERFORMANCE EVALUATION*	*IMPROVEMENT*
1 Design in project extension					
2 Commissioning new hydrocracker					
3 Operation of reactor #1					
4 Operation of reactor #2					
5 Waste disposal					
6 Asset integrity					
7 Confined space entry					
8 Desert driving					
9 CHESM (contractor management)					
10 Vapour exposure (BTX)					
11 Spillage in tanked storage					
12 Site transport					
13 Excavation					
14 Asbestos					
15 Legionella					
Overall audit opinion =					

TIP

It is important to give each main issue a 'catchy' title to capture management's imagination. These, I think, are *not* catchy:

- Housekeeping
- General Health and Safety
- Deficiencies of the Quality Policy, page 47, paragraph 2, section 3(c)(iii)
- PPE (Personal Protective Equipment)—Lack of fully documented risk assessments in accordance with the Health and Safety (Personal Protective Equipment) Regulations 1992, as amended, and the Management of Health and Safety at Work Regulations 1999, as amended

I think the following examples *are* 'catchy', since they grab the reader's attention:

- Explosion at tank farm
- Maintenance team asbestos exposure
- Product contamination—customers poisoned
- Spillage risk to Mississippi River

Four to seven carefully selected words can convey a powerful message to senior managers. They compel them to read on. Keep the title in draft at this stage, to be revised (or confirmed) later. Without trying to be too funny or too difficult to believe, I suggest you think how a tabloid, or 'red-top' newspaper might report on the potential risk area on its front page. 'Housekeeping' would not sell newspapers. 'Massive pollution' would—and has, as editors have learned when covering the *Macondo Deepwater Horizon* blowout and resulting oil spillage in the Gulf of Mexico (Figure 9.9).

FIGURE 9.9 The spillage of crude oil into the Gulf of Mexico from semi-submersible drilling rig *Deepwater Horizon* made world news from 2010 to 2017.

The questions that an audit's findings should answer for the reader:

- What are the risks emerging from the *context* (the organization's external and internal environment, and needs of interested parties aka stakeholders) that may require attention, now or in the future?
- Which business activity is being discussed?
- What are the key inherent risks in that activity, and how might they impact our objectives?
- What are the key controls expected in an effective control framework for this risk?
- Which controls (in which control elements) are present, and what is the extent of their effectiveness?
- Which controls (in which control elements) are failing, and what is the impact of this failure?
- What are the root causes of these failures?
- What needs to be done to improve the situation?
- How urgent is the corrective action/s?

Several drafts of these summary audit findings are likely to be needed, since upon them hangs much of the audit's potential for success (or otherwise).

Initial Preparation of Recommendations

Historically—and sometimes this still happens—some auditors have been castigated by auditees because they *only bring problems, not solutions*. That is not really surprising, since it is the auditee who needs to decide exactly what, when, and how they should react to audit findings. However, if the *nemawashi* ('no surprises') communication between the lead auditor and the auditee has been successful, and the auditee fully understands and accepts the control issues the audit has found, then they are more likely to take the initiative and propose what they see as the most appropriate remedial actions—even if the action necessary seems 'impossible'.

The lead auditor needs to be prepared to assist the auditee to arrive at a workable solution that is likely to be effective. So, in exactly the same way that the audit finding must be completely defensible to challenge by the auditee, the lead auditor should be able to assess the robustness of the auditee's proposed solutions based on reliable facts and information. For example, what level of competence and seniority of personnel would be required, and are such individuals available? How much time will the proposed work take, and is it acceptable for the auditee to wait that long? How much will it cost? All these facts need to have been researched as thoroughly as possible in the final part of the fieldwork.

In the best scenario, the auditee will discuss the underlying issues and provision of the necessary resources with their line manager. However, with more serious audit findings, it could take some time for management to accept the significance of a control weakness and work out what to do. Often there are knock-on effects which need to be carefully examined and external parties who need to be consulted.

CASE STUDY

TWO YEARS ON, ONE MONTH OFF

I was a member of an audit team for a drilling company in the Middle East. During our fieldwork, we identified what was, in our opinion, the most unacceptable labour practice we had seen anywhere in the world. It was a contract offered to poorly educated expatriate caterers, cleaners, and gardeners involving two years (730 days) at work followed by one month off.

A manager from the organization described the arrangements: 'They pretend to work, and we pretend to pay them.'

We reported this as an 'alert' finding in one of our business control audits, at the 'no surprises' stage as well as in our final report. The finding was immediately rejected by management as 'impossible' to improve.

We were very pleased when, after one year, the unacceptable contract was withdrawn and replaced with a fairer one.

LEARNING

Go with what feels right. It doesn't need to be unlawful or prohibited by policy or procedure to make it wrong. Auditors should always try to do the right thing.

Drafting Part 2 of the Audit Report

Chapter 10 provides my detailed guidance for writing great audit reports that capture facts in a coherent way and compel improvement. This section will help you to organize those facts ready for reporting.

An audit report is generally divided into two parts, plus appendices. Part 1 is the management (or executive) summary, and Part 2 is the full and detailed report. Part 2 is prepared first, followed by Part 1; as most managers and other professional people will know, it is usual to prepare the management summary last of all, once what needs to be summarized has become clear.

Obviously, what is required for Part 2 of the report is a readable, logically thought-through report that presents the audit team's conclusions and the basis for these.

The report deals with major weaknesses (or clusters of minor weaknesses that together constitute major weakness) in turn, recording the detailed findings in much the same way as the structured contents of the AFWPs. By using previously prepared information—mainly that written onto the AFWPs as the fieldwork progressed—this eliminates the need for a major rewrite to create Part 2 of the audit report.

The first section for each work plan item comprises a full story for each issue (or group of issues), describing the expected controls, the actual controls found, the residual risk resulting from any difference between the two, the root cause of the problem, and the impact of the residual risk on the process and/or organization objectives. The second section presents advice or recommended actions. A typical layout for this is shown in Figure 9.10.

Part 2 – Findings and Recommendations	
Assurance or Alert	Advice (Recommendation)
Work plan item and description of the risk to objectives Expected control (PDCA) Actual control system identified Results of verifications and tests Analysis of residual risk to objectives	Recommendation: • SMART • Action verbs • Compelling improvement: removing cause and reducing the residual risk

FIGURE 9.10 Typical structure of Part 2 of the audit report.

In their entirety, the contents of Part 2 of the audit report must drive the reader towards the same understanding and conclusions that led the audit team to their final audit opinion, regarding the adequacy or inadequacy of the auditee's BCF as it is currently being operated.

Evaluate Each BCF Element

A key part of *The Audit Adventure*™ methodology is that the effectiveness of the reference framework (BCF/MSS) in controlling the risks within the auditee's area of responsibility is primarily assessed by *Reviewing* and *Verifying* how well key controls within the reference framework are applied to the essential tasks occurring in the audit team's sample of significant *inherent* risk activities.

The results (facts) from auditing each of these activities are recorded on the AFWPs. Simultaneously, the control strengths and weaknesses are used to populate the control elements within the reference framework. Figure 9.11 shows an example of how this can be done in practice.

Before starting to write report Part 1, a summary statement (usually one or two paragraphs) of how each control element in the auditee's BCF is contributing to (or detracting from) its overall effectiveness should be prepared. This summary for each element should make the reader clear as to why the audit team has assessed the element as being either positive or negative in its overall contribution. Each element summary should lead inescapably to the same conclusion as the audit team arrived at, and be based on the same irrefutable and logical analysis of the facts.

The overall positive or negative assessment (or in some cases, when the facts and evidence are finely balanced, a neutral assessment) for each control element can then be added to the bottom row of the control assessment matrix *Assessment of Each BCF Element*. The audit team should compare and cross-reference these control element assessments against the weak or failing controls and root causes (and the areas of strong control) for each of the work plan items and subsequently developed summary items. Figure 9.12 shows an example of how this can be done in practice.

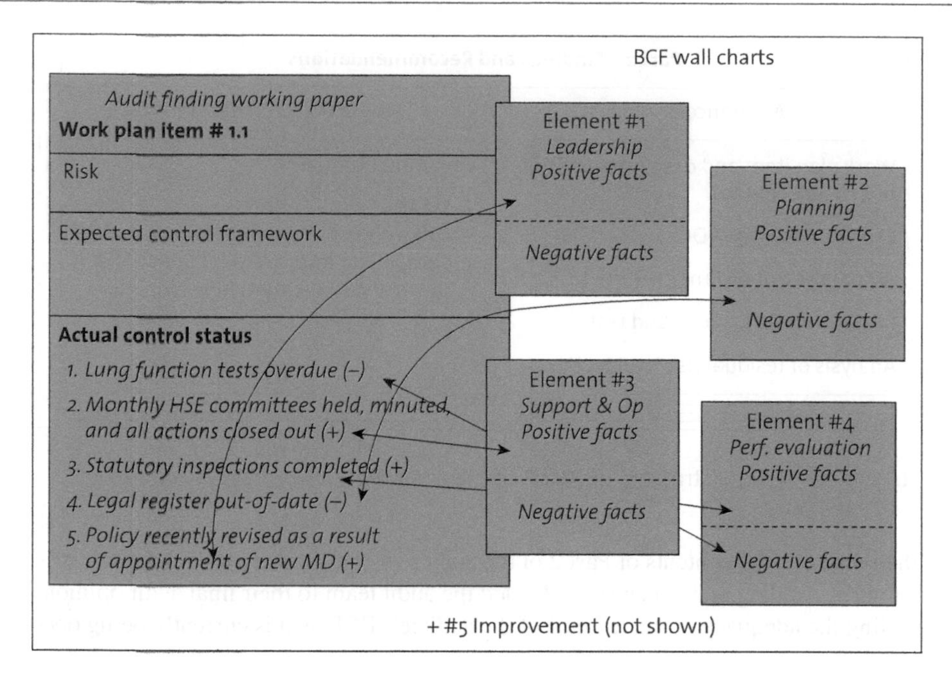

FIGURE 9.11 Adding the facts from the AFWP for work plan item 1.1 to the BCF wall charts.

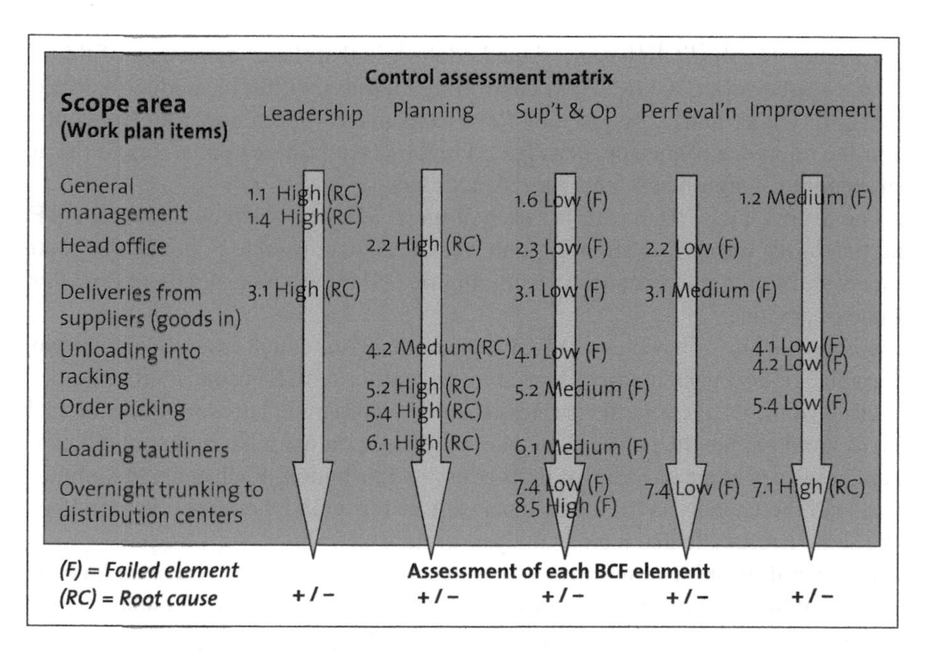

FIGURE 9.12 Cross-referencing between the results of each work plan item and the cumulative performance of each BCF element. This helps the audit team to build the 'big picture' and the overall audit opinion.

Only once this overall cross-referencing exercise has been completed can the audit team confidently say that they have arrived at their overall audit opinion on an objective basis.

OVERALL AUDIT OPINION

How to Determine Your Overall Audit Opinion

It is generally accepted practice for an organization to have an even number of graduations of overall audit *opinion* available to audit teams for reporting internal control audits. There are often four graduations, and these could be

- good,
- fair,
- poor, or
- unacceptable.

I will refer to this four-point scale here (and see Figure 9.13), but I recognize that your own organization may have its own. I like the idea that an overall opinion category, when allocated properly and consistently, can shine a bright light onto the overall status of the BCF/MSS for controlling its significant risks.

What's true is that an 'even' number of graduations reduces the opportunity for auditors to sit on the fence—'I've seen better, I've seen worse, I'll put it in the middle category'. Many large organizations have devised their own graduations, while some have dispensed with them completely. There is no definitive model.

TIP

Hours of erudite debate have been spent on the distinction between a 'review' and an 'audit'. My distinction is that an audit will result in a level of assurance being given to the auditee and the internal audit committee in the form of an audit opinion resulting from the audit work.

A review, by contrast, is primarily focused on identification of areas of strength and weakness reported individually. It is not required to aggregate these findings in an audit opinion.

It's a pretty fine distinction.

A-FACTOR 72

Overall, audit opinions should have an *even* number of gradations. The overall audit opinion should reflect the overall level of concern, based on the audit work, about the achievement of objectives.

Each gradation of audit opinion should be clearly defined in terms of an objective assessment of the level of concern that should be felt by the audit's sponsor, as shown in Figure 9.13. This will trigger the required degree of follow-up from senior management.

Caution is required in using the absolute number of findings to directly determine the audit opinion. Such a quantitative approach can provide a supporting cross-check, but, as I have said, this approach is based upon grouping weaknesses, rather than counting them! It should be the audit team's judgement of the concern arising from the weaknesses overall that provides the audit opinion.

Arriving at an audit opinion is not an art, nor can it be called a science. As professional auditors using an effective methodology, my teams strive to arrive at an opinion together, based on the objective evidence (of both weakness and strength) we have gathered. However, in the final analysis it is the lead auditor's responsibility to make the decision, and they will use their judgement to weigh all the facts available to them.

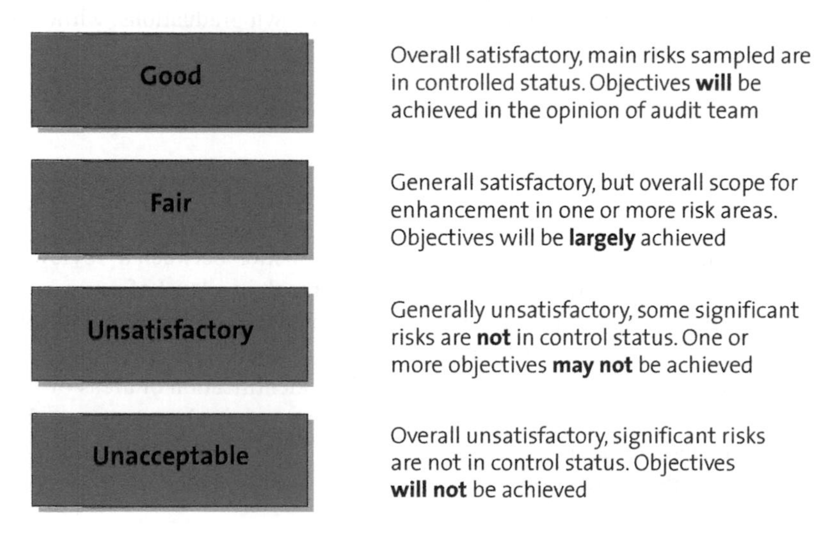

What the overall audit opinion means
Good
Overall satisfactory, main risks sampled are in controlled status. Objectives **will** be achieved in the opinion of audit team
Fair
Generall satisfactory, but overall scope for enhancement in one or more risk areas. Objectives will be **largely** achieved
Unsatisfactory
Generally unsatisfactory, some significant risks are **not** in control status. One or more objectives **may not** be achieved
Unacceptable
Overall unsatisfactory, significant risks are not in control status. Objectives **will not** be achieved

FIGURE 9.13 The audit opinion reflects the audit team's independent assessment of the organization's ability to meet its objectives, given the current level of control of their work plan sample of significant inherent risks.

A-FACTOR 73

The lead auditor is ultimately responsible for the conduct of the audit and for finalizing the overall audit opinion.

Final *nemawashi* with the Auditee

Once the audit team has arrived at this point in *Concluding* the audit, it is important that there is a final meeting with the auditee (and possibly the sponsor of the audit), to provide a final opportunity for the auditee to raise questions about any aspect of the audit, including the way it was conducted and its outcome.

If the lead auditor must convey to the auditee and the sponsor that the audit opinion is Unsatisfactory or Unacceptable, they should be prepared for a reaction. As soon as the lead auditor realizes that there is a likelihood of this outcome, they can begin preparing by briefing the auditee at early *nemawashi* meetings, stressing the depth and breadth of the control failures the audit team has found/is finding.

The lead auditor should also keep in mind what they have heard in their initial meetings with the auditee and (if applicable) the sponsor regarding their self-assessment of the effectiveness of the BCF. The wider the gap between the auditee's self-assessment and the audit opinion, the tougher the final clearance meeting should be expected to be. And, of course, however much logic there is to the conclusion, it is human nature that ambitious managers will not take kindly to criticism and will wish to avoid being judged (in their own minds) to have failed. This emotional element will be even more prominent if the audit results are part of the personal performance appraisal on which the auditee's and/or their sponsor's annual bonus is based!

Even at this late stage, the lead auditor must listen carefully to the auditee's and sponsor's contributions to the discussion. Clearly, the most important outcome from this meeting is gaining support for everything that has happened during the audit and most particularly for the results. If the lead auditor does not take the opportunity at this stage to react positively to the auditee and the sponsor, he or she may be jeopardizing a 90 per cent win at the cost of losing 100 per cent of the audit.

A-FACTOR 74

Nemawashi—remember to 'go around the roots'. Such discussions are an important informal element of major change. Successful *nemawashi* enables change to be the result of consensus among all parties.

A-FACTOR 75

The audit opinion is not negotiable once the audit team has arrived at its decision.

Prepare Part 1 of the Audit Report

The final report delivers the audit results to senior management (and sometimes, to other stakeholders). It spells out the audit opinion and the major findings. It provides an assessment of each element of the BCF. It might also describe remedial or corrective actions, particularly if these have been discussed/agreed with the auditee.

It is critical to present an audit report which senior management will want to read and believe (because of the significance of the overall level of assurance in the audited area). The painstaking work of getting hold of and documenting the detailed facts has been completed, and now the auditor's task is to explain to the auditee what it all means and, if necessary, discuss what improvement action/s is needed. An imaginative approach to this from the lead auditor will increase the likelihood of the audit's overall success.

TIP

Remember that Part 2 of the audit report provides the platform for the conclusions of Part 1, thus nothing should be reported in Part 1 unless it is linked to a weakness reported in Part 2.

In addition to declaring and formalizing the audit result, Part 1 allows the lead auditor to include an introduction to the audit. The first few paragraphs of the executive summary will describe a *context* (business environment) which is familiar to senior management. The lead auditor here needs to achieve two things: firstly, to make the point that they and their team understand the challenges faced by the leaders of this organization, and the contribution of the auditee's department to successfully achieving the corporate objectives; and, secondly, to set a scene which may be significantly disrupted or challenged when the main audit findings are read.

Appendices to the audit report should enhance the reader's understanding of the audit's coverage, process or result, and nothing else. Avoid unnecessary padding.

Figure 9.14 shows a typical structure for an audit report.

CLOSING PRESENTATION TO MANAGEMENT

Detailed guidance on how to make management presentations is beyond the scope of this book. There are numerous books and providers of presentation skills training courses, and you might consider one of these for your CPD/personal development plan if required.

Over the years, however, I have learned some techniques for how to present an audit closing presentation (as well as how not to do it!). Some of these techniques are covered in the following ten tips, four A-Factors, and my chapter-concluding case study, which describes how one well-prepared audit team dealt with a difficult situation.

Part 1 – Management Summary
Introduction
Business environment
Audit opinion
Major findings and actions

Part 2 – Findings and Recommendations	
Assurance or Alert	Advice (Recommendation)
Work plan item and description of the risk to objectives Expected control (PDCA) Actual control system identified Results of verifications and tests Analysis of residual risk to objectives	Recommendation: • SMART • Action verbs • Compelling improvement: removing cause and reducing the residual risk

Appendicies
Terms of reference
Essential materials
Action plan

FIGURE 9.14 A typical structure for a two-part-plus-appendices audit report.

> **TIP**
>
> Submit a draft report in advance, and you'll be able to refer to it. 'Please turn to the table on page 2' is a much better way of sharing data than providing lots of facts and figures in a slide presentation.

> **TIP**
>
> Prepare your own material. You are always clearer about what is meant by any model, illustration, or acronym when you are its author.

TIP

It is important that everyone knows who is doing what. A dry run before the closing meeting is a good idea, and it can serve to calm nerves. Arrive early. Check that everything works and that reference documents are to hand in sufficient numbers.

TIP

Allow the lead auditor to lead the closing presentation, as they have the 'big picture' in mind. Their role is to open and close the presentation; they will bring in individual auditors to focus on the findings related to specific risk areas they *Reviewed* and *Verified*.

TIP

If there are lots of detailed facts or figures to be provided, make notes on a card to keep in your pocket. Getting basic facts wrong is careless. The audience may wonder whether it can rely on your findings, or indeed anything else you say. Simply take out your card, comment on the facts, and slip it away in your pocket again. No one will notice. And if they do, you still look very professional and well prepared.

TIP

Present the agreed findings first, progressing to any more contentious matters later in the presentation (and see my case study at the end of this chapter).

TIP

Rehearse objections. Think as the auditee, and their team, might think. What could you have in mind or to hand if/when challenged?

TIP

Stand still when presenting, centre stage, and deliver at a measured pace. Use silence for impact, and always pause after major points.

TIP

Agree the duration of the meeting in advance if you can, so you know when to stop. It is better to leave the audience wanting more than to overstay your welcome.

TIP

The closing meeting is your final chance to make a positive impression on the auditee and their team. Be sure to take it!

A-FACTOR 76

The objective of the closing meeting is to help management arrive at the right decision regarding the audit findings. Initiating an appropriate response is essential, whether this is consolidation of existing strengths, a survival plan, or keeping the board out of jail. The strategy should be to focus management's attention on a few key messages, and the tactics should be to support these messages with factual evidence to the extent demanded by any person in the room.

A-FACTOR 77

Know who will attend the meeting and, as far as possible, ensure that the attendees are the appropriate audience and have sufficient time available. You or the auditee should send each attendee an agenda and a short description of how the audit team wishes to conduct the presentation. Requests from management for pre-meetings or discussions afterwards should be met.

A-FACTOR 78

Remember the audit is not over. The best outcome of the presentation is full support for the findings and commitment to making the appropriate response. If, however, management is not able either to accept the findings or commit to a response, then maintain a dialogue with the audited organization to give management the opportunity to express why this is not possible. Your aim is still to optimize the effect of the work done by the audit team.

CASE STUDY

HE LOVED THE GOOD NEWS

This interesting case study provides some thoughts on a powerful technique for delivering what might be perceived by some as 'bad news' audit findings and opinion. An audit team had been asked to conduct a series of health and safety audits at a large European logistics group. The organization had twenty-one locations, and a sample of seven sites was selected for the audit scope covering the three main hubs and four smaller depots.

The audit team conducted their work over several weeks. It became apparent that there was inconsistent application of group standards across the sampled sites—some were performing well, some rather less so. The auditors had taken a lot of photographs of both good and bad features of the observed performance during the audit. Some of the 'bad' issues were very bad indeed, including a failure to separate forklift and pedestrian routes, serious damage to high-bay racking, and unacceptably slow (or nil) attention to the findings of statutory lifting equipment thorough examinations.

The date for the closing presentation meeting had been set, and the audit team was advised that the logistics director would be present. The audit team had already met this director at the opening meeting, where they had noticed his tendency to praise his own work.

The team decided to do a 'best to worst' style of closing presentation. In the simplest terms, this meant showing the photographs of the better-performing sites first and suggesting that these were the practices to be shared with other distribution sites. The director was visibly inflated by these good practices and praised the work of the audit team several times for their noticing of these matters.

By the time the team had presented on the four better-performing sites, they had the director's confidence. The photographic style of reporting of the audit results, best to worst, provided a lot of indisputable facts earning early buy-in, which meant that it was harder to challenge the findings as they became more significant. Accordingly, the 'bad news' was accepted, and the improvement recommendations were agreed.

The story has a happy ending. After a 'Silver' award in 2011, the organization was awarded RoSPA 'Gold' health and safety award in the following five consecutive years from 2012, and a RoSPA Gold Medal in 2017. The audit team was delighted to have played a small but foundational part in promoting improvement in this household name organization.

LEARNING

Know (or learn about) your audience if you can. Think carefully about how you will present your findings. Feel rewarded when the findings are accepted. You might even have saved lives.

A-FACTOR 79

If, despite following this guidance, and ideally having broad support from the auditee, you do not feel that further discussion will result in management's agreement and commitment, your only course of action is to refer the matter to the head of internal audit with a view to their reporting this outcome to the chair of the internal audit committee.

Whatever you write or say, the audit conclusion should be high-level and future-focused, as shown in *The Audit Adventure*™ conclusion in Figure 9.15.

People support what they have assisted to create.

Dr Stephen Asbury

FIGURE 9.15 The conclusion is always delivered at a high level. On a good day, you can see France from here . . .

Test your knowledge with the next chapter assessment test (see Appendix 4).

Write the Audit Report and Follow Up 10

Information about the future is much more valuable and interesting than information about the past.

—Dr Stephen Asbury

I'm sorry for writing you a four-page letter. I did not have time to write you a one-page letter.

—Dr Stephen Asbury

INTRODUCTION

For a summary of this chapter, you can listen to the author's Microlearning™ presentation, which you'll find on the book's Companion Website. Then, delve deeper by reading on for further details on the Companion Website. (See Appendix 4.)

Chapter 8 concentrated on the critically important aspects of the lead auditor and auditors' abilities to connect on a professional, emotional, and psychological level with the other members of their audit team, as well as with the auditee and their staff. Chapter 9 explained how to *Conclude* the audit. This final chapter (and, by the way, thank you for following *The Audit Adventure*™ with me as far as you have!) focuses on the important ability—particularly for the lead auditor—to capture the information gathered in the *Conclusion* in a readable and accurate written form, and on how to finish your audit assignment.

If the opening meeting is the opportunity for the audit team to make a good first impression, then the report is the audit team's final opportunity to make a lasting impression. A badly written, poorly composed and punctuated report can nullify the excellent work done, while a well-written, clearly expressed report can provide the opportunities for continual improvement sought. Figure 10.1 shows the intended outcome. You'll notice that the auditee is as happy as you (the lead auditor and audit team) are!

As shown in Figure 5.5 on page 175, the *Report* stage occupies the final 20 per cent of the total audit time. That's the final two days of a two-week audit.

DOI: 10.1201/9781003364849-11

FIGURE 10.1 How a win-win audit conclusion should feel.

As you'll now completely understand, good—no, *excellent*—interpersonal relationships between the audit team (particularly the lead auditor) and the auditee are very important. This is especially true at this point in the process, when you seek to turn the time and effort you have put into the audit—its *preparation*, *conduct*, and *conclusion*—into a written report that is generally accepted and agreed upon, is easy to read, and is welcomed by the auditee's organization as a useful contribution to the future success of their organization.

One of the main reasons for writing (and three times rewriting) this book is my passionate belief that auditing is really all about facilitating and encouraging improvement. Such improvements, as I see them, are only possible if our messages—carefully worded, factually correct, and (when possible) agreed with our auditee through our *nemawashi* ('no surprises') process while the audit team is on-site—are clearly conveyed and enhanced through the written report.

A-FACTOR 80

A well-written, factually correct, timely presented written audit report is essential. It encourages, even compels, improvement when you get it right!

In this final chapter, you'll learn

- how to plan, draft, and present well-written audit reports that are easy to read and genuinely welcomed by the auditee;
- about the use of powerful words that compel action;
- how to deal with abbreviations and references;
- why it is vital to read the report after writing it and before presenting it;
- how to formally submit audit reports and recommendations; and
- about the responsibilities for audit follow-up.

HOW TO WRITE EXCELLENT AUDIT REPORTS

Let's start with a question. In your experience, when you're browsing in a bookshop, which parts of any book, prior to a purchase, do you tend to look at? I'm guessing that your answer included the cover and the summary text on the back page. Maybe the author profile? Or perhaps the first page or two? That would be my answer too.

If I now asked you which parts of a management report are likely to be read, I suspect that your answer might be the management summary at the front and the recommendations at the back. Again, that would also be my answer.

This is true in both cases, because, as we know, if these 'headlines' do not quickly engage our interest, we put the book or report back and find something else to do instead.

Well, the same is likely true for the auditee, their boss, and the members of the internal audit committee. It is very important to 'think of the reader'. All these people likely have busy days and diaries and lots of other things they could be doing. If your report is not a well-written, factually correct masterpiece from the start, it is unlikely to be read with the degree of interest you might like. To persuade a manager to do something differently, you need to understand them and to tailor your writing accordingly. Personalizing your writing to your readers satisfies their need to feel that their inputs have been heard, engages their self-interest, and probably makes your message more appealing to them.

Throughout this chapter, I'll offer tips on how to write reports that are easy to read and more likely to be welcomed. Some items that need to be 'right' are probably common sense (but lots of people still get these wrong!). If you get any of these wrong, the reader is distracted before they get started:

- The spelling of the auditee's and other managers' first and family names
- The auditee's and others' job titles
- The title of the audit
- The dates of its conduct (please, get the year right!)
- The correct expression and spelling of company names (e.g., 'plc', 'SA', 'Sdn Bhd', 'Ltd')
- The names and designatory letters of members of the audit team

Making sure these initial facts are correct gives the reader (perhaps unconsciously) feelings of comfort and confidence. To present these incorrectly is to cast doubt in the reader's mind about the audit team's ability to conduct a good audit, let alone present a coherent and useful message.

> **TIP**
>
> Your report has one chance to make a good impression. Take it!

It is very important to establish your aim when writing for other people. I like Keyser's (2012) suggestion to use 'KFC' as an aide memoir to define your purpose or objective in writing for others:

- K—What you want your reader to **K**now from the information you will share
- F—What you want your reader to **F**eel having read it
- C—What you want your reader to **C**ommit to after reading your words

I believe that 'feeling' is key; make the reader feel confident, reassured, scared, greedy, guilty, embarrassed, flattered, or angry.

> Logic makes people think, but emotion makes them act.
>
> —Dr Stephen Asbury

Structure in an Audit Report

Other important objectives to consider at an early stage include

- a clear structure,
- a simple layout, and
- a concise summary of the audit conclusion(s).

The clearest, simplest structure is

- Part 1—Management (or Executive) summary;
- Part 2—Audit findings and recommendations; and
- Appendices.

This mirrors the structure and division into parts suggested in Figure 9.14 on page 295.

> **TIP**
>
> Consider creating and using a report template as the starting point for each of your reports—you'll then add the facts and commentaries.

Report Title

So let's start writing our report—give it a snappy title on the cover. In Chapter 9, I recommended a title of four to seven words, and I do so again here.

I want you to contrast

> HSEQ Audit Report 2023—O'Hare Airport, Chicago

with:

> A report into health, safety, environment, and quality management systems, aligned to ISO 9001:2015, ISO 14001:2015 and ISO 45001:2018, at O'Hare International Airport Terminals 1, 2, 3 and 5, Chicago, Illinois, United States of America (USA) conducted between October and December 2023

I suggest you choose the former every time.

> **TIP**
>
> Don't try to be too clever. Don't try to show 'we know everything' on the cover. Compel the reader to turn the page to find out (from my example) which airport terminals are covered by your report.

Report Cover

Experienced auditors seem able to recall the necessary text for the cover from a model carried in their minds.

I recommend you try to establish your own model text—it helps a lot when you are under time pressure to prepare the final report. Or include same in your report template (see my *Tip* earlier).

My model text is presented here; this has worked well to set the scene for hundreds of audits around the world:

- [Draft/Final] audit report of a [type of audit] at [name of organization and short-form address] on [date, including year]
- Auditee: [auditee's name and job title] and Sponsor: [name and job title]
- Lead auditor and audit team members: [include your own name and the names of the audit team members, all with their pre- and/or post-nominal designatory letters; add the name of your organization if you are a contractor]

> **TIP**
>
> Designatory letters (1): Present your credentials correctly.
>
> The overall order is civil honours, military honours, KC, degrees, diplomas, certificates, and, finally, membership of academic or professional bodies in the order of their importance.

All abbreviations of degrees and other distinctions, whether all upper case or a mixture of upper and lower case, should appear without internal periods/full stops (e.g., CFIOSH, not C.F.I.O.S.H.)

If you wish to review the conventions in greater detail, Debretts maintains a helpful guide at www.debretts.com/expertise/forms-of-address/letters-after-the-name.

TIP

Designatory letters (2): If you do not have any designatory letters, it may be a good time to set about formalizing your practical skills with qualifications and/or membership of a *relevant* technical body. Chapter 3 of his book summarizes several auditor registration bodies for your consideration. What qualifications and memberships do your peers have?

Table of Contents

This is an important section of the report, since it helps readers familiarize and orientate themselves in relation to the content. In my experience, this is one of the first pages your readers will look at.

I suggest that lead auditors build a library of table of contents templates—perhaps one for each BCF/MSS/reference framework against which they conduct audits. However you present your report, the table of contents should include entries for the following appendices:

- Terms of reference (always include a copy)
- The audit work plan (always include your work plan)
- Documents that are essential to reading and understanding the report (add these selectively)
- Description or summary of the audited reference framework (optional)
- You may also list of the contributors to the audit (noting that this should be the only place audit participants/interviewees are named. Consider doing this alphabetically to preserve their confidentiality in relation to specific facts and issues).

TIP

If you do not know how to use Microsoft Word (or similar word processor software) to create a table of contents for you, find out. Page numbering in the table of contents will update throughout the document with two clicks. Finding out how to do this has saved me about 30 minutes for each report I prepare.

TIP

Similarly, an essential skill is being able to copy text using 'Ctrl+C'; cut text using 'Ctrl+X'; and paste text into its new position in the same or another document using 'Ctrl+V'.

Liability Disclaimer Statements

It seems to me that most auditors these days include some form of liability disclaimer statement in their reports. I am not a lawyer, and therefore I suggest that you discuss the pros and cons of doing this with people who can advise you as to the benefits of such a statement.

What I do suggest is avoiding putting a disclaimer on the cover or below the table of contents. I have been advised over the years that any warning given in the UK should be no less prominent than the text surrounding it, but, as a general point, I recommend auditors start with positive messages.

Let me remind you of the audit 'expectation gap' as discussed in Chapter 6:

- An auditor cannot know everything, cannot verify every record, or speak to every employee in an audit lasting two weeks, let alone one lasting only a few days.
- Any audit is based on sampling designed to give reasonable assurance to the extent that the work plan has been completed.
- An audit is not a guarantee of zero losses in those areas which are found to be adequately controlled.
- Corrective actions are the responsibility of the auditee.

However,

- audit teams follow a structured auditing process such as *The Audit Adventure*™, which is aligned to an international standard, ISO 19011:2018;
- the audit results and opinion are likely to be replicable by others following in the footsteps of the audit team (as reflected in their working papers); and
- the audit team has reported conclusions based on the facts obtained during their fieldwork (though, of course, they are reliant on the facts reported being true!).

Between these two positions lie (i) our credibility as auditors and (ii) our desire to account for and manage our own risks.

A statement I use at the end of the management summary and again at the start of the recommendations is shown later. As I too am a risk manager, I cannot recommend this particular form of words to you; rather I present it to give you a 'feel' for what your lawyers may suggest you use:

> We have considered risk factors which we were aware of at the time of the auditors' visit(s). It should be noted that there may be other relevant factors which are not reasonably identifiable or other matters which in the opinion of the auditors do not constitute risks in the context of the report. We may suggest improvements which in our opinion will reduce risks. It should not be inferred that other risks could not be reduced or further controlled, or that identified risks could not be reduced further by other measures or in other ways.

A-FACTOR 81

Discuss 'liability disclaimer statements' with competent legal advisers in your jurisdiction.

The Management Summary—Part 1 of the Audit Report

I believe that the management or executive summary is probably the most important element of your report. In my experience, it is the section that senior managers look at first, as it provides a full account of the audit work and its outcome 'in a nutshell'. They will either love it or loathe it. It is up to the lead auditor to encourage continued reading from the way this summary is written and presented.

Accordingly, preparation of this summary should not be rushed. I strongly advise you to write it last, when the main body of the report—with all its key findings, exposures (residual risks), and recommendations—has been finalized, and the appendices prepared. The lead auditor must know what the 'big picture'—the high-level and future-focused messages to be conveyed—is to be. If the *nemawashi* ('no surprises') process has been followed, they will also have a fair idea of how senior management will react to this.

> The time to begin writing an article is when you have finished it to your satisfaction. By that time, you begin to clearly and logically perceive what it is that you really want to say.
>
> Mark Twain, from his notebook for 1902–03

It is important to remember that a good management summary could (quite literally) be taken from the report to stand alone. It should have its own introduction, summary of main points, and conclusion. If possible, restrict it to between two and four pages. This distillation aims to bring transparency and clarity to the issues the audit team wishes to communicate to senior management.

The executive summary should comprise four sections. A description of each of these sections follows:

1. Introduction and *Context*
2. Audit opinion (e.g., Good/Fair/Unsatisfactory/Unacceptable)
3. Major findings and actions required
4. Overall assessment of the business control framework

Many have expressed a sentiment in the vein 'If I had had more time, I would have been briefer.' I'd like to add my own version, a quotation often incorrectly attributed to Mark Twain, which I have used on my auditor training courses for many years to remind students to write concisely:

> I am sorry for writing you a four-page letter. I did not have the time to write you a one-page letter.
>
> Dr Stephen Asbury

The point is that focus takes time. My quote also seems to echo Blaise Pascal in his *Lettres provinciales* of 1657 (quoted in O'Toole, 2012): 'I have made this longer than usual because I have not had time to make it shorter'.

1 Introduction and Context

The introduction and *Context* (aka business environment) section of the management summary should describe, in concise terms

- the type of audit performed;
- which part(s) of the organization were audited;
- aspects of the organization's business environment which are both critical to the business's future success and relevant to the audit's findings; and
- the corporate challenges, major projects and changes, and business objectives relevant to the audit's findings.

This is your opportunity to prepare the minds of senior management, most of whom will be reading the report for the first time, so that they easily recognize their own organization and subconsciously give you credit for having neatly encapsulated the key (rather than all) external and internal factors which are challenging and/or supporting their organization's potential success.

A-FACTOR 82

Top management should be able to recognize in the executive summary their organization and its challenges in a way which resonates. The summary must demonstrate that the audit team has approached their work with the aim of assuring management that they have a well-managed organization, rather than simply providing an assessment of how well the management system measures up against the specified reference framework.

Much of the information appropriate for use in the introductory paragraphs will have been obtained at the start of the audit process, during the lead auditor and audit team's familiarization with the auditee's *context* and *objectives*. It is a matter of consolidating the information gathered to give a coherent summary.

Once the audit has been completed and its conclusions arrived at, only those aspects of this familiarization work (which might include PEST, SWOT, and stakeholder analysis, along with understanding the objectives, inherent risks, and opportunities) which have clear relevance for the major findings should be mentioned in the executive summary.

TIP

The aim of this highly selective approach to the content of the introductory paragraphs of the executive summary is keeping your readers' attention focused on only those matters you need them to think about.

Either you will be commending the readers on a strong and effective BCF with respect to some of the critical issues affecting the company's future, or you will be preparing them for a 'tougher' message of *alert* which cannot be evaded.

If the latter is the case, such preparation is important as it enables reader to recognize the vital issue(s) and captures their interest before you report (in the next section) that a relevant part of the BCF/MSS requires improvement and may possibly be unacceptable in its current state.

A-FACTOR 83

Whether auditors like it or not, managers are judged by the results of audits at their facilities. Those standing in judgement include the board of directors, remuneration committees, external regulators, the legal system, and other interested parties. In such an environment, it is only human nature that 'good news' will be well received, and 'bad news' may evoke a knee-jerk rejection or pushback.

2 Audit Opinion

This section of the report is the 'newspaper headline' that really matters to senior managers. It needs to pack a punch. It must succinctly describe the auditors' assessment of each of the elements of the reference framework. Is it used as expected? Does it work as designed? Has the design been converted to practice? This description should be structured so that the reader is given specific examples of strengths and weakness within each element which are relevant to and supportive of the major findings. The totality of these facts must lead the reader along a path of irrefutable logic that ends in the reported audit opinion.

A-FACTOR 84

A well-written paragraph delivering an 'Unsatisfactory' or 'Unacceptable' audit opinion must have the effect that the logic and the facts cannot be denied, even though the reader may find these facts uncomfortable.

3 Major Findings and Actions Required

This section will be a summary of what is reported in detail in Part 2 of the report—bullet points can be a very useful way of condensing text here.

The key challenge is to demonstrate clearly what management action needs to be taken, over what suggested time frame, to avoid the identified potential unacceptable outcome. Don't try to say everything; instead, give a compelling, short version of the facts which force the reader to make the link between the control failure in the reference framework and the impact on their organization's future success.

The logical thought process used throughout the audit can provide an initial template:

- A title that evokes feelings as well as describing the issue
- The risks to business success of any control failure(s)
- The presence, absence, strength, or weakness of relevant controls
- The action and urgency required to improve the control framework

TABLE 10.1 Example Table Summarizing an Assessment of My BCF, or an MSS Aligned to ISO Annex SL (such as ISO 9001, ISO 14001, or ISO 45001)

BCF ELEMENT	*ASSESSMENT*	
	POSITIVE	*NEGATIVE*
1 Leadership	+	
2 Planning	+	
3 Support and Operation		–
4 Performance evaluation		–
5 Improvement	+	

4 Assessment of the Business Control Framework (BCF)

This information can either be retained in the body of the executive summary or attached as an appendix, with only a table of the results in the body of the report. There is an example in Table 10.1:

For each element of the BCF, there should be a clear and concise description of the most important examples of strength and weakness of control identified during the audit. If appropriate, you can make links between these specific examples and the relevant higher-risk activities in the audit work plan.

Ensure that the paragraph for each BCF element provides both a stand-alone commentary on the findings for that element and a balanced report, in terms of the quantity and significance of the control failings found, of the contribution to the assessed effectiveness of the management system overall (as shown in the table of results for the BCF).

Audit Findings and Actions—Part 2 of the Audit Report

Part 2 of the audit report presents the details behind the comments and conclusions presented in Part 1. The structuring of the detailed information will often be determined by the organization being audited. The key elements to report on:

1. The status of the control framework—comparing expected controls with actual controls
2. Exposure of the organization—identifying the extent of any unauthorized residual risk
3. Actions—as required to strengthen failing controls and address the root causes of failure

The overriding objective for an auditor—when detailing the status of the control framework needed to effectively manage specific inherent risks within activities critical to the organization—is for the reader to agree with/accept the auditor's expectation for particular controls, and to see if/where there are gaps between that expectation and the extent, type, and quality of controls found.

To the degree necessary, details of verification can be included or cross-referenced to appended documentation (for example, the list of interviewees and other contributors to the audit). The objective for the auditor when detailing the exposure resulting from the control failure is for the reader to clearly appreciate the impact of the existing situation on the organization's key objectives, both quantitative and qualitative.

A-FACTOR 85

Detailed reporting of each audit finding should examine the extent to which the organization has demonstrated 'control of sampled risks'. If the organization has failed to demonstrate structured means of control of significant risks, this should always be reported.

Without action being implemented to correct any gaps/failings identified, no improvement will occur. Irrespective of who decides on what the appropriate action is, whether the auditee or the auditor, the action must reflect a SMART approach:

- Specific—the action is clearly defined
- Measurable—the action has a defined performance level or measurement
- Achievable—the action can be carried out successfully by the actioned party
- Right—the action is appropriate considering the gap/failing/problem
- Timely—the level of urgency corresponds to the level of seriousness

From a technical standpoint, clarity of wording in recommendations is critical, especially in Part 2 of the report. Recommendations need to be given in such a way that there can be genuine commitment to them because of full understanding of what, how, and when action needs to be taken.

Clarity is a matter of balance. For example,

- balancing conciseness with detailed analysis to adequately specify the action to be taken,
- balancing corporate and process issues to improve matters impacting objectives at both levels, and
- balancing the responsibilities of the person carrying out the required actions with those of the person accountable for their effectiveness.

Using verbs is important—certainly use more verbs than nouns. Nouns (or naming words) clog up the arterial blood of your writing, but verbs (words of action, doing, and being) invigorate it. Verbs are also shorter: 'solve' is shorter than 'solution', 'manage' is shorter than 'management'. Verbs thus have two benefits: vigour and brevity.

When drafting recommendations and seeking conciseness, it is difficult but necessary to avoid vague, wishy-washy verbs such as

- consider—instead use: prepare, estimate, evaluate, measure, calculate, or compute;
- ensure—instead use: implement, check, verify, certify, justify, or investigate;

- review—instead use: eliminate, amend, adjust, correct, recreate, or rewrite; and
- monitor—instead use: analyse, investigate, revise, overhaul, or repair.

And PLEASE omit needless text—words which add no value, content, information, or impact. For example, the following:

- Don't write a page when a paragraph would do!
- Don't write a paragraph when a sentence would do!
- Don't write a sentence when a word would do!
- Instead of saying 'for the purposes of learning', say 'learn'
- Instead of saying 'this report is of a confidential nature', say 'this report is confidential'
- Instead of saying 'the audit team wishes to suggest that possibly the management should maybe consider reviewing and if necessary amending the training record-keeping system to bring it up to date', say 'the training records system should be updated'.

Keep it simple, and make every word count!

A-FACTOR 86

Using verbs is very important—certainly use more verbs than nouns—when making SMART recommendations in audit reports. Verbs bring vigour and brevity.

A-FACTOR 87

Omit needless words in audit reports—take the shortest path.

Format

An audit report should be properly formatted and presented using the conventions of professional business writing.

TIP

Number and reference tables in the text of the report as Table 1, 2, 3, etc., and pictures, photographs, graphs, and charts as Figure 1, 2, 3, etc.

Note that your word processing software is likely to be able to compile and update lists of these automatically.

TIP

For abbreviations, such as COSHH, RIDDOR, OSHA, or NASA, give the full term once when you use it for the first time, with the abbreviation in parentheses, then use just the abbreviation thereafter, for example, the Institution of Occupational Safety and Health (IOSH), then just IOSH.

TIP

Provide references to acknowledge the sources used in your writing. Including references helps you to support your claims and ensures you avoid plagiarism (using someone else's work without giving proper credit). There are many referencing styles, but they invariably consist of two things:

- A citation in your text, for example, (Asbury, 2023)
- A reference list at the end, for example, Asbury, S. (2023). *HSEQ Audits*, 4th edn. CRC Press.

READING AFTER WRITING AND BEFORE SUBMITTING

This section highlights why it is important to read the report after writing it and before submitting it. It shares some of my experiences (many of them humorous but still important) from over 30 years as an auditor.

Think of the things that might distract a reader when they are reading someone else's work:

- Spelling mistakes
- Poor punctuation and grammar
- Long and unintelligible sentences
- Obvious errors of fact
- Excessive use of 'absolute' terms ('no training' or 'no communication', for example—what, absolutely none, ever?)
- Insufficient, or too much, white space on the page

My experience is that it is difficult to proofread and review your own work. Each audit team member's work should be read by a colleague. Every part of the audit report should be read by at least two people (its author and another), looking specifically for errors of fact as well as spelling, formatting, and typographical errors.

Other reasons to check documents after writing but before submitting them include the following:

- It is good practice, and may prevent you from including something wrong or just plain silly.
- It may be part of your organization's quality system (for example an ISO 9001 procedure) to submit a sample of documents for internal review.
- Auditees can pick up on errors and use them to deflect attention from important issues; where a statement of fact or recommendation is incorrectly expressed, this may be used as an excuse not to take the action required.

If your reader is not to be distracted while reading, you need to make sure that your use of technical language familiar to them is correct. What health and safety practitioner would not be distracted by an incorrect reference to the principal health and safety legislation in their country?

Punctuation, Spelling, and Grammar

Your grammar reflects your image. Good or bad, you have made an impression.

And like all impressions, you are in total control.

—Jeffrey Gitomer; author, speaker, and trainer (1946–)

By the time you read this book, a copy editor and a proofreader will have checked my punctuation, spelling, and grammar. My production team and I want to make a good impression.

Poor punctuation, spelling, or grammar makes reading and understanding harder and can distract the reader. So check spelling carefully. A starting point is the spellchecker provided with your word processing software if its limitations are recognized. Additional proofreading is important to ensure that 'there' is not incorrectly given as 'their', and so on.

CASE STUDY

CARNAGE

I was close to submitting the following report to an insurance underwriter some years ago. Happily, it was spotted by my boss when proofreading. It said: 'the client's site has much carnage in its workshops, but this is well documented and clearly marked.'

My intention was to say, 'the client's site has much cranage [i.e., many cranes and lifting devices] in its workshops . . .', but the designers of my word processing software had pre-programmed their integral spellchecker to autocorrect selected words. Oops. But it was corrected prior to submission.

Some town, company, and surnames can be hopelessly autocorrected too, and this can be very embarrassing. If you don't believe me, try spellchecking 'Pernis' (a refinery in the Netherlands), 'Hyster' (a forklift truck manufacturer), or 'McVities' (who bake cakes and biscuits).

LEARNING

Use autocorrect and spellcheckers with care!

TIP

Peer review and proofreading of reports is a good idea. They should be read after you have finished writing them and before submitting them. Often the brain can work faster than the hand, and the report may not say quite what you wished it to!

Try this simple starter. My niece made the bookmark shown in Figure 10.2 for my wife a while back. How many errors can you spot?

It should say: 'Hands off Auntie Sue's book!'

'Hands' (not 'hand's'), 'off' (not 'of'), and 'book!' (not 'book!.').

If you feel that your punctuation may be poor, I encourage you to look at a series of exercises provided to me some years ago by the late great Dr Arthur Rothwell of De Montfort University Leicester. These certainly helped me to use apostrophes correctly and to improve my punctuation in general. You'll find these exercises to download from the companion website (https://routledgetextbooks.com/textbooks/_author/asbury/).

FIGURE 10.2 How many errors can you spot?

TIP

The plural of a single does NOT require an apostrophe.
One hand. Two hands. NOT hand's.

CASE STUDY

LEARN TO PUNCTUATE

A master's degree student submitted work to his dissertation supervisor. The supervisor returned the work, with a note which said: 'I am unable to concentrate upon the (no doubt excellent) content of your work, as I am continually distracted by your poor punctuation and grammar.'

LEARNING

Learn to punctuate if you need to!

A-FACTOR 88

Proofread your report, and have it peer-reviewed before you submit it to the client.

SUBMITTING AUDIT REPORTS

There will come a point when the report is to be submitted to your auditee or client—either because it is contractually due or because you feel that the report is ready to be read. There are two basic approaches to submitting audit reports:

- Here it is.
- Here is a draft (for comment prior to our final report).

Here It Is

Some audit reports are only presented this way. Any practitioner or auditor who has worked with a standards certification body will have prepared their audit report as they have progressed through the audit by recording the evidence of their work chronologically.

The report is then collated and, with the addition of a summary front page and recommendations at the back, handed to the auditee on the final day of work on-site.

Whilst this is a cost-effective way of audit reporting, it does not allow for issues such as

- whether the auditee can read the auditor's handwriting (I still see quite a lot of reports of this type that have been handwritten) and
- the lack of any opportunity for proofreading and peer review.

Here Is a Draft

This is my preference, and I commend this approach to you. If you can (and, of course, this depends on what has been agreed in terms of timing and budget), I suggest that a draft report is presented to the client *before* the closing presentation. If you can do this, you can refer to it during the presentation, for example to draw attention to specific facts, tables, or graphs.

I give (typically) two weeks for the receipt of the auditee's comments. The clock starts after the closing presentation and concludes just before I issue the final report. I consider all comments received, but you should note that received comments do NOT necessarily equal changes. Instead, comments provide me with an opportunity to

- resolve questions raised on the draft;
- correct errors or inaccuracies;
- adjust or moderate unnecessarily contentious statements or language, for example unfortunate references that may have slipped in to 'incompetence', or absolute negatives or positives, such as 'no', 'never', or 'always'; and
- take account of the auditee's concerns and preferences regarding actions.

Between the draft and the final report, of course, be sure to keep all your data and other work safely backed up.

A-FACTOR 89

Maintain and regularly update an off-site or cloud backup copy of all computer files to prevent irretrievable loss.

Final Check Before Issue

It is a really good idea to do a *final check* as the last thing you do before putting the final report into an envelope or attaching it to an email with, of course, an appropriately worded covering letter.

TIP

Send your invoice (if applicable) separately to the audit report.

AUDIT FOLLOW-UP AND ISO 19011:2018

The Audit Adventure™ concurs with clause 6.7 of ISO 19011:2018 in dealing with post-audit actions. These may involve corrections, corrective actions, and/or addressing appropriately opportunities for improvement.

The completion and effectiveness of these actions (advice; recommendations) should be verified. Once the final audit report has been issued, the responsibility for addressing the actions passes to the auditee, who should keep the head of internal audit informed of the status of these actions.

Formal responsibility for ensuring follow-up rests with the head of internal audit and the internal audit committee. Verification of closeout may be a part of a subsequent audit.

Meanwhile, the *Audit Adventurers* head for their next *Audit Adventure*™, taking their learning with them and sharing it.

A-FACTOR 90

The colour of company blood is green (money). If red blood is spilled, the company is going to be in trouble one way or another. However, an audit conducted throughout by using the methodology, practices, and tips provided in this book prevents losses and compels improvement.

I hope you enjoy participating in your first (or 101st, or 1001st) *Audit Adventure*™.

FIGURE 10.3 *The Audit Adventure*™—after the audit is completed, the audit team can look back on a job well done—top-down and bottom-up—prior to progressing to their next assignment.

Test your knowledge with the final chapter assessment test (see Appendix 4).

THE NATIONAL TRUST
POLDHU

Appendix 1
A-Factors

Appendix 1 brings together all ninety of the *A-Factors* which were presented throughout the book for your ease of reference. Readers of past editions have said that they have read this appendix to refresh the memories of the full text. They can be referred to as *Asbury* or *Auditing* Factors when you think or talk about them.

Together, the *A-Factors* represent a consolidation of the essential knowledge and skills for undertaking risk-based audits using *The Audit Adventure*™ auditing process.

CHAPTER 1

A-Factor 1

Organizations are concerned with transforming inputs to outputs. Inputs create outputs, and outputs create inputs.

A-Factor 2

Significant risks to organizations' objectives commonly arise in the external business environment with which they are inseparably intertwined. Senior managers should take proper account of this connection if they are to manage their enterprises successfully.

A-Factor 3

The structure of an organization is a means to an end, not an end in itself.

A-Factor 4

Recognize that, ultimately, market forces tell organizations—if they are listening carefully—what to produce (quality), when to produce it (delivery on time) and the sustainable price. These as objectives should be reflected in the business plan.

A-Factor 5

Top management should balance the influences of the competing external and internal environments to face its target market(s) with aligned and well-communicated business objectives.

A-Factor 6

Risk is the effect of uncertainty on objectives—or anything which may impact upon the achievement of objectives. It is generally quantified in terms of its *inherent* and *residual* probabilities and consequences. Value creation and value protection are the essence of an organization's success.

A-Factor 7

R = P x C (Risk = Probability x Consequences)

A-Factor 8

Where the assessed inherent risks are judged to be potentially significant, managers and auditors alike should apply the 'Four Ts' and 'E-SEAP' to effectively mitigate these.

A-Factor 9

Recognize that ultimately an audit is an independent and balanced assurance to interested parties regarding an organization's ability to meet its business objectives in increasingly volatile business environments.

CHAPTER 2

A-Factor 10

Remember Deming's PDCA. Keep things simple.

A-Factor 11

Do not permit the terminology or detail used to describe any MSS or business control framework to distract you from the structured simplicity of Plan-Do-Check-Act.

A-Factor 12

To carry out a successful and effective management system audit, an auditor needs a relevant internal control reference framework against which the auditee's performance can be assessed.

A-Factor 13

Only by using a 'structured means of control' can a manager convert high-cost *controls* into business-assuring, profit-enhancing *control*. This approach provides a higher level of assurance, as *control* is systematic, replicable, and repeatable.

A-Factor 14

Whatever the auditee's reference framework, an auditor needs to have their own 'structured means of control' in mind, which they can use to simplify the complexity of an auditee's framework. This can also be useful if there is a vacuum.

CHAPTER 3

A-Factor 15

An audit should provide a reflection, as if in a mirror, of the effectiveness of the auditee's business control framework.

A-Factor 16

The primary reasons for auditing are business improvement and providing assurance to stakeholders.

A-Factor 17

A rolling, balanced audit plan is a foundational and essential preparatory component for providing assurance to stakeholders.

A-Factor 18

The number of auditor days (the 'audit intensity') provided to deliver any single audit from the audit plan should depend upon the depth and quality of assurance the management of the organization requires.

A-Factor 19

The internal audit committee is responsible for keeping the rolling audit plan including 'the audit mix' under regular review.

A-Factor 20

Audit objectives can be referred to as *the 3As* as an aide-memoire: Assurance, Alert, Advise.

A-Factor 21

The terms of reference (ToR) are essentially the contract for the audit; the agreement between an organization and its auditors about *what* will be delivered by the end of the audit. No audit should commence without agreed terms of reference.

A-Factor 22

For internal management system audits, the team should comprise a minimum of two members (a lead auditor plus one other auditor), with access to support for peer review.

A-Factor 23

Recognize the importance to the overall audit opinion of an objective view from an independent audit team.

A-Factor 24

First impressions count. Get the highest level of professional qualification that you can, pursue CPD, and use your (applicable) designatory letters on business cards, reports, and other stationery. Don't even think of claiming qualifications to which you are not entitled.

CHAPTER 4

A-Factor 25

Time spent on reconnaissance is seldom wasted.

A-Factor 26

Request pre-audit documentation in good time. Three months before the start is an excellent start point, with a reminder at two months, and a final chase with four weeks to go.

A-Factor 27

Remember 'seek-sort-share'. Share the sorted pre-audit information with your audit team two weeks in advance of the audit.

A-Factor 28

A firm, but not crushing, handshake combined with a sincere smile is a powerful intimation of your professionalism as an auditor.

A-Factor 29

Accept that suspicion of audits and auditors is entirely normal. Do not equate it with auditees or interviewees having something to hide.

A-Factor 30

Don't drop your guard until the assignment is complete. It is essential to rehearse the possible nature of objections and to try to see things from the other person's perspective. Anticipate and have in mind answers to the types of questions auditees and others are likely to pose.

CHAPTER 5

A-Factor 31

The Audit Adventure™ comprises two simple dynamics: 'top-down' and 'bottom-up'.

A-Factor 32

A lead auditor has the authority to decide how to use the audit time. For example, if they are leading a relatively inexperienced audit team or working in an area of the business they do not know well, they might increase the time for the preparation stage above the usual 20 per cent of overall time available. This extra time for preparation will reduce the time available for the conduct of the audit, as at least 20 per cent of the overall time available must be retained for the reporting stage.

A-Factor 33

Monitoring by the lead auditor of progress against the audit timing and work plans ensures that the audit is completed on time, using those resources available to provide a level of assurance concerning the control framework within the auditee's area of responsibility.

A-Factor 34

While the Conduct stage of an audit follows a logical sequence of Reviewing and Verifying, the methods used, and the substantive content, will involve significant repetition. This is especially true of interviewing, with similar enquiries being made in different interviews at successively finer-grained levels of detail, across various lines of enquiry, and possibly across different control frameworks.

A-Factor 35

Information about the future is much more valuable and interesting than information about the past. An audit team able to set their report in the context of the future will generally be esteemed by management.

A-Factor 36

The main deliverable of *The Audit Adventure*™ is a well-presented audit report that triggers improvement in control of the most significant risks to the achievement of the organization's objectives (the *Big Rocks*).

A-Factor 37

People support what they have assisted to create.

CHAPTER 6

A-Factor 38

Lead auditors must have a clear view of the risk-based auditing process and know how to prepare for, and perform, in each stage (Prepare, Conduct, Report).

A-Factor 39

Be ready to explain the 'expectation gap' to the auditee, as well as clarifying it within the audit team. There is no expectation or possibility that you will audit 'everything'; the audit will be based on a structured sample.

A-Factor 40

The audit sample should be selected independently by the audit team (and no one else).

A-Factor 41

If the audit reference framework or audit scope are significantly changed, the head of internal audit should be advised, so that they may amend the audit plan to reflect these changes.

A-Factor 42

The audit terms of reference are generally not negotiable. They have been approved by the internal audit committee as one of their 'jigsaw pieces'. The scope as specified needs to be considered in full, without exclusions.

A-Factor 43

As a lead auditor, it is important to encourage your team to be speculative. Think ahead about the business environment (*Context*) of the audit setting and how your auditee will be managing their organization considering its foreseeable changes and challenges.

A-Factor 44

Key to successful audit is a thoroughly prepared audit team.

A-Factor 45

You only get one chance to make a positive first impression. Be sure to take it!

A-Factor 46

The audit work plan is a living document, which is continuously referred to and updated by the lead auditor. It should be adapted as necessary to account for discoveries made by the audit team in the Review stage. This may occur if the team identified a new and significant area of risk not previously considered, which may then replace an item in the plan if the remaining time remaining allows for this.

A-Factor 47

Time constraints and the need for efficiency mean that auditors should not set out to ask questions about *every* control element of the reference framework. They need to decide which of the control elements are critical as the basis for good risk management of the activity being audited. Their selections for review and testing will be shown on the work plan.

A-Factor 48

Before the time available for the *Prepare* stage expires, each auditor in the team should have prepared an agenda for each of their first interviews, together with lists of appropriate questions. This will enable them to commence the next stage of the audit—the *Conduct* stage.

A-Factor 49

Each audit has its own master file containing all the audit records, including interview notes, critical documents, and other audit evidence. This will be retained securely for an agreed period after the audit.

CHAPTER 7

A-Factor 50

The lead auditor should ensure that an audit is managed as any other project, with careful time and resource planning, including a provision for contingency.

A-Factor 51

The audit team will base its overall audit opinion on the effectiveness of the structured means of control for the sample of risks in its work plan.

A-Factor 52

The *expected* control framework should be identified before fieldwork commences. It is essential for focusing the auditor's questions about the structured means of control during the *Review* sub-stage.

A-Factor 53

A *gemba* attitude means going to the source to check the facts for yourself, to arrive at well-informed decisions (*gemba* means 'place' in Japanese).

A-Factor 54

Remember the 'ten-foot rule'. People within 10 feet (3 metres) of the operation will likely have very good knowledge of the work processes there.

A-Factor 55

The more you know about a subject, the more you can see both sides.

A-Factor 56

Remember IWA—I Will Audit (Independent, Well balanced, Appropriate) in a manner which meets the needs of the auditee's organization.

A-Factor 57

The *Verify* sub-stage involves checking that expected controls are correctly implemented and that those controls are effective at controlling the sampled risk. It can also involve verifying, in the absence of expected controls which are not considered appropriate or necessary by management, the acceptability of the residual risk exposure.

A-Factor 58

The best recommendations an audit team can ever make are those agreed in advance with the auditee. The best chance of gaining agreement comes from bringing the auditee on side at the earliest possible opportunity.

A-Factor 59

Learn about the management system by listening closely. There is more to hear in any answer than just 'yes' or 'no'!

A-Factor 60

Whatever you decide to use as the sampling strategy, record the sample size and how the sample was derived, and then report the results—what the sample told you—in Box 4 of the AFWP.

CHAPTER 8

A-Factor 61

The lead auditor's first action to promote effective teamworking is to have an early discussion with each team member—and remember the sixty-second rule!

A-Factor 62

There is no space in an effective audit team for auditors to hold back from commenting on something they are uncomfortable with. Similarly, there is no space for petulance to be shown by a team member whose work is being (constructively) challenged by another member of the team.

A-Factor 63

Avoid using absolute terms (no, never, or always) in meetings or reports, as they are seldom true.

A-Factor 64

In every questioning interaction, know specifically what you want from your question before you ask it.

A-Factor 65

The only place for the audit team to disagree is in the privacy of their team room. Outside the team room, they must present a united face. There are no minority opinions in audit reports.

A-Factor 66

It is good practice to provide the lead auditor with access to a competent colleague for peer review and to calibrate the draft audit opinion.

A-Factor 67

Rapport = building and maintaining positive engagement with other human beings. It should not be overlooked.

CHAPTER 9

A-Factor 68

Momentum accumulated 'Top-down' provides auditors with a burst of energy for the journey 'Bottom-up'.

A-Factor 69

A significant audit finding is one for which the answer to the 'So what?' question represents a significant impact on the auditee's objectives. If the auditors cannot attach this significance, then it is unlikely that senior management will either.

A-Factor 70

All areas of strength found in the application of the BCF must be recorded as fully as areas of weakness, so that there is accurate weighting of each.

A-Factor 71

The audit team must be able to *see* the balance of the emerging facts if they are to apply their 'group brain' to what those facts mean for the control of significant risks. Large wall charts are a fantastic idea because they lend great visibility.

A-Factor 72

Overall audit opinions should have an *even* number of gradations. The overall audit opinion should reflect the overall level of concern, based on the audit work, about the achievement of objectives.

A-Factor 73

The lead auditor is ultimately responsible for the conduct of the audit and for finalizing the overall audit opinion.

A-Factor 74

Nemawashi—remember to 'go around the roots'. Such discussions are an important informal element of major change. Successful *nemawashi* enables change to be the result of consensus among all parties.

A-Factor 75

The audit opinion is not negotiable once the audit team has arrived at its decision.

A-Factor 76

The objective of the closing meeting is to help management arrive at the right decision regarding the audit findings. Initiating an appropriate response is essential, whether this is consolidation of existing strengths, a survival plan, or keeping the board out of jail. The strategy should be to focus management's attention on a few key messages, and the tactics should be to support these messages with factual evidence to the extent demanded by any person in the room.

A-Factor 77

Know who will attend the meeting, and, as far as possible, ensure that the attendees are the appropriate audience and have sufficient time available. You or the auditee should send each attendee an agenda and a short description of how the audit team wishes to conduct the presentation. Requests from management for pre-meetings or discussions afterwards should be met.

A-Factor 78

Remember the audit is not over. The best outcome of the presentation is full support for the findings and commitment to making the appropriate response. If, however, management is not able to either accept the findings or commit to a response, then maintain a dialogue with the audited organization to give management the opportunity to express why this is not possible. Your aim is still to optimize the effect of the work done by the audit team.

A-Factor 79

If, despite following this guidance, and ideally having broad support from the auditee, you do not feel that further discussion will result in management's agreement and commitment, your only course of action is to refer the matter to the head of internal audit with a view to their reporting this outcome to the chair of the internal audit committee.

CHAPTER 10

A-Factor 80

A well-written, factually correct, timely-presented written audit report is essential. It encourages, even compels, improvement when you get it right!

A-Factor 81

Discuss 'liability disclaimer statements' with competent legal advisers in your jurisdiction.

A-Factor 82

Top management should be able to recognize in the executive summary their organization and its challenges in a way which resonates. The summary must demonstrate that the audit team has approached their work with the aim of assuring management that they have a well-managed organization, rather than simply providing an assessment of how well the management system measures up against the specified reference framework.

A-Factor 83

Whether auditors like it or not, managers are judged by the results of audits at their facilities. Those standing in judgement include the board of directors, remuneration committees, external regulators, the legal system, and other interested parties. In such an environment, it is only human nature that 'good news' will be well received and 'bad news' may evoke a knee-jerk rejection or pushback.

A-Factor 84

A well-written paragraph delivering an 'Unsatisfactory' or 'Unacceptable' audit opinion must have the effect that the logic and the facts cannot be denied, even though the reader may find these facts uncomfortable.

A-Factor 85

Detailed reporting of each audit finding should examine the extent to which the organization has demonstrated 'control of sampled risks'. If the organization has failed to demonstrate structured means of control of significant risks, this should always be reported.

A-Factor 86

Using verbs is very important—certainly use more verbs than nouns—when making SMART recommendations in audit reports. Verbs bring vigour and brevity.

A-Factor 87

Omit needless words in audit reports—take the shortest path.

A-Factor 88

Proofread your report, and have it peer-reviewed before you submit it to the client.

A-Factor 89

Maintain and regularly update an off-site or cloud backup copy of all computer files to prevent irretrievable loss.

A-Factor 90

The colour of company blood is green (money). If red blood is spilled, the company is going to be in trouble one way or another. However, an audit conducted throughout using the methodology, practices, and tips provided in this book prevents losses and compels improvement.

Appendix 2
Preparation, Preparation, Preparation

INTRODUCTION

This appendix was described by many individuals who corresponded with me after publication of the earlier editions as the most important and valuable component of the book. They said that it had helped them to deliver their audit assignments successfully.

Audits have failed or gone horribly wrong when proper preparation for the assignment has been skipped, or when mission-critical elements have been overlooked. Lost time cannot be recovered. A *right-first-time* approach is recommended to lead auditors and their audit teams, particularly as they will be *selling* excellence and the integrity of operations to their auditees and clients. To fail to deliver this themselves would be unforgivable.

So, just as many people (including me) have a packing list for their holidays and business trips, adjusting the list each time for *that thing* they forgot so as not to forget it again, Appendix 2 provides an audit practitioner's checklist and a guide to audit preparation and conduct. I have used it for many years, each time I have been assigned or contracted to lead or participate in an audit, and I recommend it to you. No list like this can be absolutely definitive, and you should review it prior to use according to your own needs. You will find lots of tips and suggestions here to think about, whether you are embarking on your first or your 1001st audit.

Among the suggestions, tips, and techniques, you'll find my thoughts on

- personal preparation—for example, passports and visas, immunizations and medication, data and communications, currency, credit cards, insurance;
- preparation of the audit team—for example, sharing pre-audit information;
- subject preparation—for example, context, organization's objectives, subject matter expertise, interview scheduling;
- getting there—for example, transportation, accommodation, tickets;
- conducting the audit—for example, preparation, conduct, reporting, *nemawashi*; and
- Activities after the audit—for example, issuing audit reports, and invoicing.

This checklist is provided for the guidance of all auditors, particularly lead auditors. As I have said, it is unlikely that any list of this type can be wholly complete for every user.

It does, however, aim to cover the main areas and to give a reasonable assurance that the main requirements for successfully leading audits have been considered.

I encourage lead auditors to copy this checklist for their own use. Should you wish to amend it to match your specific needs, a version is available for download from the book's companion website (https://routledgetextbooks.com/textbooks/_author/asbury/).

I also welcome constructive feedback from users of the checklist. I will incorporate and credit the best suggestions into future editions of this book and its companion materials.

How the Checklist Is Structured

The checklist (with its supporting commentary) is divided into three sections, hence the appendix title, 'Preparation, Preparation, Preparation':

- Before the audit
- During the audit
- After the audit

BEFORE THE AUDIT

---

1 Liaise with Head of Internal Audit on the Requirements for Audits

As a lead auditor, it is essential that the requirement to lead/participate in an audit is confirmed. Ideally this will be in writing, and it may be an official order. You will need the audit duration and intensity (the number of auditor days that have been allocated to the assignment).

If you are providing this audit as a consultant or contractor, this confirmation constitutes your initial work instructions. Agree budgets and the class of air travel if necessary. Pursue a work order. Without this, you may not be paid. Another important request at this time is for the draft terms of reference (ToR).

If you are being seconded from another department to the audit team, you'll be away from your own job, so cover for your absence may need to be arranged.

2 Confirm the Audit Dates

Propose and later confirm the dates for the audit with the nominated individual at the location or process to be audited (as per the ToR scope). If no specific individual is nominated, the site's manager (or similar) is likely to be the auditee. A good way of thinking about the likely auditee, if they are not named, is as the manager with the main responsibility for the scope area.

Confirm the audit dates to the auditee in writing. Add a 'read receipt' and/or ask for confirmation if it is sent by email. Mention the intention to make a pre-audit site visit if this is required (see item 9).

3 Identify/Select the Audit Team Members

My personal preference is to select my own audit team members. It allows me to nominate technically competent auditors with experience in the likely scope risks. Sometimes this is possible, sometimes not. Either way, identifying the team composition is very important.

The lead auditor should write to each nominated member of the audit team, welcoming them to the team, briefly describing their involvement, and confirming the dates and duration of the audit.

4 Develop and Send Pre-audit Requests for Information and Documents

About three months in advance of the audit, request the desired information from the auditee. If the information is not received, this gives time for follow-up at two months (*this is my second request*), and one month prior (*this is my third and final request*). I have included an example letter text, showing what it may be useful to request in advance, in Appendix 3.

Should the pre-audit documentation not be received at all, you should refer to the audit sponsor (often the head of internal audit) to advise on the situation. It is possible that the audit may have to be rescheduled.

5 Receipt of Pre-audit Documentation

When the pre-audit documentation you asked for arrives, be sure to read it. Check what you have received against the list of information you requested.

What you receive does not have to be perfectly aligned to your request, but it should cover your main requirements. If necessary, follow up with a further request for additional essential items.

6 'One-month-out' Checks

These constitute your 'final arrangements'. Check (for yourself and the audit team) that you have the appropriate documents:

- Passports—Six-months validity beyond your planned return departure date is a good standard. Check that your passport will be recognized by the territory to be visited (for example, Israeli immigration stamps invalidate passports for some Arab territories). It may be useful to have two passports. This is legal in many territories, on the basis that you may need to send one passport with a visa application at the same time as you are travelling. In the UK, your employer will need to endorse this request. Take a colour photocopy of your passport's data page with you, and leave one at home. You can apply for or renew a UK passport online at www.gov.uk/apply-renew-passport. There are also premium and *fast track* services. You'll need a credit or debit card to use those services. Alternatively, you can pick up passport application forms from a local post office and apply by post. The post office offers a *Check and Send* service which you should consider. It takes longer to apply by post than online.

- Entry visas—Find a good visa agent. I have noted the agent I have used for 15 years subsequently*. For entry to the USA, some countries can apply for an ESTA online at https://esta.cbp.dhs.gov. By contrast, I have found that some territories have particularly challenging visa conditions. For example, at the time of writing, the Angolan short-visit work visa requires that you enter the country within three days (72 hours) of its issue. The major consequence of this is that you'll likely have to book your flights prior to receiving confirmation of the issue of your visa.
- Travel arrangements—Is the team travelling together or meeting later? Make sure that everybody knows the arrangements.
- Vaccination and medication—Check the requirements. I use https://travelhealthpro.org.uk/countries as a resource. There may be requirements for anti-malarial medication, and some territories require evidence of vaccination (e.g., for yellow fever or Covid-19) at their border control. Contact a physician or GP for specific and up-to-date advice. If you are carrying prescription medication, carry a copy of your prescription or a letter from your physician.
- Flight and/or train reservations—have the six-digit locator code(s) to hand ready for when you travel.
- Car park reservations—take a copy of your booking details.
- Taxi bookings—arrange EXACTLY where you will be met and how the driver will identify themselves to you. I'm not really a fan of *jumping into airport taxis*—it's too risky these days. Pre-booking is much safer. I love Uber (download and set up the app).
- Business travel insurance—obtain a copy of your policy and details of cover to take with you when you travel.
- Accommodation for the duration of the stay—The auditee can often recommend suitable and convenient locations and may also have preferential rates agreed.
- Currency and ATM-enabled credit card—Carrying a few US dollars has always been a good fallback, as most taxi drivers (for example) will accept them.
- Directions and a map to your hotel and/or the auditee's site (including satnav and *What Three Words* location information, if required)
- Insect repellent—ideally it should contain 50 per cent DEET or greater.
- Availability of foreign-language translators, where needed.
- Initial meeting point logistics for auditors.
- Mains electrical adaptors for the electrical appliances you are taking, including for laptops. Screen/projector adaptor for your Mac (if you use one).
- Wi-Fi, mobile data, and telephone arrangements for the territories to be visited.
- Training or qualifications for specific tasks or locations—for example, helicopter escape training (HUET) for offshore locations.

- PPE (personal protective equipment)—for example, safety shoes, flameproof overalls.

**In relation to entry visas, my company has tried several visa agencies over the years, some good and some (very) bad. In recent years, we have discovered the best for us has been IJDF Worldwide Visa Services Limited, London www.ijdf.co.uk*

7 Send the Draft Terms of Reference

Send the draft ToR to the auditee approximately one month before the audit is scheduled to commence. This is an ideal way of confirming the final details.

8 Sort and Share the Pre-audit Materials with Audit Team Members

There is no need to send *everything* to your audit team. Choose wisely, and send copies of the information that is most likely to be helpful to your auditors as they prepare for their assignment. Two weeks before the audit is the ideal time frame in my experience.

Impress upon the auditors the importance of being fully prepared in advance of the audit. To do this successfully, they must find time to read what you have sent them.

TIP

Remember *Seek-Sort-Share.*

9 Arrange a Pre-audit Site Orientation Visit

Within the month before the audit, it is useful for the lead auditor to make a short orientation visit to the location for the audit.

Accept that this pre-audit visit may not be possible if the location is hundreds or thousands of miles from the auditor's home base. As an alternative, the auditee may be able to send some photographs of the site, while online tools such as *Google Earth* and *Google Maps* with *Street View* may provide useful additional information for your preparation.

10 The Day Before the Audit

Review the internet and the organization's website for last-minute news. This takes just a few minutes, and I've been surprised how often this provides information that has proven useful to my audit team.

DURING THE AUDIT

1 Your Conduct as an Auditor

Conduct yourself and the audit as per *The Audit Adventure*™ described in this book. Use the book as a resource throughout your audit.

2 Responsibilities

Lead auditors have the principal responsibility for delivering the overall audit service with an evidence-based written audit report.

The lead auditor should also take responsibility for the following:

- Acting as the principal contact between the audit team and the auditee.
- Convening and chairing all meetings of the audit team.
- Being the principal presenter at opening and closing meetings with the auditee and their team. Of course, the lead auditor may co-opt other presenters as required.
- Arranging and conducting *nemawashi* (no surprises) meetings with the auditee.
- Scheduling meetings with the auditee's staff and coordinating overall project timekeeping.
- Motivating the audit team—coaching, coordinating, and encouraging them. Also responsible for maintaining team discipline.
- Proofreading and challenging, where necessary, all documents (particularly working papers/AFWPs) produced by the team.
- Security of the audit's data and materials.
- Deciding when the team cannot. Sometimes this can mean using a 'casting vote', and sometimes leading the team to a decision—the lead auditor is responsible for the overall audit opinion.
- Coordinating production of the draft audit report, which ideally is issued prior to the closing presentation.
- Overall quality assurance of the audit and its deliverables.

AFTER THE AUDIT

The lead auditor has the following responsibilities following the conclusion of the on-site work, and after the audit team has dispersed:

- Finalizing and submitting the final audit report (having received and taken account of any comments on the draft report from the auditee), including any recommendations to the auditee and others, as agreed/required.

- Gathering, indexing, and securely archiving or destroying (depending on contractual arrangements and/or professional indemnity insurance requirements) the audit file and the working papers.
- Sending a letter of thanks to the auditee and to each member of the audit team (unless, for some reason, this is inappropriate).
- Arranging for charging or invoicing (as required) of fees and expenses incurred.

An excellent lead auditor should also do the following:

- Regularly back up all computer and mobile phone data at a secure, cloud, or other out-of-office location.
- Identify and participate in continuing professional development activities and maintain a CPD logbook of developmental training and audit experience.
- Provide one-to-one coaching, support, or training of their staff.
- Be aware of the need for professional indemnity insurance. I recommend that this is discussed with a licensed insurance broker.
- Be a member of a recognized auditing organization/auditors' register, such as the International Register of Certificated Auditors (IRCA)—see Chapter 3.

I hope you enjoy participating in your first (or 101st, or 1001st) *Audit Adventure*™.

Appendix 3
Pre-audit Letter

INTRODUCTION

This suggested text provides foundational ideas for the lead auditor's initial letter to the auditee to arrange the audit, which typically requests a set of pre-audit documents and other information.

This text is available as a document to download and, if required, to amend to suit your requirements, from the companion website (https://routledgetextbooks.com/textbooks/_author/asbury/).

Some lead auditors prefer pre-audit questionnaires, with *fill in the blanks* questions. These are sent to a location in advance of an audit to gather information. Personally, I do not favour this approach, as the answers are often too closed, too binary, or too ad hoc to be of real assistance, but as mentioned elsewhere, I would be pleased to hear from auditors who use this method.

PRE-AUDIT LETTER—SUGGESTED TEXT

[Company name]
[Address]
[Post/Zip code]

[Date]

Dear [Name]

Re: HSEQ Audit 2024/Re: HSEQ Audit 2024—Second request/Re: HSEQ Audit 2024—Third and final request [insert subject, as applicable]

I write to confirm the arrangements for the [audit subject, for example, HSEQ] audit at your site, which is scheduled to commence on [date]. My team and I plan to be on-site at [time, for example, 8 am] on the first day.

We would appreciate an opening meeting with you and senior colleagues you may wish to nominate on the first morning, and a lockable room suitable for my team to work in for the duration of the audit.

To assist with our preparations, we would like to receive [hard/electronic] copies in [preferred language, for example, English] of the documents listed, to facilitate our arrangements and pre-audit preparation.

We would also appreciate your advice on suggested hotel accommodation for the team during our assignment.

Please forward this information to [preferred street or email address], marked for the attention of [Name], lead auditor.

To allow the necessary time for our preparation, we request that this information is provided no later than [add date].

Documents:

- Directions to the site and an area map
- Site rules, pointing out any particular training, PPE, or other mandatory access requirements applicable to us during our visit
- Site plan showing perimeter, buildings, and major processes
- Comprehensive organization chart/s
- Current business plan showing the organization's major objectives
- Copies of operating licences and permits (fire, environmental, waste, fleet, etc.)
- List of laws and regulations applicable to the audit
- Table of contents for the subject [e.g., health and safety, environment, quality] manual
- HSEQ policy statement/s
- Training matrix, summarizing training requirements, training provided and scheduled
- Minutes of the most recent [subject] management review meeting, or similar review forum.

On behalf of my team, I look forward to meeting you and your colleagues.

Yours sincerely,

[Your name and designatory letters]
Lead Auditor

GUIDANCE ON PRE-AUDIT DOCUMENTATION

In my experience, less is often more. I suggest that you don't ask for complete manuals at this early stage, as the table of contents will usually suffice to provide you with an overview of coverage—you can review the whole or the sections you wish to see during the audit.

I recommend keeping your document request list tight and short—this way, you'll be more likely to receive all the items requested. Auditees have told me that they find long lists exasperating. There is no need to try to obtain *everything* in advance.

If the documents you have requested do not show up by the date you have requested, send up to two reminders. As described in Chapters 4 and 6, I suggest the initial contact at three months' range, the first follow-up at two months, and the third and final request one month from the audit start date. If the documents are not received, the audit may need to be postponed. Starting ill-prepared seldom promotes a successful audit outcome.

Please don't be too disappointed if *some* of the documents you requested are not sent—this is actually very common. But do be ready to work a little harder in the on-site preparation stage to fill in the gaps if you decide not to press further in advance. It is also common, in my experience, to receive a *standard pack* of documents, which some organizations seemingly prepare to send to various enquirers. Ultimately, you must decide whether you have enough or whether further follow-up is required.

I'll conclude Appendix 3 by reminding you of A-Factor 25: *Time spent on reconnaissance is seldom wasted.*

Appendix 4
Guide to the e-Book+ and Companion Website

INTRODUCTION

The online content for the fourth edition is presented via a companion website. The printed book is also available as an *e-Book+*. *eBook+* is the enhancement of content in Taylor & Francis eBooks with additional interactive features to provide you the reader with added value and a better user experience. eBook+ titles are designed to make reading more engaging, increase comprehension and retention of the material, and ultimately lead you and your organization to greater success.

You will find the companion website at
https://routledgetextbooks.com/textbooks/_author/asbury/
The *eBook+* is available at www.routledge.com/9781032429083
This extended content includes the following:

1 Book Microlearning™ Presentations

The companion website and the *eBook+* present *Microlearning*™ presentations by the author on the Introduction, and on each chapter—11 PowerPoint Show (.ppsx) files in total.

They are called *Microlearning*™ because they are micro-short yet provide punchy learning. The shortest is less than eight minutes, and the longest is less than 14 minutes.

Each presentation was recorded to stand alone. You can move through the introduction, and between slides, by pressing the spacebar on your keyboard.

2 Learning Assessments

There is a multiple-choice assessment test for each chapter of the book.

Successful completion of each assessment is rewarded with three hours of CPD; a total of 30 hours CPD for completion of all ten assessments. The pass mark for each assessment is 70% per cent.

You should record the achievement of your personal CPD as required by your professional body.

3 *The Audit Adventure*™ Video Tutorial

Found in Chapter 5, this film, presented by the author, takes viewers through *The Audit Adventure*™. In 30 minutes, it provides an overview of this powerful auditing methodology.

You'll also find a recording of one-hour industry webinar on management systems and organizational improvement at the end of Chapter 2. It was recorded a few years ago, introduced by Hailey Thomas of PetroSkills LLC, and led by Dr Stephen Asbury.

4 Documents

You'll also find downloadable copyright-free documents for you to use when conducting an audit aligned to *The Audit Adventure*™, including Dr Stephen's:

- Pre-audit letter
- Terms of reference (TOR)
- Audit finding working paper (AFWP)
- Audit wall charts

5 OH&S-MS Toolkits for ISO 45001:2018

Your author has been involved in establishing, operating, and auditing OH&S and other MS on over 40,000 work sites. In these 16 toolkits, found exclusively on the companion website and via links from the eBook+, he shares his copyright-free processes and documents necessary to successfully implement and certify/register to ISO 45001:2018.

1. Context of the organization toolkit
2. Mapping to the standard (preliminary review) toolkit
3. Leadership and worker participation toolkit
4. OH&S policy, Roles, Responsibilities and Authorities (RRA) toolkit
5. Risk assessment and action planning toolkit
6. Legal and Other requirements toolkit
7. Competency and awareness toolkit

7A. Training course feedback form

8. Communications toolkit
9. Documented information toolkit
10. Control of contractors toolkit
11. Emergency preparedness and response toolkit
12. Monitoring performance toolkit
13. Internal audit toolkit
14. Management review toolkit
15. Incident reporting, investigation, and continual improvement toolkit
16. Safety Self-Audit (SSA) toolkit

Please acknowledge the author (Dr Stephen Asbury) in any use.

These toolkits may also be of interest to auditors as examples of the evidence that may be presented to their team during an audit they are participating in.

6 Example MSS Frameworks

You'll find downloadable examples of MSS frameworks, including

- COSO Framework and Enterprise Risk Management summaries,
- Global Reporting Initiative Guidelines (GRI),
- London 2012 Olympic Delivery Authority HSE-MS, and
- IOGP HSE-MS 6.36/210.

7 Articles and Papers of Interest

You'll find a selection of articles and papers on HSEQ and auditing that your author anticipates you'll find to be of interest:

- Apostrophes—a brief guide to their use
- London 2012 vision—a presentation made to the International Olympic Committee on 6 July 2005
- *The Art of War* by Sun Tzu
- Report of the Commission on Auditors' Responsibilities (AICPA, 1978)
- Watchdog or bloodhound?
- What does a CFO expect from internal audit?

8 Websites and Contacts of Interest

The list of useful websites provides readers with targeted places to look for further information on management systems, auditing, and bodies associated with professional HSEQ and auditing practice.

There are millions of websites in the world, and great search engines to trawl them. The following links have been selected to be of potential interest to HSEQ practitioners and internal auditors. They are presented alphabetically in each section.

Please contact the author stephen@stephenasbury.com if you wish to suggest additions, amendments, or if broken links are identified as time passes.

H&S Regulators

Canadian Centre for Occupational Health and Safety (CCOHS)—https://www.ccohs.ca

Department of Labour (US), Occupational Safety & Health Administration (OSHA)—https://www.osha.gov

Enforcement and Inspection Service, Department of Employment & Labour, Republic of South Africa—https://www.labour.gov.za/About-Us/Ministry/Pages/IES0320-7398.aspx
Health and Safety Executive (HSE, UK)—www.hse.gov.uk
Official HSE App—https://books.hse.gov.uk/HSE-Mobile-App
Safe Work Australia—https://www.safeworkaustralia.gov.au

Environmental Regulators

Environment Agency (UK)—https://www.gov.uk/government/organisations/environment-agency
Environmental Protection Agency (EPA, US)—https://www.epa.gov

International HSE Laws

Global database on OH&S legislation (LEGOSH)—https://www.ilo.org/dyn/legosh/en/f?p=14100:1:0::NO:::
Environmental laws, Australia—https://www.environment.act.gov.au/about-us/legislation-policies-guidelines
Environmental laws, US—https://www.epa.gov/laws-regulations
International labour standards (ILO)—https://www.ilo.org/global/standards/subjects-covered-by-international-labour-standards/occupational-safety-and-health/lang--en/index.htm
UK, Scotland, Wales, and Northern Ireland Legislation (National Archives)—https://www.legislation.gov.uk

Registration and Accredited Certification Bodies (Who Also Employ Auditors)

BAB (British Assessment Bureau)—https://www.british-assessment.co.uk
British Standards Institute (BSi)—https://www.bsigroup.com
BSi worldwide country locator—https://www.bsigroup.com/en-GB/global-contact-details/
Bureau Veritas—https://certification.bureauveritas.com
DNV (Det Norske Veritas)—https://www.dnv.com
Isoqar (Alcumus)—https://www.alcumus.com/en-gb/certification/
LR (Lloyd's Register)—https://www.lr.org/en/
NQA—https://www.nqa.com/global
SGS—https://www.sgs.com/en
UKAS provides its *CertCheck* tool to search for accredited organizations and to validate its certificates—https://www.ukas.com/find-an-organisation/
In addition, the International Accreditation Forum (IAF) provides links to over 1200 accredited national certification bodies around the world—https://www.iafcertsearch.org/search/certification-bodies (IAF does not provide certification services itself)

Auditing and Auditors' Professional Organizations

American Society for Quality (ASQ)—https://asq.org/cert
Board for Global EHS Credentialing (BGC)—https://gobgc.org
Exemplar Global (formerly RABQSA)—https://exemplarglobal.org
IEMA Environmental Auditor Register—https://www.iema.net/membership/specialist-registers/environmental-auditor-register
IIA (international)—https://www.theiia.org
IIA (UK)—https://www.iia.org.uk
International Auditing and Assurance Standards Board—https://www.iaasb.org/about-iaasb Professional Evaluation and Certification Board (PECB)–https://pecb.com
International Personnel Certification Association (IPC)—http://www.ipcaweb.org
IRCA—https://www.quality.org

Environmental and Conservation Professional Organizations

British Ecological Society—https://www.britishecologicalsociety.org
Chartered Institute of Ecology and Environmental Management (CIEEM)—https://cieem.net
Chartered Institution of Water and Environmental Management (CIWEM)—https://www.ciwem.org/#
Institute of Environmental Management and Assessment (IEMA)—www.iema.net
Institution of Environmental Sciences—https://www.the-ies.org
International Society of Sustainability Professionals—https://www.sustainabilityprofessionals.org
National Trust—https://www.nationaltrust.org.uk
Society for the Environment—https://socenv.org.uk
UK Environmental Law Association—https://www.ukela.org

H&S Professional Organizations

American Society of Safety Professionals (ASSP)—https://www.assp.org
Australian Institute of Health and Safety (AIHS)—https://www.aihs.org.au
Canadian Society of Safety Engineering (CSSE)—https://www.csse.org
Institution of Occupational Safety and Health (IOSH)—www.iosh.com
International Institute of Risk and Safety Management (IIRSM)—https://www.iirsm.org
Royal Society for the Prevention of Accidents (RoSPA)—https://www.rospa.com
Safety Groups UK—https://www.safetygroupsuk.org.uk

Quality Professional Organizations

Chartered Quality Institute (CQI)—https://www.quality.org
Society of Quality Assurance—https://www.sqa.org

Publications

Fire Safety Journal (International Association of Fire Safety Science)—https://www.sciencedirect.com/journal/fire-safety-journal
IOSH Magazine—https://www.ioshmagazine.com
Journal of Safety Research (National Safety Council and Elsevier)—https://www.sciencedirect.com/journal/journal-of-safety-research
Quality Digest (QD)—https://www.qualitydigest.com
Quality World (CQI and IRCA)—https://www.quality.org/qualityworld
Safety journal (MDPI)—https://www.mdpi.com/journal/safety
Safety Science—https://www.sciencedirect.com/journal/safety-science
Transform magazine (IEMA)—https://www.iema.net/transform

Please review my References and Further reading in the text book. In addition, most of the regulatory, registration, and membership bodies noted here produce their own online and print magazines and newsletters. Contact them for further details.

LinkedIn Groups of Interest to HSEQ Practitioners and Auditors

EHS Professionals—https://www.linkedin.com/groups/46570/
EHSQ Elite (No. 1 in Safety) Environmental Health Safety Sustainability Security Quality Elite—https://www.linkedin.com/groups/113464/
Environmental, Health and Safety—https://www.linkedin.com/groups/129347/
Environmental Health and Safety Professionals—https://www.linkedin.com/groups/43169/
Food Safety & Quality Assurance—https://www.linkedin.com/groups/792737/
Good Auditing Practice—https://www.linkedin.com/groups/1893835/
Health & Safety (H&S) Professionals UK—https://www.linkedin.com/groups/2315060/
ISO 9001, ISO 14001, and ISO 45001—https://www.linkedin.com/groups/1306627/
ISO 45001, ISO 14001, ISO 22000, and ISO 20121: Health Safety and Environment Standards—https://www.linkedin.com/groups/8318823/
Jobs and Career: Auditing [etc.]—https://www.linkedin.com/groups/2064244/
Oil & Gas HSE Practitioners—https://www.linkedin.com/groups/137144/
Quality Assurance—QA Professional, Testing, Test Automation—https://www.linkedin.com/groups/60879/
Quality Assurance and Process Improvement—https://www.linkedin.com/groups/78075/
Quality & Regulatory Network—https://www.linkedin.com/groups/80327/
Safety OHS EHS HSE OSHA OSH Quality, Security, Jobs International Group—https://www.linkedin.com/groups/2877202/
The Official IOSH Group—https://www.linkedin.com/groups/3969846/

Other Websites of Interest

European Agency for Safety and Health at Work—https://osha.europa.eu/en
European Commission health and safety at work—https://ec.europa.eu/social/main.jsp?catId=148
European data on accidents at work (Eurostat)—https://cc.europa.eu/eurostat/statistics-explained/index.php?title=Category:Accidents_at_work
Financial Reporting Council (FRC) *Auditor's Responsibilities for the Audit*—https://www.frc.org.uk/auditors/audit-assurance-ethics/auditors-responsibilities-for-the-audit
Institute of Chartered Accountants (ICAEW) *What Auditors Do: The Scope of Audit*—https://www.icaew.com/technical/thought-leadership/full-collection/audit-and-assurance-thought-leadership/what-auditors-do-the-scope-of-audit
IOSH discussion forum—http://forum.iosh.co.uk
National Audit Office (UK)—https://www.nao.org.uk
National Careers Service (UK) *Auditor*—https://nationalcareers.service.gov.uk/job-profiles/auditor

Your Author's Webpages

Dr Stephen Asbury on Amazon—https://www.amazon.com/stores/Stephen-Asbury/author/B001HD3Q9G and https://www.amazon.co.uk/Stephen-Asbury/e/B001HD3Q9G
LinkedIn—https://www.linkedin.com/in/stephenasbury/
Google Scholar—https://scholar.google.co.uk/citations?user=Qdy481kAAAAJ&hl=en
ORCID—https://orcid.org/my-orcid?orcid=0000-0002-5562-7738
Taylor & Francis—https://www.routledge.com/authors/i8614-stephen-asbury# and https://routledgetextbooks.com/textbooks/_author/asbury/
IOSH—https://www.iosh.co.uk/IOSH/Home/MyIOSH/Member%20directory/Profile%20detail?Id=6515
ASSP—https://community.assp.org/profile/21198

Appendix 5
Example Examination Questions

This appendix provides six example questions related to auditing from recent international examination board papers. Questions 1–4 are at Certificate level, and Questions 5 and 6 are at Diploma level. These were kindly provided by Dr Jonathan Backhouse, a former NEBOSH examiner.

Questions 1 and 2 (Certificate), and 5 and 6 (Diploma) include examiner's commentaries and/or the suggested main points of the required answers. I have left two of the Certificate-level questions unanswered so that you can think about them!

1 A health and safety audit of an organization has identified a general lack of compliance with procedures.
   I Describe the possible reasons for procedures not being followed (10 marks).
   II Outline the practical measures that could be taken to motivate employees to comply with health and safety procedures (10 marks).

Commentary

In answering part (i) of the question, candidates suggested a range of possible reasons for health and safety procedures not being followed by employees. In the main, answers concentrated on

- inadequate supervision and enforcement of the procedures by management,
- a poor attitude toward health and safety generally (as an indication of a poor safety culture), and
- issues relating to working conditions that may hinder compliance with procedures (such as poor workstation design and inattention to ergonomic issues).

Fewer candidates included other possibilities, such as

- the procedures themselves being unrealistic or unclear;
- literacy and language issues;
- peer pressure;
- other pressures or incentives to cut corners;

- a failure by management to consult the workforce;
- a failure by management to provide the necessary information and training; and
- the repetitive, tedious, or complex nature of the tasks being performed.

For part (ii), candidates needed to outline practical measures that could be taken to motivate employees to comply with the procedures. These should have included

- a display of commitment on the part of management;
- the provision of a good working environment;
- joint consultation and the involvement of employees in drawing up and reviewing the procedures;
- the setting of personal performance targets with due recognition when these are achieved;
- the provision of information and training, including toolbox talks and the use of posters and noticeboards;
- the introduction of job rotation; and,
- finally, the taking of disciplinary action in cases where there is a deliberate failure to follow laid-down procedures.

There were some reasonable responses to this question, but many candidates did not gain all the marks that they might because they did not provide adequate depth to their answers. It was important in part (i), for instance, for candidates to provide sufficient detail to indicate why the particular issue identified might lead to a procedure not being followed.

2 Outline the reasons why an organization should monitor and review its health and safety performance (8 marks).

Commentary

Several candidates experienced problems with this question by appearing to mistake the words 'the reasons why' for 'how' and proceeding to outline the various methods by which health and safety performance can be monitored and reviewed. The need to read the question carefully and understand its purpose is advice that cannot be given too often.

Among the reasons offered by better candidates were the following:

- To identify substandard health and safety practices and conditions (perhaps by means of workplace inspections)
- To identify trends in relation to different types of incidents, or incidents in general (by analysis of relevant incident data)
- To compare actual performance with previously set targets
- To 'benchmark' the organization's performance against that of similar organizations or an industry norm
- To identify whether control measures are in use and to assess their effectiveness
- To be able to make decisions on appropriate remedial measures for any deficiencies identified
- To set priorities and establish realistic timescales

- To assess compliance with legal requirements
- To be able to provide a board of directors or safety committee with relevant information

An additional reason for monitoring and reviewing health and safety performance is quite simply because there is a legal requirement to do so under the Management of Health and Safety at Work Regulations 1999.

Weaker answers to this question never really progressed from 'so that we can see how we are doing', whereas it was the reasons 'why we should want to know how we are doing' that were being sought. The question required a little thought but allowed candidates with a good understanding of health and safety management systems to shine.

3 Identify the advantages AND disadvantages of carrying out a health and safety audit of an organization's activities by
  i an internal auditor (10 marks) and
  ii an external auditor (10 marks).
4 List the documents that are likely to be examined during a health and safety audit (8 marks).

Questions 5 and 6 are set at Diploma level for more advanced candidates.

5 As the Health and Safety Adviser to a large organization, you have decided to introduce an in-house auditing programme to assess the effectiveness of the organization's health and safety management arrangements.

Outline the issues to be addressed in the development of the audit system (20 marks).

Answer summary:

- Obtaining support and commitment from senior managers, and from other key stakeholders
- Logistics and available resources/audit intensity
- Nature, scale, and frequency of auditing (relative to level of risk)
- Terms of reference and Scope (areas to be included in each audit)
- Recognition of the need to develop audit plans and terms of reference
- ISO 45001:2018, HSG65, or other management system/reference framework
- Specific questions that might be included in the audit interviews
- Identifying the key elements of an audit process (Initiate, Prepare, Conduct, Report, Conclude)
- The types of auditing (system elements, performance standards, scope, comprehensive, horizontal, or vertical slicing, etc.)
- Use of a single auditor vs audit teams
- Personnel implications:
  - Use of audit teams
  - Auditor training

- Briefing for those affected
- Preparation and presentation of the final report

6 As the Health and Safety Adviser to a large organization, you have decided to develop and introduce an in-house auditing programme to assess the effectiveness of the organization's health and safety management arrangements.

Describe the organizational and planning issues to be addressed in the development of the audit programme. You do not need to consider the specific factors to be audited (20 marks).

Answer summary:

- A strategic approach (high-level, future-focused) is required
- Consideration of the logistics and resources required
- Obtaining the support and commitment of senior managers and other key stakeholders
- Nature, scale, and frequency of the auditing relative to the level of risk involved
- Standards against which the management arrangements are to be audited
- Identification of the key elements of the process:
  - Initiate audit planning
  - Prepare work plan
  - Conduct—review and verify
- *Nemawashi* ('no surprises') meeting
- Preparation and presentation of the final report
- Consideration of issues such as scoring systems or use of proprietary software
- Types of auditing (such as comprehensive or horizontal or vertical slicing)
- Use of a single auditor or audit teams
- Training of the auditors and briefing of those members of the organization who are likely to be affected

Appendix 6
Frequently Asked Questions

1. **What is the difference between an inspection and an audit?**

   Here are two different activities with language that often gets confused in common discourse.

   In the language of the HSEQ professions, an inspection is a verification in the present to confirm that something is in the required state. This state may be locked, unlocked, tidy, guarded, protected, undamaged, or charged at the correct pressure. It provides information on the current state to supervisors and line managers.

   An audit is a systematic review through time, often years, to verify activities, facilities, or processes against specified requirements. It leads to a high-level, future-focused audit opinion and provides a level of assurance (high or low) to the stakeholders (aka interested parties) of an organization.

2. **What is the difference between a lead auditor and an auditor?**

   A lead auditor is usually formally trained in auditing practice and has participated in numerous audits previously. They are responsible for the whole conduct of a single audit.

   An auditor may also be formally trained but works with the lead auditor as a member of an audit team to conduct a single audit.

3. **Does an auditor need formal training?**

   Ideally, yes. In some territories, this is a legal requirement.

   For an organization to maximise the benefits from its internal audits, auditor training is highly recommended.

4. **Should an auditor join an auditor registration body?**

   Ideally, yes. In my career, I have been a member of several auditor registers and registration bodies, including some of those highlighted in Chapter 3. Such bodies provide CPD opportunities, news related to auditing, and networking opportunities to their members.

I have always taken the view that individuals including auditors should pursue the highest-level qualifications and professional credentials available to them.

5. **Should an auditor complete CPD (continuing professional development)?**

Some of the auditor registration bodies make CPD a mandatory requirement for their members, and some suggest it as a best practice.

In the same way that I would expect a medical doctor or dentist to keep up to date with the latest medical and dental research, techniques, and practices, I believe that an auditor who regularly considers and keeps a record of their professional development is likely to perform better than one who does not. It is normal to keep a record of participation in training, other professional developments (such as reading this book and completing its assessments), along with the details of audits they have conducted.

Glossary of Operations Integrity Language

accident The opposite of 'on purpose'

actual incident A specific event or extended condition that results in a significant, unwanted, and unintended impact on the safety or health of people, on property, on the environment, on legal/regulatory compliance, or on security as it relates to operations integrity (see *incident* and *near-miss incident*)

aspect See *environmental aspect*

audit A process which evaluates activities, facilities, or processes against management system requirements (or expectations)

audit fieldwork Time spent during an audit—typically 60 per cent of total audit time—to determine management system effectiveness within the organization through interviews with execution resources, observations of work activities, etc. (aka Review and Verify)

bridging document A document that addresses operations integrity interfaces between functions or organizations. It describes the interfaces and ensures that responsibilities are identified, assigned, and accepted.

business environment The current and future characteristics of the place where the organization, its supply chain, and its competitors operate. Understanding the business environment sets the *context* for understanding the needs and expectations of stakeholders, and determines the necessary scope of the management system.

business objectives Objectives set by management in the context of their business environment (or *context*) to meet the requirements of their stakeholders/interested parties

commissioning The activities that relate to placing newly constructed or installed facilities or equipment in service or returning modified facilities or equipment to service. Such activities, usually carried out by the project organization in close cooperation with operators, include verification that the installation is correct with respect to drawings, filling systems with fluids, testing for leakage, test running equipment, and initial testing of entire systems to their operating conditions. Commissioning ends when the facility is accepted by the operating organization for initial operations.

competency The skills, knowledge, attitude, training, and experience ('SKATE') required of personnel to successfully perform job-related tasks to achieve business objectives

context Clause 4 of ISO Annex SL (see *business environment*)

contract A legally enforceable, documented agreement that includes purchase order and service agreement (i.e., offer, acceptance, and consideration)

controlled documents Documents managed by a system that assigns custodians who control modifications, circulation, and distribution. The system ensures removal of obsolete or superseded material.

critical See *integrity critical*

data Quantitative or qualitative information derived from testing and/or measurement

deployment The act of providing initial communication and training for a system, process, procedure, or programme

discharge The episodic or ongoing release of materials to the water or onto the land

document owner The individual assigned to provide management oversight and final approval of modifications, circulation, and distribution of a document

documented information Electronic or hard-copy manual, procedure, drawing, legal agreement, correspondence, record, or reference that contains information (i.e., documents and records)

emission The episodic or ongoing release of materials to the air

employee Any individual on the payroll of an organization

environmental aspect Element of an organization's activities, products, or services that can interact with the environment

environmental impact Any change to the environment, whether adverse or beneficial, wholly, or partly, resulting from an organization's environmental aspects

expectations The minimum requirements set out in a documented management system standard

external audit An assurance process conducted by auditors from outside the organization which determines the extent to which operations comply with the expected requirements of a specified management system. Synonymous with third-party audit and certification or registration audit.

facilities Physical equipment and/or plant, including large mobile equipment, involved in the performance of operations

function A specialized organization that supports the operating organization. Often used to describe that part of a corporate function specifically assigned to work within an asset.

gap An identified deficiency in addressing and/or complying with a management system expectation or other defined requirement

gap analysis A method or means to identify deficiencies between management system expectations and actual practices. The results of such analyses are documented.

***gemba* attitude** Going to the source of the facts to make correct decisions

guidelines A second tier of requirements developed by organizations or their regulators to further define expectations. Synonymous with approved codes of practice.

hazard Potential for harm

higher risk Combinations of probability and consequence on a risk matrix that constitute an area of significant risk as defined by management (synonymous with *Big Rocks*)

human factors Integration and application of scientific knowledge about people, facilities, and management systems to improve their interaction in the workplace, to achieve improved safety, health, and environmental and financial performance. Human factors considerations may include workplace design, equipment design, work environment, physical activities, job design, information transfer, and personal factors.

implementation A multi-step process involving allocation of resources, roles, responsibilities and authorities, competency, training and awareness, consultation, participation and communication, documentation and control of documents, system stewardship for effectiveness and continual improvement. Clauses 5–10 of ISO 9001, ISO 14001 and ISO 45001 refer to the key steps in implementation.

incident A specific event or extended condition that results (an *actual incident*) or could have resulted (a *near-miss incident*) in a significant unwanted and unintended impact on the safety or health of people, on property, on the environment, on legal/regulatory compliance or on security as it relates to operations integrity

initial training The training required before assuming job duties for a new hire, prior to performing the duties of a new position for an existing employee, or acquired during the initial performance of job duties

integrity critical Describes an activity, piece of equipment, device, process, document, or position determined to be vital to the prevention or mitigation of a major event as it relates to operations integrity. Such an event might be an uncontrolled emission, fire, or explosion that poses serious danger to people, property, or the environment.

interface A place or time at which independent systems and/or operations of two or more companies, functions, assets, departments, or groups of works should act with or communicate with each other. With respect to HSEQ management systems, the phrase 'interfaces between operations' refers to two types of interface: (i) procedural-level interfaces, typically occurring at sites with shift operations or simultaneous work activities that require special procedures (e.g., shift change procedures, simultaneous operations procedures, or manuals); and (ii) systems-level interfaces between two or more companies, functions, or assets that require system documentation (e.g., bridging documents, and service agreements).

internal audit A process conducted by members of the assessed organization operating independently which determines the extent to which operations comply with the expected requirements of a specified management system. Synonymous with first-party audit.

key positions Those positions in an organization which could have a significant impact on operations integrity through individual decision-making authority, operational control, or individual actions. Note these positions are sometimes held by third parties.

line management Management or individuals responsible for the day-to-day functioning of an organization

long-term shutdown Suspension in the operation of facilities for a duration that requires the preservation of equipment, for example, depressurization or use of corrosion inhibitors

major project A project requiring senior management endorsement, and possibly capital expenditure/Cap-Ex

management Senior management is typically made up of corporate function managers to whom operations and department managers report. Senior operations management is typically the department level of management directly responsible for operations. Operations management is typically made up of operating organization personnel responsible for direct supervision of operations. Site management is typically the most senior level of supervision working on-site.

near-miss incident An undesirable event which could have resulted, under slightly different circumstances, in an unwanted impact on the safety or health of people, property, or the environment (see *incident* and *actual incident*)

nemawashi An informal process of laying the foundations for a project or change—a Japanese term, literally 'going around the roots'

occupational health programme All activities addressing workplace health hazards and employee health. These include identification, evaluation, and control of health hazards, monitoring of worker exposure, communication of health hazard knowledge, determination of employees' medical fitness for duty, and providing or arranging for medical services necessary for the treatment of occupational illnesses or injuries.

ongoing training Training oriented toward improving an employee's performance in each position. It could include training to keep the employee current with changing technology or to provide improved skills.

operated by others (OBO) Joint venture facilities or long-term operations in which the organization has a vested business interest, which are not operated by the organization

operating organization A subset of a corporate function responsible for activity involving the production, manufacture, use, storage, movement of materials, or utilization of resources to produce an output

operation Any activity involving the production, manufacture, use, storage, or movement of material. Also, the utilization of resources by a function to produce an output.

operations integrity Operations integrity addresses all aspects of the organization's business, including security, which can impact safety, health, and environmental performance

performance measures Indicators used to determine whether a system or group of systems is meeting or progressing toward expected objectives and likely to continue the progress. Included are two types of measures: (i) active measures—utilized to measure the degree to which the execution of a system conforms to requirements (i.e., leading indicators); and (ii) reactive measures—utilized to measure the degree to which system objectives are being met (i.e., lagging indicators)

personnel A broad term generally used to describe employees, contractors (i.e., third parties) or other authorized individuals involved in a specific activity. Synonymous with *workers* or *workforce*.

practice Approved method or means of accomplishing stated tasks

procedure A documented series of steps to be carried out in a logical order for a defined operation or given situation

process A series of actions, changes, or functions that bring about an end or result

product A material that has been manufactured, refined, or treated and which is sold or exchanged

programme A plan that includes scheduled events or activities

project A planned undertaking with a specific objective and defined scope. Includes new constructions and additions or revisions to existing facilities (and see *major project*)

qualified Possessing knowledge and skills, gained through training and experience, required to successfully perform job-related tasks. The opposite of *unqualified.*

quality The ability of a *product*, service, or activity to meet or exceed requirements

quality assurance Planned and systematic actions necessary to ensure that a *product*, service, or activity can meet or exceed requirements

quality control A control that determines whether a *product*, service, or activity meets requirements through measurement, testing, and/or inspection

records Documented information that reflects evidence of past events or actions (e.g., inspections, incident investigations, training, maintenance, air emissions, risk assessments, and action plans)

refresher training Training that reinforces employees' skills and understanding, enabling them to continue to perform safely and correctly and in an environmentally sound manner

risk Impact on (business) objectives. A function of the probability (or likelihood) of an unwanted incident combined with the severity of its potential consequences. Calculated as $R = P \times C$.

risk analysis Development of a qualitative or quantitative estimate of risk

risk assessment Process of judging the significance of risk and determining whether further risk reduction is required or desirable

risk assessment matrix (RAM) A grid on which the probability of a hypothetical incident occurring is plotted against the severity of the consequences that would result if the incident did occur. A matrix on which the levels of probability and the severity of consequences are defined is usually contained in an organization's management system manuals or similar documentation. Plotting incident scenarios on this matrix is a means of prioritizing and communicating the risk level associated with assessed activities.

risk management The application of policies and practices to the tasks of identifying, assessing, and controlling risks to protect human life, the environment, physical assets, and company reputation in a cost-effective manner

root cause(s) The most fundamental or underlying reason(s) for the occurrence of an event which, when corrected, will prevent, or significantly reduce the probability or consequences of its recurrence (synonymous with TapRoot, Domino theory, or 5 Whys)

screening A methodical process to eliminate or select personnel from a large pool of candidates to develop a smaller, finite group that is ultimately used to select candidates for a position. Examples may include verification of educational credentials, residency, employment history and mercantile (credit) checks, criminal background checks, checks related to automotive and highway safety, or pending litigation.

security A set of controls or measures taken to protect personnel from injury and to protect physical assets from damage or loss as it relates to SHE, including where this could have an adverse environmental impact.

significant near-miss incident An undesirable event with significant potential consequences or high learning value that could have resulted, under slightly different circumstances, in an unwanted impact on the safety or health of people, property, or the environment (aka *high-potential near miss*)

site The place where something was, is, or will be located

stakeholders Those interested (i.e., holding a stake) in the achievement of objectives. Synonymous with 'interested parties'. Categorized into five main groups: Shareholders (or investors), Employees, Suppliers (those with whom we do business), Customers, and Society (including regulators, pressure groups, the media, and the public).

standard A documented level of requirement which must be achieved, or which is a target for achievement

stewardship The process of being accountable for an activity or function and its performance against a goal, including maintaining knowledge of its status and reporting its condition to higher management

system A structured means of control for ensuring that business objectives are achieved and sustained. The characteristics of current ISO HSEQ management systems (per Annex SL) are 4. Context of the organization, 5. Leadership, 6. Planning, 7. Support, 8. Operation, 9. Performance evaluation, and 10. Improvement.

third party A contractor, subcontractor, supplier, or vendor providing materials or services in accordance with specifications, terms, and conditions documented by a contract agreement signed by both parties

workforce A broad term generally used to describe workers, employees and contractors, or other authorized individuals involved in a specific activity. Synonymous with *personnel*.

Further Reading

In addition to my references which follow, the titles in this section are (in my view) informative sources to guide readers to learn more about corporate environments, management and management systems, and tools and techniques relating to the subject matter of business and auditing. The resources listed represent the main works which I have found useful over the years, while developing as a risk specialist and an auditor. Of course, no personal list such as this one could be exhaustive, and you may find other sources equally useful.

Ammerman, M. (1998). *The Root Cause Analysis Handbook: A Simplified Approach to Identifying, Correcting, and Reporting Workplace Errors*. Boca Raton: CRC Press.

Anderson, G., and Lorber, R. (2006). *Safety 24/7: Building an Incident-free Culture*. Aberdeen: Intertek Consulting and Training.

Barton, T., Shenkir, W., and Walker, P. (2002). *Making Enterprise Risk Management Pay-off*. Upper Saddle River, NJ: Prentice Hall/Financial Times.

Bedfordshire County Council and RSPB (Royal Society for the Protection of Birds). (1996). *A Step by Step Guide to Environmental Appraisal*. Shefford: Bedfordshire County Council.

Bendell, T., Boulter, L., and Kelly, J. (1993). *Benchmarking for Competitive Advantage*. Lanham, MD: Pitman Publishing.

Blanchard, K. and Johnson, S. (1983). *The One Minute Manager*. New York: HarperCollins Fontana.

Borge, D. (2001). *The Book of Risk*. Hoboken, NJ: Wiley.

Brooks, I., and Weatherspoon, J. (1997). *The Business Environment: Challenges and Changes*. Upper Saddle River, NJ: Prentice Hall.

Buchholz, R. (1998). *Principles of Environmental Management: The Greening of Business*. Upper Saddle River, NJ: Prentice Hall.

Campbell, D. (1997). *Organisations and the Business Environment*. Oxford: Butterworth-Heinemann.

CBI. (1990). *Narrowing the Gap: Environmental Auditing Guidelines/or Businesses*. London: Confederation of British Industry.

Chisnall, P. (1989). *Strategic Industrial Marketing*. Upper Saddle River, NJ: Prentice Hall.

Cormack, D. (1987). *Team Spirit*. Lisbon: MARC Europe.

COSLA. (1992). *Guidelines on Environmental Auditing*. Edinburgh: Convention of Scottish Local Authorities.

Council of the European Communities. (1990). Council Directive 90/313/EC on the freedom of access to information on the environment. *Official Journal*, L158.

Council of the European Communities. (1993). Council Regulation (EEC) No. 1836/93 of 29 June 1993 allowing voluntary participation by companies in the industrial sector in a community eco-management and audit scheme. *Official Journal*, L168.

Covey, S. (1989). *The 7 Habits of Highly Effective People*. New York: Simon & Schuster.

Crainer, S. (1996). *Leaders on Leadership*. Corby: The Institute of Management.

Curwin, J., and Slater, R. (1991). *Quantitative Methods for Business Decisions*. London: Chapman and Hall.

Daniels, J., and Radebough, L. (1997). *International Business: Environments and Operations*, 8th ed. Boston: Addison-Wesley.

Davies, P. (1990). *Your Total Image: How to Communicate Success*. London: Piatkus.

Department of the Environment. (1993). *A Guide to the Eco-Management and Audit Scheme for UK Local Government*. London: HMSO.
Drucker, P. (1970). *Drucker on Management*. Corby: Management Publications Ltd for British Institute of Management.
Environmental Resources Management. (1996). Environmental audit and assessment: Concepts, measures, practices, and initiatives. *SNH Review*, 46.
Finlay, P. (2000). *Strategic Management: An Introduction to Business and Corporate Strategy*. Upper Saddle River, NJ: Prentice Hall.
Friedman, T. (2005). *The World is Flat*. London: Penguin Group.
Friends of the Earth. (1989). *Environmental Charter for Local Government: Practical Recommendations*. London: Friends of the Earth.
Goldratt, E. (1988). *The Goal*. Aldershot: Gower.
Goldratt, E. (1994). *It's Not Luck*. Aldershot: Gower.
Graham, A. (1990). *Investigating Statistics*. London: Hodder and Stoughton.
Grayson, L. (1992). *Environmental Auditing: A Guide to Best Practice in the UK and Europe*. London: The British Library.
Greeno, J., Hedstrom, G., and DiBerto, M. (1988). *The Environmental, Health, and Safety Auditor's Handbook*. Boston, MA: Arthur D. Little.
Handy, C. (1995a). *Beyond Certainty*. London: Hutchinson.
Handy, C. (1995b). *Waiting for the Mountain to Move*. London: Arrow Books.
Handy, C. (1997). *The Hungry Spirit*. London: Hutchinson.
Hart, M. (1993). *Survey design and analysis using Turbostats*. London: Chapman & Hall.
Heller, R. (1998). *In Search of European Excellence*. New York, NY: HarperCollins.
Hendy, J., and Ford, M. (2004). *Redgrave, Fife, and Machin: Health and Safety*. Oxford: Butterworth.
Hill, T. (1991). *Production/Operations Management*. Upper Saddle River, NJ: Prentice Hall.
Huczynski, A., and Buchanan, D. (1991). *Organisational Behaviour*. Upper Saddle River, NJ: Prentice Hall.
James, T., and Woodsmall, W. (1988). *Time Line Therapy and the Essence of Personality*. Redwood City, CA: Meta Publications.
Jay, A. (1967). *Management and Machiavelli*. New Orleans, LA: Pelican.
Jenkins, M., Pasternak, K., and West, R. (2005). *Performance at the Limit: Business Lessons from Formula 1 Motor Racing*. Cambridge: Cambridge University Press.
Johnson, G., and Scholes, K. (1999). *Exploring Corporate Strategy*. Upper Saddle River, NJ: Prentice Hall.
Johnston, G. (2019). *Effective Root Cause Analysis: Looking at Control, Responsibility, Process Improvement and Making the Whole Activity More Effective*. Chicago: Independently published. www.amazon.co.uk/Effective-Root-Cause-Analysis-responsibility/dp/1701236419/
Jones, E., Gotts, D., and McGregor, J. (1992). Environmental auditing and its relevance to agriculture. *Farm Management*, 8(2).
Kanter, Moss-R. (1989). *When Giants Learn to Dance*. Chicago, IL: Touchstone, Simon & Schuster.
Kolk, A. (2000). *Economics of Environmental Management*. Upper Saddle River, NJ: Prentice Hall.
Kolluru, R. (1994). *Environmental Strategies Handbook: A Guide to Effective Policies and Practices*. New York: McGraw Hill.
Kolluru, R., Bartell, S., Pitblado, R., and Stricoff, S. (1996). *Risk Assessment and Management Handbook for Environmental, Health and Safety Professionals*. New York: McGraw-Hill.
LEAF. (1994). *The LEAF Environment Audit 1994*. Stoneleigh Park, UK: Linking Environment and Farming, National Agriculture Centre.
Local Government Management Board. (1991). *Environmental Auditing in Local Government: A Guide and Discussion Paper*. London: HMSO.
Lorriman, J., and Kenjo, T. (1994). *Japan's Winning Margins*. Oxford: Oxford University Press.
Magretta, J. (2002). *What Management Is*. New York, NY: HarperCollins.

Mintzberg, H., Ahlstrand, B., and Lampel, J. (1998). *Strategy Safari*. Upper Saddle River, NJ: Prentice Hall.

Morgan, G. (1986). *Images of Organisation*. Thousands Oaks, CA: Sage.

Morris, H., and Willey, B. (1996). *The Corporate Environment*. Lanham, MD: Pitman.

Moser, C., and Kalton, G. (1971). *Survey Methods in Social Investigation*. Portsmouth, NH: Heinemann.

Neale, A., and Haslam, C. (1995). *Economics in a Business Context*. London: Chapman & Hall.

Pascale, R., and Athos, A. (1986). *The Art of Japanese Management*. London: Penguin.

Peters, T. (1988). *Thriving on Chaos*. London: Pan Books.

Peters, T. (1992). *Liberation Management*. London: Pan Books.

Peters, T. (1994a). *The Pursuit of Wow!* New York: Macmillan.

Peters, T. (1994b). *The Tom Peters Seminar*. New York: Macmillan.

Peters, T. (1997). *The Circle of Innovation*. London: Hodder and Stoughton.

Peters, T., and Waterman, R. (1982). *In Search of Excellence*. New York: HarperCollins.

Porteous, A. (1996). *Dictionary of Environmental Science and Technology*. Hoboken, NJ: Wiley.

Pritchard, P. (2000). *Environmental Risk Management*. Oxford: Earthscan.

Raemaekers, J. (1993). Corporate environmental management in local government: the quality and management of action programmes, internal audits and State of the Environment Reports. *Research Paper No*. 48, School of Planning and Housing, Edinburgh College of Art/Heriot-Watt University, Edinburgh.

Raemaekers, J., Cowie, L., and Wilson, L. (1991). An index of local authority Green Plans. *Research Paper* 37, School of Planning and Housing, Edinburgh College of Art/Heriot-Watt University, Edinburgh.

Roderich, O. (2021). *5 Whys: The Effective Root Cause Analysis*. Independently published. www.amazon.co.uk/Whys-Effective-Root-Cause-Analysis/dp/B08WJY53FS/

Steiner, G., and Steiner, G. (1994). *Business, Government and Society: A Managerial Perspective*, 7th ed. New York: McGraw Hill.

Syed, M. (2015). *Black Box Thinking: Marginal Gains and the Secrets of High Performance*. London: John Murray.

System Improvements Inc. (2023). *TapRooT ® Root Cause Analysis*. www.taproot.com

Therivel, R. (1994). *Environmental Appraisal of Development Plans in Practice. School of Planning*. Oxford: Oxford Brookes University.

Thompson, S., and Therivel, R. (1991). Environmental auditing. *Working Paper* No. 130, Schools of Biological and Molecular Sciences, and Planning, Oxford Brookes University, Oxford.

Welford, R., and Gouldson, A. (1993). *Environmental Management and Business Strategy*. Lanham, MD: Pitman Publishing.

Whitmore, J. (2011). *Coaching for Performance: Growing Human Potential and Purpose: The Principles and Practice of Coaching and Leadership*. London: Nicholas Brealey Publishing.

Witte, R., and Witte, J. (2017). *Statistics*, 11th ed. Hoboken: Wiley.

Worthington, I., and Britton, C. (2000). *The Business Environment*, 3rd ed. Upper Saddle River, NJ: Prentice Hall.

References

AICPA. (1978). *The Commission on Auditors' Responsibilities: Report, Conclusions and Recommendations*. Durham: American Institute of Certified Public Accountants.

Alsop, P., and LeCouteur, M. (1999). Measurable successes from implementing an integrated OHS management system at Manningham City Council. *Journal of Occupational Health and Safety—Australia and New Zealand*, 10, 275–78.

ANSI. (2019). *Occupational Health and Safety Management Systems*. ANSI/ASSP Z10.0–2019. https://webstore.ansi.org/standards/asse/ansiasspz102019. Accessed 10 February 2023.

Antle, R. (1981). In Simunic, D. (1984), Auditing, Consulting and Auditor Independence, *Journal of Accounting Research*, 22(2), 679–702.

Asbury, S. (2005). A risk-based approach to auditing. *The Environmentalist*, 29 June.

Asbury, S., and Ball, R. (2016). *The Practical Guide to Corporate Social Responsibility*. Abingdon and New York: Routledge an imprint of Taylor & Francis.

ASQ. (2023). *About ASQ*. https://asq.org/about-asq. Accessed 14 February 2023.

Atherton, J., and Gil, F. (2008). *Incidents that Define Process Safety*. Hoboken, NJ: Wiley Interscience.

BAB. (2015). www.british-assessment.co.uk/guides/iso-9001-opens-doors-for-uk-businesses. Accessed 11 January 2017.

Baird, D. (2005). The implementation of a health and safety management system and its interaction with organisational/safety culture: An industrial case study. In *Policy and Practice in Health and Safety* (vol. 3(1), 17–39). Leicester: IOSH Services Limited.

Ball, D., and Ball-King, L. (2011). *Public Safety and Risk Assessment—Implementing Decision Making*. London and New York: Routledge.

Bandura, A. (1997). *Self Efficacy—The Exercise of Control*. Derby: Worth Publishers (Macmillan).

Bennett, J., and Foster, P. (2007). Developing an industry-specific approach to a safety management system. In *Policy and Practice in Health and Safety* (vol. 5(1), 37–59). Leicester: IOSH Services Limited.

Bernstein, P. (1996). *Against the Gods: The Remarkable Story of Risk*. Hoboken, NJ: Wiley.

BGC. (2023). *About Us*. https://gobgc.org/about_us/. Accessed 10 February 2023.

Bhattacharjee, S., and Moreno, K. (2013). The Role of Auditors' emotions and moods on audit judgements: A research summary with suggested practice implications. *Current Issues in Auditing*, 7(2), 1–8.

Bird, L. (2013). *Quotations Book*. http://quotationsbook.com/quote/22788/#sthash.fAPWsVFF.L9hhwTtT.dpbs. Accessed 9 June 2013.

Blanpain, R., and Inston, R. (1996). *The Bosman Case*. London and Edinburgh, UK: Sweet & Maxwell.

Bottani, E., Monica, L, and Vignali, G. (2009). Safety management systems: Performance differences between adopters and non-adopters. *Safety Science*, 47(2), 155–162.

Boyle, T. (2002). *Health and Safety: Risk Management*. Leicester, UK: The Institution of Occupational Safety and Health (IOSH).

Brivot, M., Roussy, M., and Mayer, M. (2018). Conventions of audit quality: The perspective of public and private company audit partners. *Auditing: A Journal of Practice and Theory*, 37, 51–71.

Broberg, P. (2013). What do auditors do? *Mercury Magazine*, 5–6, 102–107.

Bryce, L. (1991). *The Influential Manager*. London, UK: Piatkus.

Bunn III, W., Pikelny, D., Slavin, T., and Paralkar, S. (2001). Health, safety, and productivity in a manufacturing environment. *Journal of Occupational and Environmental Medicine*, 43, 47–55.

Burns-Warren, A. (2006) *Personal communication* with the author.
Carter, A. (1986). Management aspects of reliability. In *Mechanical Reliability*, 354–366. London: Palgrave.
CCPS. (2007). *Guidelines for Risk-based Process Safety*. Hoboken, NJ: Wiley and the Centre for Chemical Process Safety.
CQI. (2023). *About Us*. www.quality.org/knowledge/about-us. Accessed 10 February 2023.
Christman, P., and Taylor, G. (2002). Globalization and the environment: Strategies for international voluntary environmental initiatives. *Academy of Management Perspectives*, 16(3), 121–135.
CNN. (2010). *BP Chief to Gulf Residents: "I'm Sorry"*. http://edition.cnn.com/2010/US/05/30/gulf.oil.spill/index.html. Accessed 8 June 2013.
Coca-Cola Company [The]. (2023). *Purpose and Vision*. www.coca-colacompany.com/company/purpose-and-vision. Accessed 17 March 2023.
Cohen, A. (1977). Factors in successful occupational safety programs. *Journal of Safety Research*, 9, 168–178.
Collins, J. (2001). *Good to Great*. New York: Random House.
Cooper, D. (1998). *Improving Safety Culture—A Practical Guide*. London and Hoboken, NJ: John Wiley.
Corbett, C., Montes-Sancho, M., and Kirsch, D. (2005). The financial impact of ISO 9000 certification in the United States: An empirical analysis. *Management Science*, 51(7), 1046–1059.
COSO. (2013). *Internal Control: Integrated Framework*. www.aicpa.org/cpe-learning/publication/internal-control-integrated-framework-executive-summary-framework-and-appendices-and-illustrative-tools-for-assessing-effectiveness-of-a-system-of-internal-control-3-volume-set. Accessed 7 September 2023.
COSO. (2017). *Enterprise Risk Management: Integrating with Strategy and Performance*. www.coso.org/SitePages/Guidance-on-Enterprise-Risk-Management.aspx?web=1. Accessed 17 March 2023.
CSB. (2007). *Chief Executive's Testimony*. www.csb.gov/assets/1/19/MerrittEnergyCommerceTestimony5_16_07.pdf. Accessed 21 October 2022.
da Silva, S., and Amaral, F. (2019). Critical factors of success and barriers to the implementation of occupational health and safety management systems: A systematic review of literature. *Safety Science*, 117, 123–132.
Dalton, A. (1998). *Safety, Health and Environmental Hazards at the Workplace*. London and New York: Cassell.
Darabont, D. C., Badea, D. O., and Trifu, A. (2020), Comparison of four major industrial disasters from the perspective of human error factor. In *MATEC Web of Conferences* (vol. 305(17)). EDP Sciences.
Dasgupta, S., Hettige H., and Wheeler, D. (2000). What improves environmental compliance? evidence from mexican industry. *Journal of Environmental Economics and Management*, 39(1), 39–66.
Dekker, S. (2014). *Safety Differently—Human Factors for a New Era*. Boca Raton, FL: CRC Press, Taylor & Francis.
Dekker, S. (2017). *The Safety Anarchist: Relying on Human Expertise and Innovation, Reducing Bureaucracy and Compliance*. London: Routledge.
Deming, W. (1982). *Out of the Crisis*. Cambridge: Massachusetts Institute of Technology (MIT).
Deming, W. (1993). *The New Economics for Industry, Government, Education*. Cambridge, MA: MIT Press.
Drucker, P. (2013). Quotations by Author: Peter Drucker. *The Quotations Page*. www.quotationspage.com/quotes/Peter_Drucker. Accessed 24 May 2013.
Edkins, G. (1998). The Indicate safety program: Evaluation of a method to proactively improve airline safety performance. *Safety Science*, 30, 27595.
EFQM. (2023). *The EFQM Model*. https://efqm.org/the-efqm-model/. Accessed 10 March 2023.
Eichenwald, K. (2005). *Conspiracy of Fools: A True Story*. New York: Random House.

EMAS. (2023). *Eco Management and Audit Scheme*. https://green-business.ec.europa.eu/eco-management-and-audit-scheme-emas_en. Accessed 11 February 2023.

Evans, C. (2006). *Personal Communication with the author at the time of preparations for the first edition.*

Eves, D., and Gummer, J. (2005). *Questioning Performance: The Director's Essential Guide to Health, Safety, and the Environment. Cover Text*. Leicester, UK: IOSH.

Fadzil, F., Haron, H., and Jantan, M. (2005). Internal auditing practices and internal control systems. *Management Auditing Journal*, 20(8), 844–866.

Frick, K., and Wren, J. (2000). Reviewing occupational health and safety management: Multiple roots, diverse perspectives, and ambiguous outcomes. In K. Frick, P. L. Jensen, M. Quinlan and T. Wilthagen (Eds.), *In Systematic Occupational Health and Safety Management: Perspectives on an International Development* (17–42). Amsterdam: Pergamon.

FSA. (2023). *Food Hygiene Rating*. https://ratings.food.gov.uk. Accessed 10 March 2023.

FT. (2012). *The Ultimate Lesson of Watergate*. www.ft.com/content/f1874392-b644-11e1-8ad0-00144feabdc0. Accessed 13 February 2023.

Fuller, C., and Vassie, L. (2004). *Health and Safety Management: Principles and Best Practice*. Upper Saddle River, NJ: Prentice Hall.

Gallagher, C., Underhill, E., and Rimmer, M. (2003). Occupational safety and health management systems in Australia: Barriers to success. In *Policy and Practice in Health and Safety* (vol. 1(2), 67–81). Leicester: IOSH Services Limited.

Gardner, D. (2009). *Risk*. London, UK: Virgin Books.

Gardner, D. (2000). Barriers to the implementation of management systems: Lessons from the past. *Quality Assurance*, 8, 3–10.

Ghosh, B. (2012). The agents of outrage. *Time*, 13 September. http://world.time.com/2012/09/13/the-agents-of-outrage/. Accessed 8 June 2013.

Gray, J., Anand, G., and Roth, A. (2015). The influence of ISO 9000 certification on process compliance. *Production and Operations Management,* 24(3), 369–82.

HACCP. (2017). *Hazard Analysis and Critical Control Point*. www.food.gov.uk/business-guidance/hazard-analysis-and-critical-control-point-haccp. Accessed 10 February 2023.

Hafey, R. (2009). *Lean Safety: Transforming Your Safety Culture with Lean Management*. Boca Raton, FL: CRC Press, Taylor & Francis.

Handy, C. (1994). *The Empty Raincoat*. London: Arrow Business Books.

Hansell (2012), In Saujani, M. (2016). World-class safety culture: Applying the five pillars of safety. *Professional Safety*, 61(2), 37–41.

Heinrich, P. H. (1931). *Industrial Accident Prevention: A Scientific Approach*. Maidenhead: McGraw-Hill.

Heras, I., Dick, G., and Casadesus, M. (2002). ISO 9000 registration's impact on sales and profitability: A longitudinal analysis of performance before and after accreditation. *International Journal of Quality and Reliability Management*, 19(6), 774.

Hian Chye Koh, and E-Sah Woo. (1998). The Expectation Gap in Auditing, In *Managerial Auditing Journal*, 13(3), 147–154.

Higson, A. (2002). *Corporate Financial Reporting: Theory and Practice*. Thousand Oaks, CA: Sage Publications.

Hollnagel, E. (2014). *Safety-I and Safety-II*. Oxford and New York: Routledge.

Hopkins, A. (2009). *Failure to Learn: The BP Texas City Refinery Disaster*. North Ryde NSW, Australia: CCH Australia Limited.

HSE. (1991). *Successful Health and Safety Management* (HSG65), 1st ed. Liverpool, UK: HSE Books.

HSE. (1997). *Successful Health and Safety Management* (HSG65), 2nd ed. Liverpool, UK: HSE Books.

HSE. (2013). *Successful Health and Safety Management* (HSG65), 3rd ed. Liverpool, UK: HSE Books. www.hse.gov.uk/pubns/priced/hsg65.pdf. Accessed 10 February 2023.

HSE. (2015). *Historical Picture—Trends in Work-related Injuries and Ill-Health in Great Britain since the Introduction of the Health and Safety at Work Act 1974*. www.hse.gov.uk/statistics/history/historical-picture.pdf. Accessed 24 September 2017.

HSE. (2016). Personal correspondence with Helen Wilson, HSE Science Directorate—Statistics and Epidemiology Unit, 22 January 2016.

Humphrey, C. (1997), Debating audit expectations. In M. Sherer and S. Turley (Eds.), *Current Issues in Auditing*. London: Sage.

IAQG. (2016). *Quality Management Systems—Requirements or Aviation, Space, and Defence Organizations*. https://iaqg.org. Accessed 12 September 2023.

IATF. (2016). International standard for automotive quality management systems. *International Automotive Task Force*. www.iatfglobaloversight.org. Accessed 10 February 2023.

IEMA. (2022). *IEMA Sustainability Policy, Version February 2022*. www.iema.net/resources. Accessed 10 February 2023.

IIA. (2004). *Practice Advisory 2060–2. Relationship with the Audit Committee*, 12 February.

IIA. (2023a). *Is It Mandatory to Have an Internal Audit?* www.theiia.org/en/about-us/advocacy/about-the-profession/faq/. Accessed 9 February 2023.

IIA. (2023b). *What is Internal Auditing?* www.theiia.org/en/about-us/advocacy/about-the-profession/faq/. Accessed 9 February 2023.

IIA. (2023c). *About Us*. www.theiia.org/en/about-us/. Accessed 12 February 2023.

ILO. (2009). *ILO-OSH 2001 Guidelines on Occupational Safety and Health Management Systems*. www.ilo.org/wcmsp5/groups/public/—ed_protect/—protrav/—safework/documents/normativeinstrument/wcms_107727.pdf. Accessed 8 February 2023.

ILO. (2016). Safety in Numbers. *IOGP website*. www.iogp.org/blog/2016/08/02/safety-in-numbers/, Accessed 7 February 2023.

IOGP. (2014). *Operating Management System Framework*. www.iogp.org/bookstore/product/operating-management-system-framework-for-controlling-risk-and-delivering-high-performance-in-the-oil-and-gas-industry/. Accessed 8 February 2023.

IOSH, (2023). *Disciplinary and Appeals*. https://iosh.com/about-iosh/who-we-are/governance-structure/board-of-trustees/board-committees/disciplinary-and-appeals/. Accessed 5 March 2023.

IPC. (2023). *About Us*. www.ipcaweb.org/index?p=MjEEQUALS. Accessed 14 February 2023.

IRCA. (2023). *Personal Communication with Ian Howe, head of membership via email*, 10 January.

ISO. (1990). *ISO 10011–1:1990 Guidelines for Auditing Quality Systems. Part 1—Auditing*. Geneva: International Organization for Standardization.

ISO. (2002). *ISO 19011:2002 Guidelines for Auditing Management Systems*. Geneva: International Organization for Standardization.

ISO. (2012a). *Annex SL (previously ISO Guide 83) of the Consolidated ISO Supplement of the ISO/IEC Directives*. Geneva: International Organization for Standardization.

ISO. (2012b). *ISO/IEC 17024:2012 Conformity Assessment: General Requirements for Bodies Operating Certification of Persons*. www.iso.org/iso/catalogue_detail?csnumber=52993. Accessed 17 March 2023.

ISO. (2012c). *ISO 39001:2012 Road Traffic Safety (RTS) Management Systems—Requirements with Guidance for Use*. Geneva: International Organization for Standardization.

ISO. (2012d). *ISO 20121:2012 Event Sustainability Management Systems—Requirements with Guidance for Use*. Geneva: International Organization for Standardization.

ISO. (2015a). *ISO 9001:2015 Quality Management Systems: Requirements*. Geneva: International Organization for Standardization.

ISO. (2015b). *ISO 14001:2015 Environmental Management Systems: Requirements with Guidance for Use*. Geneva: International Organization for Standardization.

ISO. (2016a). *ISO 13485:2016 Medical Devices—Quality Management Systems—Requirements for Regulatory Purposes*. Geneva: International Organization for Standardization.

ISO. (2016b). *ISO 37001:2016 Anti-bribery Management Systems*. Geneva: International Organization for Standardization.

ISO. (2017a). *ISO/IEC 17021:2015 Conformity Assessment: Requirements for Bodies Providing Audit and Certification of Management Systems.* www.iso.org/standard/61651.html. Accessed 26 January 2023.

ISO. (2017b). *ISO 44001:2017 Collaborative Business Relationship Management Systems—Requirements and Framework.* Geneva: International Organization for Standardization.

ISO. (2018a). *ISO 45001:2018 Occupational Health and Safety Management Systems—Requirements with Guidance for Use.* Geneva: International Organization for Standardization.

ISO. (2018b). *ISO 19011:2018 Guidelines for Auditing Management Systems*, 3rd ed (1st edn. 2002; 2nd edn. 2011). Geneva: International Organization for Standardization.

ISO. (2018c). *ISO 31000:2018 Principles and General Guidelines on Risk Management.* Geneva: International Organization for Standardization.

ISO. (2018d). *ISO 22000:2018 Food Safety Management Systems.* Geneva: International Organization for Standardization.

ISO. (2018e). *Energy Management Systems—Requirements with Guidance for Use.* Geneva: International Organization for Standardization.

ISO. (2018f). *ISO/IEC 20000–1 Information Technology—Service Management—Part 1: Service Management System Requirements.* Geneva: International Organization for Standardization.

ISO. (2019). *ISO 22301:2019 Security and Resilience- Business Continuity Management Systems—Requirements.* Geneva: International Organization for Standardization.

ISO. (2020). *ISO 29001:2020—Petroleum, Petrochemical, and Natural Gas Industries—Sector-Specific Quality Management Systems.* Geneva: International Organization for Standardization.

ISO. (2022b). *ISO 22301:2019 Security and Resilience—Business Continuity Management Systems.* Geneva: International Organization for Standardization.

ISO. (2022c). *ISO 28000:2022 Security and Resilience—Security Management Systems—Requirements.* Geneva: International Organization for Standardization.

ISO. (2023a). *The ISO Survey.* https://www.iso.org/the-iso-survey.html. Accessed 18 September 2023.

ISO. (2023b). *International Harmonized Stage Codes.* www.iso.org/files/live/sites/isoorg/files/developing_standards/docs/en/stage_codes.pdf. Accessed 26 January 2023.

ISO. (2023c). *ISO 45002:2023 Occupational Health and Safety Management Systems—General Guidelines for the Implementation of ISO 45001:2018.* Geneva: International Organization for Standardization.

James, T. and James, A. (2009). *The Intensive NLP Practitioner Certification Training Manual, version 6.53.* Quotation used with permission of Dr Adriana James and The Tad James Company. Nevada: Tad James Company.

JCB. (2023). *Our Mission.* www.jcb-finance.co.uk/about-us/our-mission. Accessed 17 March 2023.

Jennings, M., Kneer, D., and Reckers, P. (1993). The significance of audit decision aids and precase jurists' attitudes on perceptions of audit firm culpability and liability. *Contemporary Accounting Research*, 9(2), 489–507.

Jilcha, K., and Kitaw, D. (2016). A literature review on global occupational safety and health practice and accidents severity. *International Journal for Quality Research*, 10(2).

Johnson, S. (1999). *Who Moved My Cheese?* London, UK: Vermilion.

Joyston-Bechal, S. (2022). *Retained EU Law (Revocation and Reform) Bill: UK H&S Legislation on the Edge?* www.shponline.co.uk/legislation-and-guidance/retained-eu-law-revocation-and-reform-bill-uk-hs-regulation-on-the-edge/. Accessed 30 January 2023.

Kao, K. Y., Thomas, C. L., Spitzmueller, C., and Huang, Y. H. (2019). Being present in enhancing safety: Examining the effects of workplace mindfulness, safety behaviors, and safety climate on safety outcomes. *Journal of Business and Psychology*, 1–15.

Karapetrovic, S., and Willborn, W. (1998). Integrated audit of management systems. *International Journal of Quality and Reliability Management*, 15(7), 694–711.

Katz, R. (2007). *Political Institutions in the United States.* Oxford: Oxford University Press.

Keyser, S. (2012). The write stuff. *Business Life*, 14 October.

Knechel, W., Krishnan, G., Pevzner, M., Shefchik, L., and Velury, U. (2013), Audit Quality: Insights from the Academic Literature, *Auditing: A Journal of Practice and Theory*, 32, 385–421.

Knutt, E. (2016). *What Exactly is "Safety Differently"*. www.healthandsafetyatwork.com/safe-systems-of-work/what-is-safety-differently. Accessed 16 February 2017.

Kolluru, R., Bartell, S., Pitblado, R., and Stricoff, S. (1996). *Risk Assessment and Management Handbook for Environmental, Health and Safety Professionals*. New York: McGraw-Hill.

Krause, T. (1990). *The Behavior-based Safety Process*. New York: Van Nostrand Reinhold.

Lane, M. (2011). *Plausible Denial: Was the CIA Involved in the Assassination of JFK?* New York, NY: Skyhorse Publishing.

Levine, D., and Toffel, M. (2010). Quality management and job quality: How the ISO 9001 standard for quality management systems affects employees and employers. *Management Science*, 56(6), 978–96.

Lorriman, J., and Kenjo, T. (1994). *Japan's Winning Margins*. Oxford: Oxford University Press.

Lucius, T. (2002). *Department of Defence Quality Management Systems and ISO 9001:2000* (Naval Postgraduate School, Monterey, CA). http://hdl.handle.net/10945/5983. Accessed 8 February 2023.

Luscombe, B. (2012). 10 Questions for Sir Tim Berners-Lee. *Time*, 5 September. http://techland.time.com/2012/09/05/10-questions-for-sir-tim-berners-lee/. Accessed 23 February 2023.

Maltby, J. (2007). There is no such thing as an audit society. *Working Paper*, Department of Management Studies, University of York.

Manuele, F. (2002). *Heinrich Revisited: Truisms or Myths*. Itasca, IL: National Safety Council.

Manuele, F. (2011). Reviewing Heinrich: Dislodging two myths from the practice of safety. *Professional Safety*, 56(10).

Marsh, T. (2013). *Talking Safety: A User's Guide to World-class Safety Conversation*. Aldershot: Gower.

Marson, B. (1993). Building customer-focused organizations in British Columbia. *Public Administration Quarterly*, 17(1), 30–41.

McCormick, J. (2015). *European Union Politics*. Basingstoke: Palgrave Macmillan.

McDonald's. (2023). *Our Mission and Values*. https://corporate.mcdonalds.com/corpmcd/our-company/who-we-are/our-values.html#:~:text=We%20serve%20delicious%20food%20people,and%20personalization%20our%20customers%20expect. Accessed 17 March 2023.

Mearns, K., Whitaker, S., and Flin, R. (2003). Safety climate, safety management practice and safety performance in offshore environments. *Safety Science*, 41, 641–680.

Merritt, C. (2005). *Statement to the BP Independent Safety Review Panel. US Chemical Safety and Hazard Investigation Board (CSB)*. www.csb.gov/assets/1/20/carolyn_statement_3.pdf?13854. Accessed 10 February 2023.

Michaels, A., Hoyos, C., and Parker, A. (2004). Retired Shell engineer played central role. *Financial Times*, 25 August.

Moiser, P. (1997). Independence. In M. Sherer and S. Turley (Eds.), *Current Issues in Auditing*. London: Sage.

Monroe, G., and Woodliffe, D. (1993). The effect of education on the audit expectation gap. *Accounting and Finance*, 33(1), 1–91.

Morrin, D. (2016). We are much more than watchdogs: The dual identity of auditors at the UK national audit office. *Journal of Accounting and Organisational Change*, 12(4), 569–589.

Morton, C. (2016). *Becoming World Class*. New York, NY: Springer.

Nair, G., and Tauseef, S. (2018). Predicting effectiveness of management systems: Measuring successes against failures. *International Journal of Engineering Technology Science and Research*, 5(1).

National Trust. (2023). *Poldhu Cove > Cornwall*. www.nationaltrust.org.uk/visit/cornwall/poldhu-cove. Accessed 14 February 2023.

OGP. (1994). *HSE-MS 6.36/210 Guidelines for the Development and Application of Health, Safety and Environmental Management Systems*.

ONS. (2016) *Labour Force Report*. www.ons.gov.uk/employmentandlabourmarket/peopleinwork/employmentandemployeetypes/bulletins/uklabourmarket/dec2016. Accessed 30 December 2016.

O'Toole, G. (2012). If i had more time, i would have written a shorter letter. *Quote Investigator, blog posting*. http://quoteinvestigator.com/2012/04/28/shorter-letter/. Accessed 22 February 2023.

Oxebridge. (2022). *Annual Analysis of the ISO Survey*. www.oxebridge.com/emma/iso-survey-2021/. Accessed 7 February 2023.

Oxford University Press. (2008). *Concise Oxford English Dictionary*, 11th ed. Oxford: Oxford University Press.

PECB. (2023). *About Us*. https://pecb.com/en/about. Accessed 10 February 2023.

Porter, B. (1993). An empirical study of the audit expectation-performance gap. *Accounting and Business Research*, 24(93).

Potoski, M., and Prakash, A. (2005a). Covenants with weak swords: ISO 14001 and facilities' environmental performance. *Journal of Policy Analysis and Management*, 24(4), 745–769.

Potoski, M., and Prakash, A. (2005b). Green clubs and voluntary governance: ISO 14001 and firms' regulatory compliance. *American Journal of Political Science*, 49(2), 235–248.

Power, M. (1994). *The Audit Explosion*. London: Demos.

Prasad, S. (2022). *China to Dominate 95% of Solar Panel Supply Chain*. www.downtoearth.org.in/news/energy/china-to-dominate-95-of-solar-panel-supply-chain-83651. Accessed 3 February 2023.

Quick, R. (2020). The audit expectation gap: A review of the academic literature. *Maandblad Voor Accountancy en Bedrijfseconomie*, 94, 5.

Rae, A., and Provan, D. (2019). Safety work versus the safety of work. *Safety Science*, 111, 119–127.

Reason, J. (1990). *Human Error*. Cambridge: Cambridge University Press.

Reason, J. (2013). *A Life in Error*. Boca Raton, FL: CRC Press, Taylor & Francis.

Redinger, C., and Levine, S. (1998). Development and evaluation of the Michigan occupational health and safety management system assessment instrument: A universal OHSMS performance measurement tool. *American Industrial Hygiene Association Journal*, 59, 572–581.

Robson, L. (2017). Personal discussion with Dr Lynda Robson, Institute for Work and Health Toronto, lead researcher/author of Robson et al. (2007), on 12 December 2017.

Robson, L., Clarke, J., Cullen, K., Bielecky, A., Severin, C., Bigelow, P., Irvin, E., Culyer, A., and Mahmood, Q. (2005). *The Effectiveness of Occupational Health and Safety Management Systems: A Systematic Review*. Toronto: Institute for Work and Health.

Robson, L., Clarke, J., Cullen, K., Bielecky, A., Severin, C., Bigelow, P., Irvin, E., Culyer, A., and Mahood, Q. (2007). *The Effectiveness of Occupational Health and Safety Management System Interventions: A Systematic Review, in Safety Sciences* (vol. 45(3), 329–53). London: Elsevier.

Robson, L., Macdonald, S., Gray, G., Van Eerd, D., and Bigelow, L. (2012). A descriptive study of the OHS management and methods used by public sector organisations conducting audits of workplaces: Implications for audit reliability and validity, *Safety Science*, 50(2), 181–189.

Rodway, J. (2023). *Risk*. www.safec.co.uk/services/risk-management/. Accessed 6 February 2023.

Saksvik, P. O., and Quinlan, M. (2003). Regulating systematic occupational health and safety management: comparing the Norwegian and Australian experience. *Relations Industrielles*, 58, 721–738.

Salehi, M., Mansoury, A., and Azary, Z. (2009). Audit independence and expectation gap: Empirical evidence from iran. *International Journal of Economics and Finance*, 1(1).

Saraf, A. (2019). *Efficacy of ISO 9001: 2015 to Support Operational Performance*. Doctoral dissertation, Metropolitan State University.

Sass, R., and Crook, G. (1981), Accident proneness: Science or non-science? *International Journal of Health Services*, 11(2), 175–190.

Saujani, M. (2016). World-class safety culture: Applying the five pillars of safety. *Professional Safety*, 61(2), 37–41.

Schuman, M. (2012a). Why China must push reset. *Time*, 18 June. www.time.com/time/magazine/article/0,9171,2116604-2,00.html. Accessed 24 May 2013.
Schuman, M. (2012b). The new great wall of China. *Time*, 24 September. www.time.com/time/magazine/article/0,9171,2124406-1,00.html. Accessed 24 May 2013.
Siegel, G. (2002). Business partner and corporate cop: do the roles conflict? *Strategic Finance* 82(3), 89–90 (Institute of Management Accountants).
Silberman, B. (2013). The year in highlights. *Time*, 7 January.
Simunic, D. (1984). Auditing, consulting and auditor independence. *Journal of Accounting Research*, 22(2), 679–702.
Sky. (2023). *Culture of Silence as NHS Staff are Scared to Report Problems*. https://news.sky.com/story/culture-of-silence-as-nhs-staff-are-scared-to-report-problems-fearing-repercussions-12798287. Accessed 30 January 2023.
Statista. (2023a). *Number of Internet and Social Media Users*. www.statista.com/statistics/617136/digital-population-worldwide/. Accessed 20 March 2023.
Statista. (2023b). *United Kingdom Employment from 2016 to 2023*. www.statista.com/statistics/275311/employment-in-the-united-kingdom/. Accessed 6 February 2023.
Stewart, J., and Stewart, J. (2002). *Managing for World Class Safety*. New York, NY: John Wiley & Sons.
Suan, A. (2017). A mini review on efficacy of safety management systems in construction. *International Journal of Engineering Science and Computing*. https://ijesc.org/upload/4f44f8a2ec9ac3f6c469bfd04ba8d758.A%20Mini%20Review%20on%20Efficacy%20of%20Safety%20Management%20Systems%20in%20Construction.pdf. Accessed 6 July 2023.
Taleb, N. M. (2010). *The Black Swan: The Impact of the Highly Improbable*, 2nd ed. New York and London: Random House, Allen Lane/Penguin.
Taylor, J. (2004). What do we know about audit quality? *The British Accounting Review*, 36(4), 345–368.
Terlaak, A., and King, A. (2006). The effects of certification with the ISO 9000 quality management standard: S signalling approach. *Journal of Economic Behaviour & Organization*, 60(4), 579–602.
The Law Times. (1896). *The Law Times* (vol. LXXIV). London: Court of Appeal.
Tiwari, P., and Shukla, V. (2018). An innovating methodology for measuring the effective implementation of OHSMS (Occupational Health and Safety Management System) in small and medium scale industries. *International Journal of Scientific Research & Engineering Trends*, 4(4).
Toone, B. (2004). *Protect Your People-and Your Business*. Leicester, UK: IOSH.
Tzu, S. (2009, first published 1910). *The Art of War* (trans. Giles, L). Pax Librorum. https://www.paxlibrorum.com/books/taowde/. Accessed 6 July 2023.
Viswanathan, K., Johnson, S., and Toffel, M. (2021). Do management system standards indicate superior performance? *Working Paper No*. 22–042, Harvard Business School.
Willis Corroon. (1996). *Environmental Management Manual*. London: Willis Corroon Environmental Forum.
World Bank. (2022). *Risk of Global Recession in 2023*. www.worldbank.org/en/news/press-release/2022/09/15/risk-of-global-recession-in-2023-rises-amid-simultaneous-rate-hikes. Accessed 25 January 2023.
Yassi, A. (1998). Utilising data systems to develop and monitor occupational health programs in a large Canadian hospital. *Methods of Information in Medicine*, 37, 125–129.
Zakaria, F. (2006). Voices. *Newsweek*, 147(22), 29.
Zakaria, F. (2012). Tax and spend. *Time*, 23 July. www.time.com/time/magazine/article/0,9171,2119336,00.html. Accessed 24 May 2013.
Zukav, G. (2013). Gary Zukav > Quotes > Quotable Quote. *Goodreads*. www.goodreads.com/quotes/149697-reality-is-what-we-take-to-be-true-what-we. Accessed 9 June 2013.

Index

Page numbers in *italics* indicate figures; page numbers in **bold** indicate tables.

B

C

G

H

I

J

K

L

T

U

V

9781032427577